OXFORD MATHEMATICAL MONOGRAPHS

Series Editors

OXFORD MATHEMATICAL MONOGRAPHS

The Fourth Janko Group

A. A. IVANOV

CLARENDON PRESS · OXFORD

2004

OXFORD

UNIVERSITY PRESS

Great Clarendon Street, Oxford OX2 6DP

Oxford University Press is a department of the University of Oxford.
It furthers the University's objective of excellence in research, scholarship,
and education by publishing worldwide in

Oxford New York

Auckland Bangkok Buenos Aires Cape Town Chennai
Dar es Salaam Delhi Hong Kong Istanbul Karachi Kolkata
Kuala Lumpur Madrid Melbourne Mexico City Mumbai Nairobi
São Paulo Shanghai Taipei Tokyo Toronto

Oxford is a registered trade mark of Oxford University Press
in the UK and in certain other countries

Published in the United States
by Oxford University Press Inc., New York

British Library Cataloguing in Publication Data

(Data available)

Library of Congress Cataloging in Publication Data

(Data available)

ISBN 0-19-852759-4

1 3 5 7 9 10 8 6 4 2

Typeset by Newgen Imaging Systems (P) Ltd., Chennai, India
Printed in Great Britain
on acid-free paper by
Biddles Ltd., King's Lynn, Norfolk

To Lena

- Позволь, сударыня, мне сделать то же точно,
В чем упражнялись те, кто делали тебя,
Авось и мне удастся ненарочно
Сделать такую ж, хоть не для себя.

Иван Барков "Требование"

PREFACE

We start with the following classical situation. Let V be a 10-dimensional vector space over the field $GF(2)$ of two elements. Let q be a non-singular quadratic form of Witt index 5 on V. Let $H \cong O_{10}^+(2)$ be the group of invertible linear transformations of V that preserve q. Let $\Omega = D^+(10, 2)$ be the dual polar graph associated with the pair (V, q), that is the graph on the set of maximal subspaces of V which are totally singular with respect to q (these subspaces are 5-dimensional); two such subspaces are adjacent in Ω if and only if their intersection is of codimension 1 in each of the two subspaces. Then $(\Omega, H) = (D^+(10, 2), O_{10}^+(2))$ belongs to the class of pairs (Ξ, X), where Ξ is a graph and X is a group of automorphisms of Ξ satisfying the following conditions (C1) to (C3):

(C1) Ξ is connected of valency $31 = 2^5 - 1$;
(C2) the group X acts transitively on the set of incident vertex-edge pairs in Ξ;
(C3) the stabilizer in X of a vertex of Ξ is the semi-direct product with respect to the natural action of the general linear group in dimension 5 over the field of two elements and the exterior square of the natural module of the linear group.

The constrains imposed by the conditions (C1) to (C3) concern only 'local' properties of the action of X on Ξ and these properties remain unchanged when one takes suitable 'coverings'. The constrains are encoded in the structure of the stabilizers in X of a vertex of Ξ and of an edge containing this vertex, and also in the way these two stabilizers intersect. Denote the vertex and edge stabilizers by $X^{[0]}$ and $X^{[1]}$, respectively and 'cut them out' of X to obtain what is called the *amalgam*

$$\mathcal{X} = \{X^{[0]}, X^{[1]}\}$$

(the union of the element-sets of the two groups with group operations coinciding on the intersection $X^{[01]} = X^{[0]} \cap X^{[1]}$). Because of (C2) the isomorphism type of $\mathcal{X}$ is independent of the choice of the incident vertex–edge pair. If the pair (Ξ, X) is simply connected which means that Ξ is a tree, then X is the universal completion of $\mathcal{X}$ which is known to be the free product of $X^{[0]}$ and $X^{[1]}$ amalgamated over the common subgroup $X^{[01]}$. A reasonable question to ask is about the possibilities for the isomorphism type of $\mathcal{X}$. The following lemma gives the answer.

Lemma A. *Let (Ξ, X) be a pair satisfying (C1) to (C3) and let $\mathcal{X}$ be the amalgam formed by the stabilizers in X of a vertex of Ξ and of an edge incident to*

*this vertex. Then $\mathcal{X}$ is either the classical amalgam $\mathcal{H}$ contained in $H = O_{10}^{+}(2)$
or one extra amalgam $\mathcal{G} = \{G^{[0]}, G^{[1]}\}$.*

In a certain sense the existence of the additional amalgam $\mathcal{G}$ in Lemma A
is due to the famous isomorphism of the general linear group in dimension four
over the field of two elements and the alternating group of degree eight.

In order to obtain all the pairs satisfying (C1) to (C3) we should consider the
quotients of the universal completions of the amalgams $\mathcal{H}$ and $\mathcal{G}$ over suitable
normal subgroups. However, there are far too many possibilities for choosing
that normal subgroup and the project of finding them all appears fairly hopeless.
Nevertheless, we may still try to find particular examples which are 'small and
nice' in one sense or another.

It can be shown that for every pair (Ξ, X) satisfying (C1) to (C3) there is
a 'nice' family of cubic (that is valency 3) subgraphs in Ξ as described in the
following lemma.

Lemma B. *Let (Ξ, X) be a pair satisfying (C1) to (C3). Then Ξ contains a
family $\mathcal{S}$ of connected subgraphs of valency 3. This family is unique subject to
the condition that it is stabilized by X and whenever two subgraphs from $\mathcal{S}$ share
a vertex and if the neighbours of this vertex in both subgraphs coincide, the whole
graphs are equal.*

The subgraphs forming the family $\mathcal{S}$ in Lemma B are called *geometric
cubic subgraphs*. If Ξ is a tree then every geometric cubic subgraph is a
cubic tree which is 'large', even infinite. On the other hand, in the classical
example $(D^{+}(10, 2), O_{10}^{+}(2))$ the geometric cubic subgraphs correspond to the
3-dimensional totally singular subspaces in the 10-dimensional orthogonal space
V. The subgraph corresponding to such a subspace U is formed by the maximal
totally singular subspaces in V containing U. This subgraph is complete bipartite on 6 vertices denoted by $K_{3,3}$. Thus here the geometric cubic subgraphs are
small and nice.

Let (Ξ, X) be a pair satisfying (C1) to (C3). Let Σ be a geometric cubic
subgraph in Ξ, let S and T be the global and the vertexwise stabilizers of Σ in
X. Let $\widehat{\Sigma}$ be the graph on the set of orbits of the centralizer $C_S(T)$ of T in S on
the vertex set of Σ in which two orbits are adjacent if there is at least one edge
of Σ which joins them. Then the natural mapping $\psi : \Sigma \to \widehat{\Sigma}$ turns out to be a
covering of graphs commuting with the action of S. Put

$$\widehat{S} = S/(TC_S(T))$$

which is the image of S in the outer automorphism group of T.

Direct but somewhat tricky calculation in the amalgams $\mathcal{H}$ and $\mathcal{G}$ give the
following.

Lemma C. *In the above terms the following hold:*

(i) *if $\mathcal{X} = \mathcal{H}$ then $\widehat{S} \cong \mathrm{Sym}_3 \wr \mathrm{Sym}_2$ and $\widehat{\Sigma}$ is the complete bipartite graph $K_{3,3}$:*

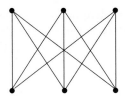

(ii) *if $\mathcal{X} = \mathcal{G}$ then $\widehat{S} \cong \mathrm{Sym}_5$ and $\widehat{\Sigma}$ is the Petersen graph:*

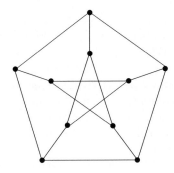

By the remark after Lemma B if the pair (Ξ, X) is $(D^+(10, 2), O_{10}^+(2))$ then the mapping $\psi : \Sigma \to \widehat{\Sigma}$ is an isomorphism. The following characterization has been established by P. J. Cameron and C. E. Praeger in 1982.

Proposition D. *Let (Ξ, X) be a pair satisfying $(C1)$ to $(C3)$ with $\mathcal{X} = \mathcal{H}$ and suppose that $\psi : \Sigma \to \widehat{\Sigma}$ is an isomorphism. Then $(\Xi, X) = (D^+(10, 2), O_{10}^+(2))$.*

It is natural to ask what happens when Σ attains the other minimal possibility in Lemma C, that is the Petersen graph. The main purpose of this book is to answer this questing by proving the following

Main Theorem. *Let (Ξ, X) be a pair satisfying $(C1)$ to $(C3)$ with $\mathcal{X} = \mathcal{G}$ and suppose that $\psi : \Sigma \to \widehat{\Sigma}$ is an isomorphism. Then the pair (Ξ, X) is determined uniquely up to isomorphism. Furthermore*

(i) *X is non-abelian simple;*
(ii) *$|X| = 2^{21} \cdot 3^3 \cdot 5 \cdot 7 \cdot 11^3 \cdot 23 \cdot 29 \cdot 31 \cdot 37 \cdot 43 =$*
     $86,775,571,046,077,562,880;$
(iii) *X contains an involution z such that $C_X(z) \cong 2_+^{1+12} \cdot 3 \cdot \mathrm{Aut}\ (M_{22})$.*

The group X in the Main Theorem is the fourth Janko sporadic simple group J_4 discovered by Zvonimir Janko in 1976 and constructed in Cambridge in 1980 by D. J. Benson, J. H. Conway, S. P. Norton, R. A Parker, J. G. Thackray as a subgroup of $L_{112}(2)$.

The book was derived from research which originated almost exactly twenty years ago. This was the golden age of the research seminar on algebra and geometry at the Institute for System Studies (VNIISI) in Moscow. Some of its members were examining a special case of the quasithin groups' classification (and J_4 is a quasithin group); others were searching for new distance-transitive graphs. The latter activity led me to the discovery of the geometry $\mathcal{F}(J_4)$. At this stage our quasithin experts suggested that the discovery might eventually emerge into a geometric construction and uniqueness proof for J_4. The project that seemed preposterous at that time has been fully realized here.

My foremost gratitude goes to Vladimir L'vovich Arlazarov and Igor Alexandrovich Faradjev, who both founded and maintained the VNIISI seminar. The crucial ingredient in the characterization is the simple connectedness of the geometry $\mathcal{F}(J_4)$ which I proved in the summer of 1989 within the joint project with Sergey Shpectorov on the classification of the Petersen and tilde geometries. I am very thankful to Sergey for his long term friendship and cooperation and hope this will last. When the proof was discussed in (then still West) Berlin in June 1990, Geoff Mason noticed that the simple connectedness formed a basis for the uniqueness proof for J_4. At the Durham Symposium in July 1990, together with Ulrich Meierfrankenfeld, we found a way of transferring the simple connectedness into a computer-free construction. It took a further ten years to see the construction published. Many original ideas here and indeed almost the whole of chapter 8 are due to Ulrich.

The main layout of the book was designed when I gave a series of lectures at the University of Tokyo in autumn 2002. I am very thankful to Atsushi Matsuo who organized these lectures and to the audience for many stimulating questions and discussions. My special thanks go to Satoshi Yoshiara and Hiroki Shimakura who took the notes of the lectures; I used these notes as a draft for the first two chapters of the book. Hiroki wrote the notes in Japanese and now they are published as (Ivanov 2003). In order to proceed with the project I needed a transparent description of the pentad subgroup in J_4. This was achieved in Oberwolfach in summer of 2003, again, thanks to a fruitful cooperation with Sergey Shpectorov. The writing up began in September 2003 and took about six months. Half way through I received useful comments on the draft from Ernie Shult. His note: 'nice' against Lemma 4.5.5 was particularly inspiring. Thank you, Ernie. Corinna Wiedorn is one of the very few who read through our construction paper with Ulrich. So even during the early stages of working on the book, I was sure that it will have at least one dedicated reader. Her incisive comments on the final draft exceeded my best expectations. Antonio Pasini not only kindly offered to read the final draft but also assured me that this was not a favour at all and that this was solely for his own pleasure. His thoughtful comments are of great value to me.

I am pleased to acknowledge that my the insight in J_4 was gradually built up through discussions with experts including Michael Aschbacher, John Conway, Wolfgang Lempken, Simon Norton, Richard Parker, Gernot Stroth and Richard Weiss. Whenever I got stuck, Dima Pasechnik was always ready to help by performing another ingenious computer calculation. I am thankful to my colleagues and friends including Colin Atkinson, Michael Deza, Martin Liebeck, Jan Saxl and Leonard Soicher for encouragement to write a book whose title will be just 'J_4'. The reader can judge to what extent I have achieved this goal. Last, but certainly not least, my sincere gratitude goes to my wife Lena, my son Denis and my daughter Nina for support, understanding and immense patience over the years.

CONTENTS

1

CONCRETE GROUP THEORY

According to the current state of art, the existence of large sporadic simple groups (and J_4 is certainly one of them) is explained by exceptional properties of concrete small groups. We invite the reader to join the devious trail to J_4, which wends its way through peculiar features of common first members of otherwise disjoint families of finite simple groups. As a basic unifying concept we have chosen the class of symplectic and orthogonal forms over the field of two elements. For basic definitions, notation and background results in this and further chapters the reader is advised to consult the Appendices and well as standard literature.

1.1 Symplectic and orthogonal $GF(2)$-forms

Let V be an n-dimensional $GF(2)$-vector space. A mapping

$$f : V \times V \rightarrow GF(2)$$

is a *symplectic form* on V if it is bilinear and for all $u, v \in V$ we have $f(u,v) = f(v,u)$ and $f(u,u) = 0$. For a symplectic form f a mapping

$$q : V \rightarrow GF(2)$$

is an *orthogonal form* on V *associated with* f if

$$f(u,v) = q(u) + q(v) + q(u+v)$$

for all $u, v \in V$.

With V as above let f be a symplectic form on V and let q be an orthogonal form associated with f. Then the pairs (V, f) and (V, f, q) are called *symplectic* and *orthogonal* spaces, respectively (although f determines q in the latter case, we still prefer to keep the explicit reference to f).

For $S \subseteq V$ by $f|_S$ and $q|_S$ we denote the restrictions of f and q to $S \times S$ and S, respectively. It is useful to keep in mind the following easy observation.

Lemma 1.1.1 *Let (V, f) be an n-dimensional symplectic space and B be a basis of V. Then for every function $h : B \rightarrow GF(2)$ there is a unique orthogonal space (V, f, q), such that $q|_B = h$. In particular there are exactly 2^n orthogonal forms associated with a given symplectic form.*

Let (V, f) be a symplectic space. Two vectors $u, v \in V$ are said to be *perpendicular* if $f(u,v) = 0$ (particularly every vector is perpendicular to itself). The set of vectors perpendicular to a given vector $u \in V$ is called the *perp* of u denoted by $u^\perp$. If $S \subseteq V$ then $S^\perp$ consists of the vectors perpendicular to every vector of S. It is clear that $S^\perp$ is a subspace and $S^\perp = \langle S \rangle^\perp$ (here and elsewhere $\langle S \rangle$ is the subspace in V spanned by S). The symplectic space (V, f) is *non-singular* if $V^\perp = 0$ and *singular* otherwise.

Lemma 1.1.2 *Let (V, f) be a non-singular symplectic space and U be a subspace of V. Then*

$$\dim U + \dim U^\perp = \dim V.$$

Proof For $w \in V$ the map $l_w : u \mapsto f(u, w)$ is a linear function on U, that is an element of the dual space U^* of U. Since (V, f) is non-singular, the mapping $\nu : V \to U^*$, where $\nu(w) = l_w$ is a surjective homomorphism of vector spaces, and by the definition, $U^\perp$ is the kernel of ν. Now the result is immediate from elementary linear algebra. ∎

Two symplectic spaces (V, f) and (V', f') are isomorphic if there is an isomorphism $\alpha : V \to V'$ of vector spaces such that $f(u, v) = f'(\alpha(u), \alpha(v))$. The isomorphism of orthogonal spaces is defined in the obvious similar way. Let (U, g) and (W, h) be symplectic spaces. The *direct sum* of these spaces is a symplectic space (V, f), where V is the direct sum of U and W; $f|_U = g$ and $f|_W = h$, and (with respect to f) every vector from U is perpendicular to every vector from W. In the above terms if p and r are orthogonal forms associated with g and h, respectively, then the direct sum of (U, g, p) and (W, h, r) is an orthogonal space (V, f, q) such that (V, f) is the direct sum of (U, g) and (W, h), and q the only orthogonal form associated with f such that $q|_U = p$ and $q|_W = r$.

We use the symbol $\oplus$ to denote the direct sum operation on symplectic and orthogonal spaces. When the spaces are considered up to isomorphism, this operation is commutative and associative. The following two lemmas are rather straightforward.

Lemma 1.1.3 *The direct sum $(U, g) \oplus (W, h)$ is non-singular if and only if both the summands are such.*

Lemma 1.1.4 *Let (V, f, q) be an orthogonal space. Suppose that V is the direct sum of its subspaces U and W with $U \leq W^\perp$ and $W \leq U^\perp$. Then $(V, f, q) \cong (U, f|_U, q|_U) \oplus (W, f|_W, q|_W)$.*

Now we are well prepared to present the well-known classification of the non-singular symplectic and associated orthogonal spaces.

Lemma 1.1.5 *Let (V, f) be an n-dimensional non-singular symplectic space. Then*

    (i) *n is even, say $n = 2m$;*
    (ii) *the isomorphism type of (V, f) is uniquely determined;*
    (iii) *(V, f) is the direct sum of m hyperbolic planes.*

Proof We proceed by induction on n. If $n = 1$, then the only non-zero vector is perpendicular to itself and $V = V^\perp$. Thus $n \geq 2$ and there must be a pair of non-perpendicular vectors. For $n = 2$ this implies any two non-zero vectors are non-perpendicular and hence (V, f) is uniquely determined and called the *hyperbolic plane.* Let us turn to the case $n \geq 3$. Suppose that $f(u, v) = 1$ and set $U = \langle u, v \rangle$. Then $(U, f|_U)$ is a hyperbolic plane; $U \cap U^\perp = 0$ and by (1.1.2) V is the direct sum of U and $U^\perp$. By (1.1.4)

$$(V, f) \cong (U, f|_U) \oplus (U^\perp, f|_{U^\perp})$$

and by (1.1.3) the second summand is a non-singular $(n - 2)$-dimensional symplectic space to which the induction hypothesis applies. ∎

The direct sum decomposition in (1.1.5 (iii)) will be called a *hyperbolic decomposition.* Let (V, f) be a non-singular n-dimensional symplectic space. By (1.1.5) $n = 2m$ and we can chose a basis $(e_1, e_2, \ldots, e_n)$ called the *hyperbolic basis* such that for $1 \leq i < j \leq n$, we have $f(e_i, e_j) = 1$ if $j - i = m$ and $f(e_i, e_j) = 0$ otherwise. Then the subspaces $\langle e_i, e_{i+m} \rangle$ taken for $i = 1, 2, \ldots, m$ are pairwise perpendicular hyperbolic planes.

Let q be an orthogonal form associated with f. A vector $u \in V$ is said to be *singular* or *non-singular* depending on whether $q(u) = 0$ or $q(u) = 1$. A subspace U of V is said to be *totally singular* if $q(u) = 0$ for every $u \in U$. A subspace U is said to be *totally isotropic* if $f(u, v) = 0$ for every $u, v \in U$.

In the case $n = 2$ it is immediate that the isomorphism type of an orthogonal space (V, f, q) is determined by the number of non-singular vectors which is one or three. These orthogonal spaces of dimension 2 we denote by V_2^+ or V_2^-, respectively. If (e_1, e_2) is a hyperbolic basis of V then V_2^+ is obtained by setting $q(e_1) = q(e_2) = 0$, while V_2^- by setting $q(e_1) = q(e_2) = 1$.

Lemma 1.1.6 *Up to isomorphism there are exactly two orthogonal spaces associated with a non-singular 2-dimensional symplectic space, which are V_2^+ and V_2^-.*

By (1.1.1), (1.1.5), (1.1.6) given a non-singular symplectic space of dimension $n = 2m$ in order to obtain an orthogonal space it is sufficient to define the orthogonal form on each plane in a hyperbolic decomposition by turning this plane into V_2^+ or V_2^-. This immediately shows that the number of isomorphism types of orthogonal spaces is at most m. This number is dramatically reduced by the following.

Lemma 1.1.7 $V_2^- \oplus V_2^- \cong V_2^+ \oplus V_2^+$.

Proof Let (e_1, e_2, e_3, e_4) be a hyperbolic basis of $V_2^- \oplus V_2^-$, so that $\langle e_1, e_2 \rangle$ and $\langle e_3, e_4 \rangle$ are perpendicular hyperbolic planes and $q(e_i) = 1$ for $1 \leq i \leq 4$. Put $d_1 = e_1 + e_3$, $d_2 = e_1 + e_4$, $d_3 = e_2 + e_3 + e_4$, $d_4 = e_1 + e_2 + e_3 + e_4$. Then (d_1, d_2, d_3, d_4) is another hyperbolic basis and $q(d_i) = 0$ for $1 \leq i \leq 4$. Therefore

$$\alpha : e_i \to d_i$$

defines the required isomorphism. ∎

Define V_{2m}^+ to be the direct sum of m copies of V_2^+ and V_{2m}^- to be the direct sum of $m - 1$ copies of V_2^+ and one copy of V_2^-. By (1.1.7) and the paragraph before that lemma we obtain the following.

Lemma 1.1.8 *An orthogonal space associated to a non-singular symplectic space of dimension $n = 2m$ is isomorphic to either V_{2m}^+ or V_{2m}^-.*

In the next section among other things we will see that V_{2m}^+ and V_{2m}^- are non-isomorphic.

1.2 $S_{2m}(2)$, $O_{2m}^+(2)$, and $O_{2m}^-(2)$

Let $n = 2m$ be a positive even integer and (V, f) be a non-singular n-dimensional symplectic space which is uniquely determined up to isomorphism by (1.1.5 (ii)). Define the symplectic group $S_{2m}(2)$ to be the group of invertible linear transformations α of V which preserve f in the sense that

$$f(u, v) = f(\alpha(u), \alpha(v))$$

for all $u, v \in V$.

Lemma 1.2.1 *The following assertions hold:*

   (i) $S_2(2) \cong L_2(2) \cong \mathrm{Sym}_3$;
   (ii) $S_{2m}(2)$ *acts transitively on the set of ordered hyperbolic decompositions of (V, f);*
   (iii)

$$|S_{2m}(2)| = 2^{m^2} \cdot \prod_{i=1}^{m} (2^{2i} - 1).$$

Proof (i) is immediate from the description of the hyperbolic plane while (ii) follows from commutativity and associativity of the direct sum operation. Let u be a non-zero vector of V. Then by (1.1.2) $u^\perp$ is a hyperplane in V and hence there are 2^{n-1} vectors generating with u a hyperbolic plane U. The stabilizer in $S_{2m}(2)$ of the ordered decomposition

$$(V, f) = (U, f|_U) \oplus (U^\perp, f|_{U^\perp})$$

is $S_2(2) \times S_{2m-2}(2)$ and (iii) comes by inductive calculations. ∎

Now we turn to the orthogonal groups and start with the following.

Lemma 1.2.2 *Let $(V, f, q) \cong V_{2m}^\varepsilon$ and*

$$k_\delta^\varepsilon(m) = |\{v \mid v \in V, q(v) = \delta\}|.$$

Then

$$k_0^+(m) = k_1^-(m) = 2^{2m-1} + 2^{m-1}, \quad k_1^+(m) = k_0^-(m) = 2^{2m-1} - 2^{m-1}.$$

Proof We know that $k_0^+(1) = k_1^-(1) = 3$, $k_1^+(1) = k_0^-(1) = 1$. For $m \geq 2$ we represent V_{2m}^ε as direct sum $U \oplus W$, where $U \cong V_{2m-2}^\varepsilon$, $W \cong V_2^+$. Then a vector from V (which might or might not be zero) is singular if and only if its projections into U and W are of the same type (both singular or both non-singular). This provides us with a recurrence relation for the $k_\delta^\varepsilon(m)$'s which in turn gives the result. ∎

By (1.2.2) the spaces V_{2m}^+ and V_{2m}^- are indeed non-isomorphic.

Let $(V, f, q) \cong V_{2m}^\varepsilon$ be an orthogonal space. The *orthogonal group* associated with this space is the group of invertible linear transformations of V which preserve q. This group is denoted by $O_{2m}^\varepsilon(2)$. Since f is uniquely determined by q, both $O_{2m}^+(2)$ and $O_{2m}^-(2)$ are contained in $S_{2m}(2)$.

Lemma 1.2.3 *The following assertions hold:*

(i) $O_2^+(2) \cong \mathrm{Sym}_2$, $O_2^-(2) \cong S_2(2) \cong L_2(2) \cong \mathrm{Sym}_3$;
(ii) *if $m \geq 2$ then $O_{2m}^\varepsilon(2)$ acts transitively on the set of hyperbolic pairs (the pairs $u, v \in V$ such that $q(u) = q(v) = 0$, $f(u,v) = 1$);*
(iii)

$$|O_{2m}^+(2)| = 2^{m^2 - m + 1}(2^m - 1) \cdot \prod_{i=1}^{m-1}(2^{2i} - 1)$$

$$|O_{2m}^-(2)| = 2^{m^2 - m + 1}(2^m + 1) \cdot \prod_{i=1}^{m-1}(2^{2i} - 1).$$

Proof (i) is immediate. If (u, v) is a hyperbolic pair and $U = \langle u, v \rangle$, then

$$(V, f, q) = (U, f|_U, q|_U) \oplus (U^\perp, f|_{U^\perp}, q|_{U^\perp}) \cong V_2^+ \oplus V_{2m-2}^\varepsilon$$

and (ii) follows. The stabilizer of the above decomposition in $O_{2m}^\varepsilon(2)$ is $\mathrm{Sym}_2 \times O_{2m-2}^\varepsilon(2)$. Arguing as in (1.2.1) and making use of (1.2.2) we perform inductive calculations to obtain (iii). ∎

Next we would like to discuss all the orthogonal forms q associated with a given non-singular symplectic form (V, f). Such a form is said to be of *plus* or *minus* type depending on whether (V, f, q) is isomorphic to V_n^+ or to V_n^-.

Lemma 1.2.4 *Let $\mathcal{Q}^+$ and $\mathcal{Q}^-$ be the sets of quadratic forms of plus and minus type respectively associated with f. Then*

(i) *the orthogonal group $O_{2m}^\varepsilon(2)$ permutes transitively both the non-zero singular and the non-singular vectors;*
(ii) *the symplectic group $S_{2m}(2)$ acts transitively both on $\mathcal{Q}^+$ and $\mathcal{Q}^-$;*
(iii) *$|\mathcal{Q}^+| = 2^{2m-1} + 2^{m-1}$ and $|\mathcal{Q}^-| = 2^{2m-1} - 2^{m-1}$.*

Proof (i) and (ii) are easy consequences of (1.2.1 (ii)) and (1.2.3 (ii)). We obtain (iii) comparing the orders of $S_{2m}(2)$ and $O_{2m}^\varepsilon(2)$ given in (1.2.1 (iii)) and (1.2.3 (iii)). ∎

By (1.2.4 (iii)) we have $|\mathcal{Q}^+| + |\mathcal{Q}^-| = 2^{2m} = 2^n$ which perfectly agrees with (1.1.1).

The following well-known result is also easy to establish using induction. Recall that a subspace U of an orthogonal space (V, f, q) is totally singular if $q(u) = 0$ for every $u \in U$.

Lemma 1.2.5 *Let* $(V, f, q) \cong V_{2m}^\varepsilon$ *be an orthogonal space and let* U *be a maximal totally singular subspace in this space. Then* $\dim U = m$ *if* $\varepsilon = +$ *and* $\dim U = m - 1$ *if* $\varepsilon = -$.

1.3   Transvections and Siegel transformations

In this section we discuss some specific automorphisms of orthogonal spaces. Recall some basic terminology. If U is a $GF(2)$-vector space, u is a non-zero vector in U and W is a hyperplane containing u. Then the *transvection* with centre u and axis W is the element of the general linear group of U which fixes W vectorwise and maps $v \in U \setminus W$ onto $v + u$. It is well-known that $L_n(2)$ is generated by its transvections for $n \geq 2$.

Lemma 1.3.1 *Let* $(V, f, q) \cong V_{2m}^\varepsilon$. *Then the following hold:*

(i) *If* u *is a non-singular vector in* V, *then the orthogonal transvection*

$$t_u : v \mapsto v + f(u, v)u$$

(which is the transvection with centre u *and axis* $u^\perp$*) is an element of* $O_{2m}^\varepsilon(2)$;

(ii) *If* $u, v \in V$ *and* $T = \langle u, v \rangle$ *is a totally singular 2-space then the Siegel transformation*

$$s_T : w \mapsto w + f(u, w)v + f(v, w)u$$

is an element of $O_{2m}^\varepsilon(2)$.

Proof It can be checked directly that t_u and s_T preserve the orthogonal form. ∎

Notice that the Siegel transformation s_T does not depend on the choice of the generating vectors u, v. Whenever $T \subseteq w^\perp$ the vector w is fixed by s_T and if $T \not\subseteq w^\perp$ then s_T shifts w by the unique non-zero vector in T which is perpendicular to w.

If $m \geq 3$, then $O_{2m}^\varepsilon(2)$ is generated by the transvections t_u taken for all the non-singular vectors u, while the Siegel transformations s_T taken for all totally singular 2-dimensional subspaces, generate the commutator subgroup $\Omega_{2m}^\varepsilon(2)$ which has index 2 in $O_{2m}^\varepsilon(2)$ (cf. Taylor (1992) for details). It is well known and easy to check that the orthogonal transvections form a class of 3-transpositions in $O_{2m}^\varepsilon(2)$. More precisely, the following lemma holds.

Lemma 1.3.2 *Let u and v be non-singular vectors in V. Then t_u and t_v commute whenever u and v are perpendicular; otherwise their product is of order 3.*

Now we turn to the Siegel transformations.

Lemma 1.3.3 *Let U be a totally singular subspace in V of dimension $k \geq 2$. Then*

(i) *if w is a singular vector outside $U^\perp$ and $u \in w^\perp \cap U$, then $s_{\langle u,w \rangle}$ induces on U the transvection whose centre is u and whose axis is $w^\perp \cap U$;*

(ii) *if W is a totally singular subspace which complements $U^\perp$ in V then the Siegel transformations $s_{\langle u,w \rangle}$, taken for all perpendicular $u \in U$ and $w \in W$, generate in $O_{2m}^\varepsilon(2)$ a subgroup $L \cong L_k(2)$ for which U and W are the natural and the dual natural modules.*

Proof Statement (i) follows directly by the definition. Notice that L acts trivially on $\langle U, W \rangle^\perp$, so it is a subgroup of the stabilizer in $O_{2k}^+(2)$ of two disjoint maximal totally singular subspaces in the orthogonal $2k$-space. Since there is a duality between U and W this stabilizer acts faithfully on each of U and W. Therefore $L \leq L_k(2)$. Now (ii) follows from (i) and the fact that $L_k(2)$ is generated by its transvections. ∎

In order to establish one further property of Siegel transformations we need to recall a well-known characterization of the exterior squares (cf. Lemma 3.1.3 in Ivanov and Shpectorov (2002)).

Lemma 1.3.4 *Let U be a k-dimensional $GF(2)$-space, $k \geq 2$, let $L \cong L_k(2)$ be the general linear group of U and let $\begin{bmatrix} U \\ 2 \end{bmatrix}$ be the set of 2-subspaces in U. Let W be a non-zero $GF(2)$-module for L which satisfies the following:*

(i) *there is a mapping $d : \begin{bmatrix} U \\ 2 \end{bmatrix} \to W$ which commutes with the action of L and the image of d spans W;*

(ii) *whenever $T_1, T_2, T_3 \in \begin{bmatrix} U \\ 2 \end{bmatrix}$ are pairwise distinct with $\dim \langle T_1, T_2, T_3 \rangle = 3$ and $\dim (T_1 \cap T_2 \cap T_3) = 1$, the equality $d(T_1) + d(T_2) + d(T_3) = 0$ holds.*

Then $\dim W = k(k-1)/2$, W is irreducible and W is isomorphic to the exterior square $\bigwedge^2 U$ of U.

Lemma 1.3.5 *Let U be a totally singular subspace in V of dimension $k \geq 2$. Then the Siegel transformations s_T taken for all 2-subspaces T in U generate in $O_{2n}^\varepsilon(2)$ an elementary abelian subgroup Q of rank $k(k-1)/2$. If $L \cong L_k(2)$ as in (1.3.3) then Q and the exterior square of U are isomorphic as L-modules.*

Proof We apply (1.3.4) for $d : T \to s_T$. It is an easy exercise to check that the equality in (1.3.4 (ii)) holds, hence the result. ∎

1.4   Exterior square via symplectic forms

Let U be a $GF(2)$-space and let f be a symplectic form on U which is no longer assumed to be non-singular. Then

$$U^\perp = \{u \mid f(u,v) = 0 \text{ for every } v \in U\}$$

is the *radical* of the symplectic space (U, f). Consider the factor space $\bar{U} = U/U^\perp$ and the *induced form* $\bar{f}$ on $\bar{U}$ defined by

$$\bar{f}(\bar{u}, \bar{v}) = f(u, v),$$

where $\bar{u} = u + U^\perp$, $\bar{v} = v + U^\perp$.

Lemma 1.4.1 *The following assertions hold:*

(i) *$f(u_1, v_1) = f(u, v)$ whenever $u_1 \in \bar{u}$, $v_1 \in \bar{v}$, so that $\bar{f}$ is well defined;*
(ii) *$(\bar{U}, \bar{f})$ is a non-singular symplectic space;*
(iii) *(U, f) is uniquely determined by $U^\perp$ and $(\bar{U}, \bar{f})$.*

Proof By bilinearity and since $u + u_1, v + v_1 \in U^\perp$ we obtain (i). The form $\bar{f}$ is symplectic because f is such. Finally (iii) is immediate from (i).   ∎

The codimension of the radical is called the *rank* of the form. When the non-singularity condition is dropped the set of all the symplectic forms on a given space U is closed under pointwise addition. This is stated in the following lemma whose proof is immediate.

Lemma 1.4.2 *Let f_1 and f_2 be symplectic forms on U. Then $f_1 + f_2$ defined by*

$$f_1 + f_2 : (u, v) \mapsto f_1(u, v) + f_2(u, v)$$

is a symplectic form on U.

The space $\mathcal{S}(U)$ of all the symplectic forms on a k-dimensional space U is $\binom{k}{2}$-dimensional since having a fixed basis B of U in order to define such a form it is necessary and sufficient to assign the value 0 or 1 to every pair of distinct vectors in B.

Lemma 1.4.3 *The space $\mathcal{S}(U)$ as a module for $GL(U)$ is isomorphic to the exterior square of the dual U^* of U.*

Proof By (1.1.5) the radical of a symplectic form has even codimension. Furthermore by the proof of (1.1.5) there is only one non-singular 2-dimensional symplectic space. Therefore the symplectic spaces of rank 2 are indexed by the codimension 2 subspaces in U (that is by the 2-dimensional subspaces in U^*). Let U^1 be a hyperplane in U, let U^3 be a codimension 3 subspace in U and let f_1, f_2, f_3 be the symplectic forms of rank 2 whose radicals contain U^3 and are contained in U^1. Then it is straightforward to check that

$$f_1 + f_2 + f_3$$

is the zero form and the result follows from the characterization of the exterior squares in (1.3.4), since the forms of rank 2 span the space of all the symplectic forms. ∎

We conclude this section with another well known result.

Lemma 1.4.4 *The spaces $\bigwedge^2 U$ and $\bigwedge^2 U^*$ are dual to each other as $GL(U)$-modules.*

Proof By (1.3.4) there are mappings

$$d_1 : \begin{bmatrix} U \\ 2 \end{bmatrix} \to \bigwedge {}^2U, \quad d_2 : \begin{bmatrix} U^* \\ 2 \end{bmatrix} \to \bigwedge {}^2U^*$$

With a 2-dimensional subspace V^* in U^* we bijectively associate a codimension 2 subspace $K(V^*)$ in U which is the intersection of the kernels of the linear forms in V^*. It is easy to check that there is a unique bilinear mapping

$$\delta : \bigwedge {}^2U \times \bigwedge {}^2U^* \to GF(2)$$

such that $\delta(d_1(W), d_2(V^*)) = 0$ if and only if $W \cap K(V^*) \neq 0$. Then δ provides us the required $GL(U)$-invariant duality. ∎

1.5 Witt's theorem

Let U be an n-dimensional $GF(2)$-space and let f be a symplectic form on U (which might or might not be non-singular). Let S_f be the *symplectic group* of the space (U, f), which is the group of invertible linear transformations of U preserving the form f. Then $S_f \cong S_{2m}(2)$ if $n = 2m$ and f is non-singular. Let W be a complement to $U^\perp$ in U (by (1.1.5) the dimension of W is even, say $2k$). Then $(W, f|_W)$ is a non-singular symplectic space and

$$(U, f) = (W, f|_W) \oplus (U^\perp, f|_{U^\perp}),$$

where the form $f|_{U^\perp}$ is identically zero. Since the isomorphism type of (U, f) is uniquely determined by the dimensions of the subspaces in the above direct sum decomposition, we have the following.

Lemma 1.5.1 *In terms introduced in the paragraph before the lemma the following assertions hold:*

(i) *S_f induces the full linear group of $U^\perp$;*

(ii) *S_f permutes transitively the complements W to $U^\perp$ in U;*

(iii) *the stabilizer in S_f of a complement W to $U^\perp$ induces on W the symplectic group $S_{2k}(2)$, where $2k = \dim W$.*

The following two lemmas are special cases of what is called Witt's theorem (compare Aschbacher (1986) and Taylor (1992)).

Lemma 1.5.2 *Let (U, f) be a non-singular symplectic space of dimension $n = 2m$. Then two subspaces $U^{(1)}$ and $U^{(2)}$ of U are contained in the same orbit of the symplectic group $S_{2m}(2)$ if and only if $(U^{(1)}, f|_{U^{(1)}})$ and $(U^{(2)}, f|_{U^{(2)}})$ are isomorphic.*

Proof The 'only if' part is obvious. Suppose that the restricted symplectic spaces are isomorphic, let l be the dimension and let $2k$ be the rank of each of them. We claim that for $i = 1$ and 2 there exists a decomposition of U into pairwise perpendicular hyperbolic planes such that (a) k planes are contained in $U^{(i)}$; (b) $l - 2k$ planes intersect the radical of $(U^{(i)}, f|_{U^{(i)}})$ in 1-dimensional subspaces; (c) the remaining pairs are disjoint from $U^{(i)}$. The claim is easy to establish arguing inductively as in the proof of (1.2.1). It is also clear that the claim implies the assertion.   ■

Combining (1.5.1) and (1.5.2) it is easy to establish the following.

Lemma 1.5.3 *Let (U, f) be an arbitrary symplectic space. Then the orbit of a subspace W of U under the group S_f is uniquely determined by the isomorphism type of $(W, f|_W)$ and by $\dim (W \cap U^\perp)$.*

1.6   Extraspecial 2-groups

Let E be a 2-group with centre Z and put $\bar{E} = E/Z$. Then E is said to be an *extraspecial 2-group* if $|Z| = 2$ and $\bar{E}$ is elementary abelian.

A beautiful relationship between the extraspecial 2-groups and the orthogonal spaces over $GF(2)$ is established by the following result.

Proposition 1.6.1 *Let E be an extraspecial 2-group with centre Z and $\bar{E} = E/Z$. Consider Z as the field $GF(2)$ of two elements and $\bar{E}$ as a $GF(2)$-vector space. Define the mappings $f : \bar{E} \times \bar{E} \to Z$ and $q : \bar{E} \to Z$ by*

$$f(uZ, vZ) = [u, v], \quad q(uZ) = u^2.$$

Then

(i) *f and q are well defined;*
(ii) *f is a non-singular symplectic form;*
(iii) *q is an orthogonal form associated with f.*

Proof Since Z is the centre of E we have (i). The basic relation for the commutator of a product reads as follows:

$$[uv, w] = [u, w]^v [v, w] = [u, w][v, w]$$

and therefore f is linear in the first coordinate. Clearly $[u, u] = 1$ and $[u, v] = [v, u]$ since Z has order 2. Therefore (ii) is established. Finally

$$(uv)^2 = u^2 v^2 [u, v]$$

and (iii) follows. ∎

By (1.6.1) $s(E) := (\bar{E}, f, q)$ is a non-singular orthogonal space and by (1.1.5) it is isomorphic to V_{2m}^{ε}, where $\varepsilon \in \{+, -\}$ and m is a positive integer (notice that $|E| = 2^{2m+1}$). It turns out that the isomorphism type of $s(E)$ determines that of E. One way of proving this is to show that $s(E)$ uniquely determines the homology class of 2-cocycles which specifies E as an extension of Z by $\bar{E}$ (cf. chapter 5 in Frenkel *et al.* (1988)). Here we follow a more geometric approach and refer to section 23 in Aschbacher (1986) for the missing proofs.

Let E_1 and E_2 be extraspecial 2-groups whose centres Z_1 and Z_2 are generated by elements z_1 and z_2, respectively. Then the *central product* $E_1 * E_2$ of these groups is the quotient of the direct product

$$E_1 \times E_2 = \{(e_1, e_2) \mid e_1 \in E_1, e_2 \in E_2\}$$

over the normal subgroup of order 2 generated by the element (z_1, z_2). The central product operation is commutative and associative (provided that the result is considered up to isomorphism) and therefore one can consider the central product of any number of extraspecial groups.

Lemma 1.6.2 *If E_1 and E_2 are extraspecial 2-groups, then $E_1 * E_2$ is also extraspecial and*

$$s(E_1 * E_2) \cong s(E_1) \oplus s(E_2),$$

where $\oplus$ stands for the orthogonal direct sum.

Lemma 1.6.3 *Let E be an extraspecial group with centre Z, let $Z \leq U \leq E$, let $\bar{U}$ be the image of U in $\bar{E}$, and let $s(E) = (\bar{E}, f, q)$. Then*

(i) *U is abelian if and only if $\bar{U}$ is totally isotropic with respect to f;*
(ii) *U is elementary abelian if and only if $\bar{U}$ is totally singular with respect to q;*
(iii) *U is extraspecial if and only if $(\bar{U}, f|_{\bar{U}})$ is non-singular.*

The following lemma supplies us with the classification of the smallest extraspecial 2-groups.

Lemma 1.6.4 *An extraspecial group of order 8 is isomorphic either to the dihedral group D_8 or to the quaternion group Q_8.*

The next lemma forms the inductive basis for the general classification.

Lemma 1.6.5 *Let E be an extraspecial group and suppose that $s(E)$ is the orthogonal sum of two non-singular subspaces $\bar{V}$ and $\bar{U}$. Let V and U be the preimages in E of $\bar{V}$ and $\bar{U}$, respectively. Then $E \cong V * U$.*

Now combining (1.6.2), (1.6.3), (1.6.4), (1.6.5) with (1.1.6), (1.1.7) and (1.1.8) we obtain the main result of the section.

Proposition 1.6.6 *Let E be an extraspecial 2-group and $s(E) = (\bar{E}, f, q)$. Then one of the following holds.*

(i) $s(E) \cong V_{2m}^+$ *for some positive integer m, E is isomorphic to the central product of m copies of D_8, and $|E| = 2^{2m+1}$;*

(ii) $s(E) \cong V_{2m}^-$ *for some positive integer m, E is isomorphic to the central product of $m - 1$ copies of D_8 together with one copy of Q_8, and $|E| = 2^{2m+1}$.*

As a by-product of the above proposition we obtain the isomorphism $D_8 * D_8 \cong Q_8 * Q_8$.

The extraspecial 2-groups of order 2^{2m+1} corresponding to the possibilities (i) and (ii) in (1.6.6) are denoted by 2_+^{1+2m} and 2_-^{1+2m}, respectively and are said to be of type $+$ and $-$, respectively. Thus for every odd power of 2 greater than 2, up to isomorphism there are exactly two extraspecial groups of that order.

Lemma 1.6.7 *Let $E \cong 2_\varepsilon^{1+2m}$ be an extraspecial group of type $\varepsilon \in \{+, -\}$ with centre Z and U be a maximal elementary abelian subgroup in E. Then*

(i) $U \cong 2^{m+\delta}$, *where $\delta = 1$ if $\varepsilon = +$ and $\delta = 0$ otherwise;*

(ii) *U contains Z;*

(iii) *U is normal in E and if $\varepsilon = +$ then $C_E(U) = U$;*

(iv) *the group E acting on U by conjugation induces the elementary abelian group of order $2^{m-1+\delta}$ generated by the transvections with centre z (which is the generator of Z).*

Proof (i) follows from (1.2.5) and (1.6.3 (ii)) while (ii) follows from the definition of U. Since U contains Z, which is also the commutator subgroup of E, it is normal in E. This and (1.1.2) give (iii). Let $x \in E \setminus U$. Then by (iii), x acts on U as a transvection with axis z and (iv) follows by the order consideration. $\blacksquare$

Lemma 1.6.8 *Let $E \cong 2_\varepsilon^{1+2m}$. Then $\mathrm{Aut}\,(E)$ is an extension of $\mathrm{Inn}\,(E) \cong E/Z = \bar{E} \cong 2^{2m}$ by the automorphism group $O_{2m}^\varepsilon(2)$ of the orthogonal space $(\bar{E}, f, q)$, where f and q are defined as in (1.6.1).*

Proof It is clear that an automorphism of E preserves f and q. On the other hand every automorphism of V_2^+ or V_2^- is realized as an automorphism of D_8 and Q_8, respectively. $\blacksquare$

The extension of 2^{2m} by O_{2m}^ε in (1.6.8) does not split when $m \geq 3$ (Griess 1973).

1.7   The space of forms

As in the previous section let (V, f) be a non-singular symplectic space of dimension $n = 2m$, $m \geq 1$, let $\mathcal{Q}^+$ and $\mathcal{Q}^-$ be the sets of orthogonal forms of plus and minus types associated with f.

Lemma 1.7.1 *Let V^* be the dual space of V (which is the space of linear functions on V). Then $V^* \cup \mathcal{Q}^+ \cup \mathcal{Q}^-$ is closed under the pointwise addition.*

Proof A $GF(2)$-valued function p on V belongs to V^* if and only if

$$p(u) + p(v) + p(u + v) = 0$$

for all $u, v \in V$. Comparing this with the definition of an orthogonal form we immediately get the result. ∎

For every $w \in V$ the mapping $l_w : v \mapsto f(v, w)$ is an element of V^* and since f is non-singular, the mapping $\nu : w \mapsto l_w$ is an isomorphism of V onto V^*.

Lemma 1.7.2 *Let* $q \in \mathcal{Q}^+ \cup \mathcal{Q}^-$ *and* $w \in V$. *Then* $q + l_w \in \mathcal{Q}^+ \cup \mathcal{Q}^-$ *and the types (which are plus or minus) of* q *and of* $q + l_w$ *coincide if and only if* $q(w) = 0$.

Proof Let $q \in \mathcal{Q}^-$ and $F \cong O_{2m}^-(2)$ be the corresponding orthogonal group. By (1.2.4 (i)) F permutes transitively the $2^{2m-1} + 2^{m-1}$ vectors w in V such that $q(w) = 1$. Therefore the type of $q + l_w$ is independent of the choice of such w. Since by (1.2.4 (iii)) there are only $2^{2m-1} - 2^{m-1}$ forms of minus type, $q + l_w$ is of plus type and the result follows. ∎

1.8 $S_4(2)$ and Sym_6

There is a simple way to associate a non-singular $2m$-dimensional symplectic space $s(P_{2m+2}) = (V, f)$ with a set P_{2m+2} of $2m + 2$ elements. The vectors in V are the unordered partitions of P_{2m+2} into two even subsets (with $\{P_{2m+2}, \emptyset\}$ being the zero vector); the sum of two partitions $\{A_1, A_2\}$ and $\{B_1, B_2\}$ is the partition $\{C_1, C_2\}$, where

$$C_1 = (A_i \cup B_i) \setminus (A_i \cap B_i) \text{ for } i \in \{1, 2\},$$
$$C_2 = (A_i \cup B_j) \setminus (A_i \cap B_j) \text{ for } \{i, j\} = \{1, 2\}.$$

It is straightforward to check that this operation turns V into a $GF(2)$-vector space of dimension $2m$. Define $f(\{A_1, A_2\}, \{B_1, B_2\})$ to be the size of $|A_i \cap B_j|$ taken modulo 2. This definition is independent on the choice of $1 \le i, j \le 2$ and provides us a non-singular symplectic form. The symmetric group Sym_{2m+2} of P_{2m+2} acts naturally on V preserving both the vector space structure and the symplectic form f which gives the following.

Lemma 1.8.1 $\mathrm{Sym}_{2m+2} \le S_{2m}(2)$.

We check using (1.2.1) that the order of $S_4(2)$ is exactly 6!, which is of course the order of Sym_6 and the above inclusion gives the famous isomorphism.

Lemma 1.8.2 $S_4(2) \cong \mathrm{Sym}_6$.

The above consideration also provides us with a useful combinatorial model $s(P_6)$ of the non-singular 4-dimensional symplectic space (V, f). We modify this description to obtain a new one in terms of the group Sym_6. The non-trivial even partitions of P_6 are identified with the set of 2-element subsets of P_6 and in turn the latter set is identified with the set T of transpositions in Sym_6 (with respect to the action on P_6). Then for $t_1, t_2 \in T$ with $t_1 \neq t_2$ we have

(∗) $f(t_1, t_2) = 0$ if and only if t_1 and t_2 commute and $t_3 = t_1 + t_2$ is defined
so that $\langle t_1, t_2, t_3 \rangle = \langle t_1, t_2 \rangle$.

This gives a model $s(T)$ of the non-singular 4-dimensional symplectic space in terms of the transpositions in Sym_6. This model is merely the reformulation of the structure of $s(P_6)$ along with the canonical correspondence between the transpositions and the non-trivial even partitions of P_6.

We know that Sym_6 possesses an outer automorphism which maps T onto the set D of products of three pairwise commuting elements of T. Therefore there is a model $s(D)$ of the non-singular 4-dimensional symplectic space where the symplectic form and the vector space structure on $D \cup 0$ are defined by the rule (∗). By (1.1.5) we have the isomorphism between $s(T)$ and $s(D)$ which one can use to reprove the existence of the outer automorphism of Sym_6. Notice that $s(D)$ is canonically isomorphic to $s(R_6)$, where R_6 is a 6-element set on which Sym_6 acts in such a way that the elements from D are transpositions. Consider Sym_6 as an abstract group and let P_6 and R_6 be two sets of size 6 on which Sym_6 acts in non-equivalent ways permuted by an outer automorphism. Then we have two non-singular symplectic 4-spaces

$$s(P_6) = s(T) \text{ and } s(R_6) = s(D)$$

with Sym_6 inducing the full symplectic group on each of the spaces.

The following lemma describes the orthogonal spaces associated with $s(P_6) = s(T)$.

Lemma 1.8.3 *Let $(V, f) = s(P_6) = s(T)$ and let q be an orthogonal form associated with f. Then one of the following holds:*

(i) *q is of plus type and there is a unique partition $\{A, C\}$ of P_6 into two triples such that $q(t) = 1$ for $t \in T$ if and only if t stabilizes the partition;*
(ii) *q is of minus type and there is a unique element $a \in P_6$ such that $q(t) = 1$ for $t \in T$ if and only if t stabilizes a.*

The orthogonal spaces in (1.8.3 (i)) and (1.8.3 (ii)) will be denoted by $Q(P_6, \{A, C\})$ and $Q(P_6, a)$, respectively. By elementary calculations one can see that there are ten forms of plus type, each having exactly six non-singular vectors and six forms of minus type, each having exactly ten non-singular vectors. This is certainly consistent with (1.2.4 (iii)). The stabilizer in Sym_6 of an element $a \in P_6$ is the symmetric group Sym_5 while the stabilizer of a partition of P_6 into two triples is the wreath product $\mathrm{Sym}_3 \wr \mathrm{Sym}_2$. This gives two further exceptional isomorphisms.

Lemma 1.8.4 $O_4^+(2) \cong \mathrm{Sym}_3 \wr \mathrm{Sym}_2$ *and* $O_4^-(2) \cong \mathrm{Sym}_5$.

The stabilizer in Sym_6 of an element $a \in P_6$ is not conjugate to the stabilizer of an element $b \in R_6$ (they are conjugate in the automorphism group of Sym_6).

On other hand the stabilizer of a partition of P_6 into two triples is the normalizer of a Sylow 3-subgroup and hence it is conjugate to the stabilizer of a partition of R_6 into two triples.

Lemma 1.8.5 *Let $X \cong \mathrm{Sym}_6$. Then*

(i) *X contains two classes of subgroups Sym_5 with representatives F_a and F_b which are the orthogonal groups of minus type for the symplectic spaces $s(P_6)$ and $s(R_6)$, respectively;*

(ii) *X contains a unique class of subgroups $\mathrm{Sym}_3 \wr \mathrm{Sym}_2$ with representative E which is the orthogonal group of plus type for both the symplectic spaces $s(P_6)$ and $s(R_6)$;*

(iii) *F_a acting on the non-zero vectors of $s(P_6)$ has two orbits with lengths 5 and 10 formed by vectors which are singular and non-singular with respect to the quadratic form stabilized by F_a;*

(iv) *the action of F_a on the set of non-zero vectors in $s(R_6)$ is transitive.*

There is an outer automorphism α of Sym_6 which permutes F_a and F_b and an outer automorphism β which normalizes $E \cong \mathrm{Sym}_3 \wr \mathrm{Sym}_2$.

1.9 $\Omega_6^+(2)$, Alt$_8$, and $L_4(2)$

In the case when $2m+2$ is divisible by 4 (that is when m is odd, say $m = 2k+1$) the inclusion in (1.8.1) can be refined. Let $V(P_{4k+4}) = (V, f)$ be the symplectic space defined in the beginning of Section 1.8 and put

$$q(\{A_1, A_2\}) = \frac{|A_1|}{2}$$

taken modulo 2. It is easy to check that q is an orthogonal form associated with f. Furthermore, q is of plus or minus type depending on whether k is odd or even, respectively. This form is clearly preserved by the symmetric group Sym_{4k+4} and we have the following.

Lemma 1.9.1 *For every positive integer k the inclusion*

$$\mathrm{Sym}_{4k+4} \leq O_{4k+2}^{\varepsilon}(2)$$

holds, where $\varepsilon = +$ or $-$ according to whether k is odd or even.

By (1.2.3 (iii)) the order of $O_6^+(2)$ is 8! which is of course the order of Sym_8 and hence

Lemma 1.9.2 *$O_6^+(2) \cong \mathrm{Sym}_8$.*

There is yet another model of the orthogonal 6-dimensional space of plus type which leads to the next exceptional isomorphism. Let U be a 4-dimensional $GF(2)$-space, let V be the exterior square of U and

$$(u, v) \mapsto u \wedge v$$

be the natural bilinear mapping of $U \times U$ onto V. Define a $GF(2)$-valued function q on V by setting $q(w) = 0$ if and only if $w = u \wedge v$ for some $u, v \in U$. The following result is well-known and easy to check.

Lemma 1.9.3 *In the above terms q is an orthogonal form of plus type on V associated with a non-singular symplectic form.*

Since U is 4-dimensional, V is isomorphic to the exterior square of the dual of U. Hence along with $L_4(2)$ the contragredient automorphism τ of that group acts on V and preserves the above defined orthogonal form q. Since the order of $L_4(2).\langle \tau \rangle$ is also 8! we have the next isomorphism.

Lemma 1.9.4 $O_6^+(2) \cong L_4(2).\langle \tau \rangle$.

Since $\Omega_6^+(2)$, Alt_8, and $L_4(2)$ are the only index 2 subgroups in $O_6^+(2)$, Sym_8, and $L_4(2).\langle \tau \rangle$, respectively, we also have the following.

Lemma 1.9.5 $\Omega_6^+(2) \cong \mathrm{Alt}_8 \cong L_4(2)$.

1.10 $L_2(7)$ and $L_3(2)$

Let $P_8 = \{\infty, 0, 1, \ldots, 6\}$ be the projective line over the field $GF(7)$ of seven elements. The stabilizer of the projective line structure in the symmetric group of P_8 is $PGL_2(7)$ and the latter contains the group $L_2(7)$ with index 2. The model $Q(P_8)$ of 6-dimensional orthogonal space in Section 1.9 provides us with an embedding

$$PGL_2(7) \leq O_6^+(2).$$

In this section we refine this embedding geometrically.

Lemma 1.10.1 *Let $(V, f, q) = Q(P_8)$ and let $O_6^+(2)$ be the corresponding orthogonal group. Then there is a pair I_1, I_2 of disjoint maximal totally singular subspaces in V such that*

   (i) *$L_2(7)$ is the joint stabilizer in $O_6^+(2)$ of I_1 and I_2;*
   (ii) *$PGL_2(7)$ is the stabilizer in $O_6^+(2)$ of the unordered pair $\{I_1, I_2\}$.*

In particular

   (iii) *$L_2(7) \cong L_3(2)$;*
   (iv) *$PGL_2(7) \cong L_3(2).\langle \tau \rangle$, where τ is the contragredient automorphism.*

Proof Let $Q = \{1, 2, 4\}$ and $N = \{3, 5, 6\}$ be the set of non-zero squares and the set of non-squares in $GF(7)$, so that

$$P_8 = \{\infty\} \cup \{0\} \cup Q \cup N.$$

Put $u_1 = \{\{\infty\} \cup Q, \{0\} \cup N\}$ and $u_2 = \{\{\infty\} \cup N, \{0\} \cup Q\}$ and define I_1 and I_2 to be the subspaces in V generated by the images under $L_2(7)$ of u_1

and u_2, respectively. Then direct check shows that I_1 and I_2 are as required. Furthermore an element from $PGL_2(7) \setminus L_2(7)$ permutes I_1 and I_2. Standard calculations show that

$$|L_2(7)| = |L_3(2)| = 2^3 \cdot 3 \cdot 7.$$

Since $L_3(2)$ is the stabilizer in $O_6^+(2)$ of the (ordered) direct sum vector space decomposition $V = I_1 \oplus I_2$, we obtain (i), (ii), and (iii). Since $I_1 \cap u_2^\perp$ is a hyperplane in I_1, there is a duality between I_1 and I_2, which gives (iv).   ∎

Exercises

1. Calculate the number of totally singular k-dimensional subspaces in V_{2m}^ε.
2. Show that the actions of $S_{2m}(2)$ on $\mathcal{O}^+$ and $\mathcal{O}^-$ are doubly transitive.
3. Check that the orthogonal transvections in $O_{2m}^\varepsilon(2)$ form a class of 3-transpositions.
4. Generalize (1.3.4) to the case of an arbitrary exterior power of U.
5. Show that the orthogonal form q in (1.6.1) determines the homology class of 2-cocycles which specifies E as an extension of Z by $\bar{E}$.
6. Apply the isomorphism $\mathrm{Sym}_6 \cong S_4(2)$ to recover an outer automorphism of Sym_6.

2

$O_{10}^+(2)$ AS A PROTOTYPE

We will construct J_4 as a completion group which is constrained at level 2 of a certain amalgam $\mathcal{G} = \{G^{[0]}, G^{[1]}\}$. In fact J_4 is the only such completion, which proves the uniqueness of J_4. The amalgam $\mathcal{G}$ is a modification of a classical amalgam $\mathcal{H} = \{H^{[0]}, H^{[1]}\}$ contained in the orthogonal group $O_{10}^+(2)$. In this chapter we illustrate our strategy by showing that $O_{10}^+(2)$ is the only completion group of $\mathcal{H}$ which is constrained at level 2. This reproves a special case of a classical result by P. J. Cameron and C. E. Praeger.

2.1 Dual polar graph

Let $(V, f, q) \cong V_{10}^+$ be a 10-dimensional orthogonal space of plus type. This means that V is a 10-dimensional $GF(2)$-space, f is a non-singular symplectic form on V and q is an orthogonal form of plus type associated with f. Let $H \cong O_{10}^+(2)$ be the automorphism group of (V, f, q). We have mentioned (referring to Taylor (1992) for a proof) that H is generated by the transvections

$$t_u : v \mapsto v + f(u, v)u$$

taken for all non-singular vectors $u \in V$, while the commutator subgroup $H' \cong \Omega_{10}^+(2)$ (which is non-abelian simple with index 2 in H) is generated by the Siegel transformations

$$s_U : w \mapsto w + f(u, w)v + f(v, w)u$$

taken for all totally singular planes (2-subspaces) $U = \langle u, v \rangle$ in V.

Let $\mathcal{O} = \mathcal{O}^+(10, 2)$ be the dual polar space associated with the orthogonal space under consideration. Then for $0 \leq i \leq 4$ the elements of type i in $\mathcal{O}$ are the $(5 - i)$-dimensional totally singular subspaces in V (with respect to q) with two such subspaces being incident if one of them contains the other one. The geometry $\mathcal{O}$ belongs to the following diagram:

As usual above a node on the diagram we put the *type* of the corresponding elements of the geometry. The leftmost edge is the geometry of the vertices and edges of the complete bipartite graph $K_{3,3}$.

Let $\Omega = D^+(10, 2)$ be the dual polar graph. The vertices of Ω are the maximal (5-dimensional) totally singular subspaces, the edges are the premaximal (4-dimensional) totally singular subspaces in V with the incidence relation defined via inclusion. If U is an edge of Ω then $U^\perp/U \cong V_2^+$ and the latter contains exactly two 1-dimensional totally isotropic subspaces whose preimages are precisely the vertices incident to U. Therefore Ω is indeed an undirected graph. Clearly the intersection of two distinct 5-dimensional subspaces is at most 4-dimensional, therefore Ω contains no multiple edges.

It is well-known and easy to prove (cf. Brouwer et al. (1989)) that Ω is distance-transitive with respect to the action of H with the following intersection array:

$$i(\Omega) = \{31, 30, 28, 24, 16; 1, 3, 7, 15, 31\}.$$

In particular Ω is connected bipartite of valency 31 (the stabilizer of the bipartition is exactly $H' \cong \Omega_{10}^+(2)$).

Let $x \in \Omega$ be a vertex and $\Omega(x)$ be the set of neighbours of x in Ω. Then $\Omega(x)$ is in a canonical correspondence with the set of hyperplanes of x (treated as a 5-subspace) that is with the non-zero vectors of the dual of x. Therefore there is a natural structure Π_x of the projective space of rank 4 over $GF(2)$ defined on $\Omega(x)$. This space is canonically isomorphic to the residue of x in $\mathcal{O}$.

The geometry $\mathcal{O}$ possesses a nice reformulation in terms of specific subgraphs in Ω called *geometric subgraphs*. For an element U of type i in $\mathcal{O}$ let $\Omega^{[i]}$ denote the subgraph in Ω induced by the vertices which contain U (as subspaces of V). For $0 \leq i \leq 4$ let $\mathcal{S}^{[i]}$ denote the set of all the subgraphs $\Omega^{[i]}$. The following statement is an easy consequence of the theory of locally projective graphs exposed in chapter 9 in Ivanov (1999). In the considered situation everything can be checked directly by a straightforward calculation in the underlying orthogonal space.

Lemma 2.1.1 *The following assertions hold:*

  (i) *$\mathcal{O}$ is canonically isomorphic to the geometry in which $\mathcal{S}^{[i]}$ is the set of elements of type i for $0 \leq i \leq 4$ and the incidence is by the symmetric inclusion;*
  (ii) *$\mathcal{S}^{[0]}$ is the set of vertices while $\mathcal{S}^{[1]}$ is the set of edges of Ω;*
  (iii) *if $2 \leq i \leq 4$ then*
      (a) *the subgraph $\Omega^{[i]}$ is connected of valency $2^i - 1$;*
      (b) *$\Omega^{[i]}$ isomorphic to the dual polar graph $D^+(2i, 2)$;*
      (c) *if $x \in \Omega^{[i]}$ then $\Omega^{[i]}(x)$ (the set of neighbours of x in $\Omega^{[i]}$) corresponds to a subspace in Π_x of projective dimension i;*
  (iv) *whenever $\Omega^{[i]}, \Xi^{[i]} \in \mathcal{S}^i$, $x \in \Omega^{[i]} \cap \Xi^{[i]}$ and $\Omega^{[i]}(x) = \Xi^{[i]}(x)$, the equality $\Omega^{[i]} = \Xi^{[i]}$ holds;*
  (v) *if $x, y \in \Omega$ and $d_\Omega(x, y) = i$, for $1 \leq i \leq 3$, then there is a unique subgraph $\Omega^{[i]}$ in $\mathcal{S}^{[i]}$ which contains both x and y;*
  (vi) *if $\Omega^{[2]} \in \mathcal{S}^{[2]}$ and $y \in \Omega$ then there is a unique vertex in $\Omega^{[2]}$ closest to y in Ω.*

Let Φ be a maximal flag in $\mathcal{O}$. Then Φ can be treated as a chain

$$0 < U_1 < U_2 < U_3 < U_4 < U_5$$

of totally singular subspaces, where U_j is j-dimensional, or as a chain

$$\Omega^{[0]} = \{x\} \subset \Omega^{[1]} = \{x, y\} \subset \cdots \subset \Omega^{[4]}$$

of the geometric subgraphs in Ω where $\Omega^{[i]}$ is induced by the vertices containing U_{5-i}. Notice that the geometric subgraphs are *convex* in the sense that whenever it contains a pair of vertices it contains all the shortest paths joining them.

Let $H^{[i]}$ be the stabilizer of U_{5-i} (equivalently of $\Omega^{[i]}$) in H, which is a maximal parabolic subgroup associated with Φ. Let $Q^{[i]}$ be the kernel of the action of $H^{[i]}$ on V/U_{5-i} and $Z^{[i]}$ be the kernel of the action of $Q^{[i]}$ on U_{5-i}. Let W_{2i} be a complement to U_{5-i} in $U_{5-i}^\perp$, so that dim $W_{2i} = 2i$ and

$$(W_{2i}, f|_{W_{2i}}, q|_{W_{2i}}) \cong V_{2i}^+$$

and let T_{5-i} be a $(5-i)$-dimensional totally singular subspace in $W_{2i}^\perp$ disjoint from $U_{5-i}^\perp$. We assume that $U_5 \cap W_{2i}$ is i-dimensional and that $U_5 \cap W_{2i} < U_5 \cap W_{2j}$ for $1 \le i < j \le 4$.

Then

$$V = U_{5-i} \oplus W_{2i} \oplus T_{5-i}.$$

Let $C^{[i]}$ be the stabilizer of T_{5-i} in $H^{[i]}$ (since W_{2i} is the perp of $\langle U_{5-i}, T_{5-i} \rangle$ the subgroup $C^{[i]}$ stabilizes every term in the above direct sum decomposition), finally let $L^{[i]}$ and $R^{[i]}$ be the kernels of the actions of $C^{[i]}$ on W_{2i} and U_{5-i}, respectively.

The following proposition is a specialization of the structure of maximal parabolics in groups of Lie type (Azad et al. 1990).

Lemma 2.1.2 *For $0 \le i \le 4$ the following assertions hold (with obvious interpretation of the degenerate cases):*

   (i) *if $i \ne 1$ then $Q^{[i]} = O_2(H^{[i]})$ and $O_2(H^{[1]}) = Q^{[1]}R^{[1]}$;*
   (ii) *both $Q^{[i]}/Z^{[i]}$ and $Z^{[i]}$ are elementary abelian 2-groups and unless one of them is trivial, $Z^{[i]}$ is the centre of $Q^{[i]}$;*
   (iii) *$H^{[i]}$ is the semidirect product of $Q^{[i]}$ and $C^{[i]}$;*
   (iv) *$C^{[i]}$ is the direct product of $L^{[i]}$ and $R^{[i]}$;*
   (v) *$L^{[i]} \cong L_{5-i}(2)$ and $R^{[i]} \cong O_{2i}^+(2)$;*
   (vi) *$Z^{[i]}$ is generated by the Siegel transformations s_W taken for all 2-subspaces W contained in U_{5-i} and $Z^{[i]}$ is isomorphic to the exterior square of U_{5-i};*
   (vii) *$Q^{[i]}/Z^{[i]}$ is generated by the images of the Siegel transformations s_W taken for all totally singular 2-subspaces $W = \langle u, v \rangle$ such that $u \in U_{5-i}$, $v \in W_{2i}$ and (as a module for $L^{[i]} \times R^{[i]}$) it is isomorphic to the tensor product*

$$U_{5-i} \otimes W_{2i}.$$

Proof By (1.2.3 (iii)) we know the order of $O_{10}^+(2)$ and it is easy to get from (1.2.2) the number of totally singular subspaces in $V_{10}^+(2)$ of dimension i for

$1 \leq i \leq 5$. Thus the order of $H^{[i]}$ is known and all we have to do is to produce enough elements of the parabolics. The assertion (vi) as well as the whole structure of $H^{[0]}$ are immediate from (1.3.3) and (1.3.5). ∎

Therefore the structure of the parabolics is as follows:

$$H^{[0]} \cong 2^{10} : L_5(2), \quad H^{[1]} \cong 2^{6+8} : (L_4(2) \times 2),$$

$$H^{[2]} \cong 2^{3+12} : (L_3(2) \times O_4^+(2)), H^{[3]} \cong 2_+^{1+12} : (L_2(2) \times O_6^+(2))$$

$$H^{[4]} \cong 2^8 : O_8^+(2).$$

Now we are ready to establish some important properties of the action of H on Ω. As usual for a vertex $x \in \Omega$, by $\Omega(x)$ we denote the set of neighbours of x in Ω, by $\Omega_2(x)$, the set of vertices at distance 2 from x in Ω. Furthermore, $H(x)$ is the stabilizer of x in H and $H_1(x)$ is the kernel of the action of $H(x)$ on $\Omega(x)$.

Lemma 2.1.3 *The following assertions hold:*

(i) *the action of H on Ω is arc-transitive, that is transitive on the incident vertex–edge pairs;*

(ii) *$H(x) = H^{[0]} \cong 2^{10} : L_5(2)$, $H_1(x) = Q^{[0]} \cong 2^{10}$ and $H(x)$ induces the full automorphism group $L_5(2)$ of the projective space structure Π_x on $\Omega(x)$;*

(iii) *for $y \in \Omega(x)$ there is a unique bijection $\psi_{x,y}$ of the set $L_x(y)$ of lines in Π_x containing y onto the set $L_y(x)$ of lines in Π_y containing x;*

(iv) *$Q^{[0]}$ acts faithfully on $\Omega_2(x)$; for $y \in \Omega(x)$ and for $l \in L_y(x)$ the set $l\setminus\{x\}$ is an orbit of $H_1(x)$ on $\Omega(y)\setminus\{x\}$;*

(v) *H acts transitively on the set $S^{[i]}$ of geometric subgraphs of valency $2^i - 1$ for every $2 \leq i \leq 4$;*

(vi) *for $\Omega^{[i]} \in S^{[i]}$ we have $H^{[i]} = H\{\Omega^{[i]}\}$ and $H^{[i]}$ induces on $\Omega^{[i]}$ the orthogonal group $O_{2i}^+(2) \cong R^{[i]}$ and the kernel of the action is $K^{[i]} := Q^{[i]}L^{[i]}$.*

Proof The assertion (i) can be seen inductively using (1.2.3 (ii)), while (ii) follows from (2.1.2). The group $H^{[01]} := H(x, y) = H^{[0]} \cap H^{[1]}$ induces on each of the sets $\Omega(x)\setminus\{y\}$ and $\Omega(y)\setminus\{x\}$ the maximal parabolic subgroup $2^4 : L_4(2)$ of $L_5(2)$ (the kernels of the actions are subgroups $O_2(H(x))$ and $O_2(H(y))$ intersecting in $Z^{[1]} \cong 2^6$). In terms of (2.1.2) $R^{[1]} \cong O_2^+(2)$ is of order 2 generated by the orthogonal transvection t_u, where u is the only non-singular vector in W_2. Then $t_u \in H^{[1]}\setminus H^{[01]}$ and t_u commutes with the quotient $H^{[01]}/Q^{[1]}$ (it even commutes with a complement $L^{[1]} \cong L_4(2)$ to $Q^{[1]}$ in $H^{[01]}$). The action of t_u on the set $L_x(y) \cup L_y(x)$ establishes the required bijection $\psi_{x,y}$. The assertion (iv) possesses a direct check and also follows from the irreducibility of the action of $L_5(2)$ on the exterior square of its natural module. The assertions (v) and (vi) about the geometric subgraphs are rather straightforward. ∎

2.2 Geometric cubic subgraphs

We continue with the notation from the previous section, in particular $H \cong O_{10}^+(2)$ and $\Omega = D^+(10, 2)$. In the next section we will see that much of the combinatorial structure of Ω possesses a natural reformulation in terms of the group H. Here we are making a shortcut to this reformulation for the cubic geometric subgraphs. This provides us with a very simple way to refer to such subgraphs as we have done in the Preface.

We start with the following.

Lemma 2.2.1 Π_x *is the unique projective space structure over* $GF(2)$ *preserved by the action of* $H(x)$ *on* $\Omega(x)$.

Proof Let y and z be distinct vertices from $\Omega(x)$. Then the elementwise stabilizer of $\{y, z\}$ in $H(x)$ stabilizes a unique further point which we denote by u. Then $\{y, z, u\}$ is the unique line in Π_x containing y and z. It is well known that a projective space is uniquely determined by its line-set, hence the result. ∎

Let l be a line in Π_x. Then by the definition of Π_x we know that l is formed by the maximal totally singular subspaces intersecting $U_5 = x$ in hyperplanes containing a given 3-dimensional subspace W in U_5. Therefore (compare (2.1.1 (iv))) there is a unique $\Omega^{[2]}$ in $\mathcal{S}^{[2]}$ containing x such that $\Omega^{[2]}(x) = l$. Suppose that the edge $\{x, y\}$ is contained in $\Omega^{[2]}$. By (2.1.3 (ii)), $H(x, y)$ induces on $L_x(y)$ and $L_y(x)$ two isomorphic actions of $L_4(2)$ of degree 15. Therefore the stabilizer S of l in $H(x, y)$ stabilizes a unique line m in $L_y(x)$ which is $m = \psi_{x,y}(l)$. Since S stabilizes $\Omega^{[2]}$, by the uniqueness condition (2.1.1 (iv)) we conclude that $\Omega^{[2]}(y) = m$. Since any two distinct points in Π_x determine a unique line, we arrive with at following result.

Lemma 2.2.2 *Let* $\Omega^{[2]} \in \mathcal{S}^{[2]}$ *and let* (u, v, w) *be a 2-path in* $\Omega^{[2]}$. *Then* $\Omega^{[2]}(w) = \psi_{v,w}(l)$ *where* l *is the unique line in* Π_v *containing* u *and* w.

We are ready to prove the following characterization.

Lemma 2.2.3 *Let* $\mathcal{S}$ *be a non-empty family of connected cubic subgraphs in* Ω *such that*

(i) $\mathcal{S}$ *is stable under* H;
(ii) *whenever* $\Omega^{[2]}, \Xi^{[2]} \in \mathcal{S}$, $x \in \Omega^{[2]} \cap \Xi^{[2]}$ *and* $\Omega^{[2]}(x) = \Xi^{[2]}(x)$, *the equality* $\Omega^{[2]} = \Xi^{[2]}$ *holds.*

Then $\mathcal{S} = \mathcal{S}^{[2]}$.

Proof Let $\Sigma \in \mathcal{S}$. Let $\{x, y\}$ be an edge of Σ. Since $H_1(x)$ fixes $\Sigma(x)$, by hypothesis (ii) Σ must be stable under $H_1(x)$. Hence $\Sigma(y) \backslash \{x\}$ is a union of $H_1(x)$-orbits on $\Omega(y) \backslash \{x\}$. By (2.1.3) this means that $\Sigma(y)$ is a line l from $L_y(x)$. By the obvious symmetry $\Sigma(x)$ is a line m from $L_x(y)$. We claim that $\psi_{x,y}(m) = l$. This must be true, since $\psi_{x,y}(m)$ is the only line in $L_y(x)$ which is stabilized by the stabilizer of m in $H(x, y)$. Now we apply induction on the distance from x to show that Σ is uniquely determined by $\Sigma(x)$ and that Σ coincides with the subgraph $\Omega^{[2]}$ from $\mathcal{S}^{[2]}$ which contains x and such that $\Omega^{[2]}(x) = m$.   ∎

2.3   Amalgam $\mathcal{H} = \{H^{[0]}, H^{[1]}\}$

We would like to characterize the pair (Ω, H) where $H \cong O_{10}^+(2)$ and $\Omega = D^+(10, 2)$ in terms of certain combinatorial properties of Ω and of the action of H on this graph. Towards this end we follow the standard procedure of redefining Ω in terms of the group H and certain of its subgroups. The procedure is very general and only requires the arc-transitivity of the action of H on Ω.

As above $H^{[0]} = H(x)$ is the stabilizer in H of the vertex x and $H^{[1]} = H\{x, y\}$ is the stabilizer of the edge $\{x, y\}$ which contains x. Recall that $V(\Omega)$ is the vertex-set of Ω while $E(\Omega)$ is the edge-set.

Since H acts transitively on $V(\Omega)$, for every $z \in V(\Omega)$ there is an element h_z in H which maps x onto z and the set of all such elements forms a left coset $h_z H^{[0]}$ of $H^{[0]}$ in H (we adopt the convention that elements acts from the left). Thus there is a mapping

$$\vartheta^{[0]} : V(\Omega) \rightarrow H/H^{[0]}$$

defined by $\vartheta^{[0]}(z) = h_z H^{[0]}$. It is immediate that different cosets correspond to different vertices; therefore $\vartheta^{[0]}$ is a bijection and clearly it commutes with the natural action of H. Similarly we define a bijection

$$\vartheta^{[1]} : E(\Omega) \rightarrow H/H^{[1]}$$

commuting with the action of H. This provides us with a bijection ϑ between the unions $V(\Omega) \cup E(\Omega)$ and $H/H^{[0]} \cup H/H^{[1]}$ whose restrictions to $V(\Omega)$ and $E(\Omega)$ are $\vartheta^{[0]}$ and $\vartheta^{[1]}$, respectively. The only question left is how the incidence between vertices and edges is formulated in terms of the cosets. The answer is the rather simple one given in the following lemma.

Lemma 2.3.1 *Let* $z \in V(\Omega)$, $\{s, t\} \in E(\Omega)$. *Then* z *and* $\{s, t\}$ *are incident (i.e.* $z \in \{s, t\}$) *if and only if*

$$\vartheta(z) \cap \vartheta(\{s, t\}) \neq \emptyset.$$

Proof Suppose first that $z \in \{s, t\}$, say $z = s$. By arc-transitivity there is $h \in H$ which maps x onto s and y onto t. Then $h \in h_z H^{[0]} \cap h_{\{s,t\}} H^{[1]}$. On the other hand if h is in the intersection then it sends the incident pair $(x, \{x, y\})$ onto the pair $(z, \{s, t\})$. Since h is an automorphism of Ω the latter pair also must be incident.   ∎

The next step is to consider $\mathcal{H} := \{H^{[0]}, H^{[1]}\}$ as an abstract amalgam of two independent groups $(H^{[0]}, *_0)$ and $(H^{[1]}, *_1)$ whose element sets intersect in $H^{[01]}$ and the group operations $*_0$ and $*_1$ are the restrictions of the group operation in H to $H^{[0]}$ and $H^{[1]}$, respectively. Clearly in this case $*_0$ and $*_1$ coincide when restricted to $H^{[01]}$.

We introduce the following construction. Let $\mathcal{F} = \{F^{[0]}, F^{[1]}\}$ be an amalgam of rank 2 such that $[F^{[1]} : F^{[01]}] = 2$ and $[F^{[0]} : F^{[01]}] \geq 2$

$$\varphi : \mathcal{F} \to F$$

be a faithful generating completion of $\mathcal{F}$ in F. Recall that by this we mean that φ is an injection; its restrictions to $F^{[0]}$ and $F^{[1]}$ are homomorphisms and F is generated by the image of φ. Define $\Lambda = \Lambda(\mathcal{F}, \varphi, F)$ to be a graph such that $V(\Lambda) = F/\varphi(F^{[0]})$, $E(\Lambda) = F/\varphi(F^{[1]})$ with $v \in V(\Lambda)$ and $e \in E(\Lambda)$ incident if and only if their intersection (as cosets) is non-empty. By (2.3.1) we have the following

Lemma 2.3.2 *Let* id *be the identity mapping of* $\mathcal{H} = \{H^{[0]}, H^{[1]}\}$ *into* H. *Then* Ω *is isomorphic to* $\Lambda(\mathcal{H}, \mathrm{id}, H)$.

It turns out (cf. Table 1) that quite a few properties of Ω can be already seen inside the amalgam $\mathcal{H}$.

Let $\alpha : \mathcal{H} \to A$ be an arbitrary faithful generating completion of $\mathcal{H}$. Then it can easily be checked (cf. Exercises 3 and 4 at the end of the chapter) that $\Lambda(\mathcal{H}, \alpha, A)$ has no multiple edges. Furthermore, the properties in the second column of Table 1 hold with $(A, \Lambda(\mathcal{H}, \alpha, A))$ in place of (H, Ω).

There is an important class of morphisms in the category of faithful generating completions of $\mathcal{H}$. Let $\alpha : \mathcal{H} \to A$ and $\beta : \mathcal{H} \to B$ be two such completions and $\chi : A \to B$ be a homomorphism. Then χ is called a *morphism of completions* if

$$\chi(\alpha(h)) = \beta(h)$$

for every $h \in H^{[0]} \cup H^{[1]}$.

TABLE 1. Properties of (H, Ω) and $\mathcal{H}$

N	Properties of (H, Ω)	Properties of $\mathcal{H}$
1	Ω is an undirected graph	$[H^{[1]} : H^{[01]}] = 2$
2	Ω is connected	$H = \langle H^{[0]}, H^{[1]} \rangle$
3	H acts faithfully on Ω	If $N \leq H^{[01]}$ is normal in $H^{[0]}$ and $H^{[1]}$ then $N = 1$
4	The valency of Ω is 31	$[H^{[0]} : H^{[01]}] = 31$
5	$H(x)^{\Omega(x)} \cong L_5(2)$	$H^{[0]}$ induces $L_5(2)$ acting on the cosets of $H^{[01]}$
6	$H_1(x) \cong 2^{10}$	$\bigcap_{h \in H^{[0]}/H^{[01]}} (H^{[01]})^h \cong 2^{10}$
7	There is a canonical bijection $\psi_{x,y}$ of $L_x(y)$ onto $L_y(x)$	$H^{[1]}/O_2(H^{[01]}) \cong L_4(2) \times 2$

Lemma 2.3.3 *Let* $\alpha : \mathcal{H} \to A$ *and* $\beta : \mathcal{H} \to B$ *be two faithful generating completions and* $\chi : A \to B$ *be a morphism of completions. Then* χ *induces a covering*

$$\nu : \Lambda(\mathcal{H}, \alpha, A) \to \Lambda(\mathcal{H}, \beta, B)$$

of graphs.

Proof We define $\nu(\alpha(aH^{[i]})) = \chi(a)\beta(H^{[i]})$ for every $a \in A$ and $i = 0, 1$. This mapping clearly preserves the incidence relation defined in terms of non-emptiness of intersections. Furthermore, since both α and β are faithful

$$|\alpha(H^{[0]})/\alpha(H^{[01]})| = |H^{[0]}/H^{[01]}| = |\beta(H^{[0]})/\beta(H^{[01]})|$$

and therefore ν is a covering of graphs. ∎

2.4 The universal completion

As is common, I ignore certain technicalities concerning universal covers and universal completions. The standard reference for a detailed treatment of them is Serre (1977).

Let $\rho : \mathcal{H} \to \widetilde{H}$ be the universal completion of $\mathcal{H}$. Then ρ is generating, $\widetilde{H}$ possesses a homomorphism onto any other generating completion group of $\mathcal{H}$, in particular it possesses such a homomorphism

$$\mu : \widetilde{H} \to H$$

onto $H \cong O_{10}^+(2)$. The universal completion group is the free product of the groups $H^{[0]}$ and $H^{[1]}$ amalgamated over their common subgroup $H^{[01]}$ while the graph

$$\widetilde{\Omega} = \Lambda(\mathcal{H}, \rho, \widetilde{H})$$

is the regular infinite tree of valency 31.

Let $\nu : \widetilde{\Omega} \to \Omega$ be the covering of graphs as in (2.3.3) induced by the completion homomorphism μ. Since ν is a covering of graphs, for every $\widetilde{u} \in \widetilde{\Omega}$ the restriction of ν to $\{\widetilde{u}\} \cup \widetilde{\Omega}(\widetilde{u})$ is a bijection onto $\{\nu(\widetilde{u})\} \cup \Omega(\nu(\widetilde{u}))$.

Referreing to the definition of $\Lambda(\mathcal{H}, \rho, \widetilde{H})$ in the paragraph before (2.3.2) let $\widetilde{x}$ be the vertex of $\widetilde{\Omega}$ which is the cosets of $\rho(H^{[0]})$ containing the identity and let $\{\widetilde{x}, \widetilde{y}\}$ be the edge which is the coset of $\rho(H^{[1]})$ containing the identity. Then

$$\nu(\widetilde{x}) = x \text{ and } \nu(\widetilde{y}) = y$$

where $H^{[0]} = H(x)$ and $H^{[1]} = H\{x, y\}$. The natural action of $\widetilde{H}$ on $\widetilde{\Omega}$ is locally projective of type $(5, 2)$ and this action clearly commutes with ν. Furthermore, according to our notation

$$\widetilde{H}(\widetilde{x}) = \rho(H^{[0]}) \text{ and } \widetilde{H}\{\widetilde{x}, \widetilde{y}\} = \rho(H^{[1]}).$$

There is a standard way to describe $\widetilde{\Omega}$ and ν in terms of arcs in Ω originating at x which works as follows.

Let $\pi = (x_0 = x, x_1, \ldots, x_n)$ be an n-arc originating at x and terminating at x_n. Then (because of the covering property of ν) there is a unique n-arc $\widetilde{\pi}$ in $\widetilde{\Omega}$ originating at $\widetilde{x}$ which maps onto π. Namely

$$\widetilde{\pi} = (\widetilde{x}_0 = \widetilde{x}, \widetilde{x}_1, \ldots, \widetilde{x}_n),$$

where for $1 \leq i \leq n$ the vertex $\widetilde{x}_i$ is the unique preimage of x_i in $\widetilde{\Omega}(\widetilde{x}_{i-1})$. Since $\widetilde{\Omega}$ is cycleless, $\widetilde{\pi}$ is the only arc which joins $\widetilde{x}_0$ and $\widetilde{x}_n$. On the other hand, since $\widetilde{\Omega}$ is connected, its every vertex is joint with $\widetilde{x}$ by an arc of some length. Therefore we can identify the vertex-set of $\widetilde{\Omega}$ with the set of arcs in Ω originating at x. In these terms two vertices in $\widetilde{\Omega}$ are adjacent if one of the corresponding arcs is a continuation of the other one by a single edge. The covering ν sends an arc onto its terminal vertex.

Now let us discuss what happens with cubic geometric subgraphs under the covering ν. Arguing as in Section 2.2 it is easy to show that $\widetilde{\Omega}$ contains a family of such subgraphs. Using the above construction we can provide a more explicit description.

Let $\Omega^{[2]}$ be the geometric cubic subgraph as in (2.1.3) which is stabilized by $H^{[2]}$. Recall that $\Omega^{[2]}$ is the complete bipartite graph $K_{3,3}$, $K^{[2]} \cong 2^{3+12} : L_3(2)$ is the vertexwise stabilizer of $\Omega^{[2]}$ in H and

$$H^{[2]}/K^{[2]} \cong R^{[2]} \cong O_4^+(2) \cong \mathrm{Sym}_3 \wr \mathrm{Sym}_2$$

is the action induced by $H^{[2]}$ on $\Omega^{[2]}$ (this action coincides with the automorphism group of $\Omega^{[2]}$). The vertex x and the edge $\{x, y\}$ are contained in $\Omega^{[2]}$ and

$$\mathcal{H}^{[2]} = \{H^{[02]}, H^{[12]}\}$$

is the amalgam formed by the vertex and edge stabilizers in the action of $H^{[2]}$ on $\Omega^{[2]}$. Notice that $K^{[2]}$ is the largest subgroup in $H^{[012]} = H^{[02]} \cap H^{[12]}$ which is normal in both $H^{[02]}$ and $H^{[12]}$.

Let $\widetilde{\Omega}^{[2]}$ be the geometric cubic subgraph in $\widetilde{\Omega}$ which contains $\widetilde{x}$ and which maps onto $\Omega^{[2]}$ under the covering ν. We can describe $\widetilde{\Omega}^{[2]}$ in two different ways. First, when the vertices of $\widetilde{\Omega}$ are treated as the arcs in Ω originating at x, the vertices in $\widetilde{\Omega}^{[2]}$ are precisely the arcs all whose vertices are contained in $\Omega^{[2]}$. Alternatively, if we put

$$\widetilde{H}^{[2]} = \langle \rho(H^{[02]}), \rho(H^{[12]}) \rangle$$

then $\widetilde{\Omega}^{[2]}$ is induced by the images of $\widetilde{x}$ under $\widetilde{H}^{[2]}$.

The group $\widetilde{H}^{[2]}$ is the free product of its subgroups $\rho(H^{[02]})$ and $\rho(H^{[12]})$ amalgamated over $\rho(H^{[012]})$; the graph $\widetilde{\Omega}^{[2]}$ is the infinite cubic tree. The restriction ν_r of ν to $\widetilde{\Omega}^{[2]}$ is a covering onto $\Omega^{[2]}$ and the restriction of μ to $\widetilde{H}^{[2]}$ is clearly a homomorphism onto $H^{[2]}$. Let M be the kernel of μ and put

$$M^{[2]} = M \cap \widetilde{H}^{[2]},$$

so that $M^{[2]}$ is the kernel of the restriction of μ to $\widetilde{H}^{[2]}$. Since μ maps the amalgam $\{\rho(H^{[0]}), \rho(H^{[1]})\}$ isomorphically onto $\mathcal{H} = \{H^{[0]}, H^{[1]}\}$, the intersection $M \cap \rho(K^{[2]})$ is trivial. Thus $M^{[2]}$ and $\rho(K^{[2]})$ are two normal subgroups in $\widetilde{H}^{[2]}$ which have trivial intersection. Hence $M^{[2]}$ is contained in the centralizer of $\rho(K^{[2]})$ in $\widetilde{H}^{[2]}$. On the other hand, it is easy to deduce from (2.1.2) that $C_{H^{[2]}}(K^{[2]}) = 1$. This gives the following.

Lemma 2.4.1 *Let M be the kernel of the completion homomorphism $\mu : \widetilde{H} \to H$ and $M^{[2]} = M \cap \widetilde{H}^{[2]}$. Then $M^{[2]} = C_{\widetilde{H}^{[2]}}(\rho(K^{[2]}))$.*

Since $\rho(K^{[2]})$ is the vertexwise stabilizer of $\widetilde{\Omega}^{[2]}$ in $\widetilde{H}$ we have $M^{[2]} \cap \rho(K^{[2]}) = 1$, since the action of $M^{[2]}$ on $\widetilde{\Omega}^{[2]}$ is faithful. On the other hand, $M^{[2]}$ is the group of deck automorphisms with respect to the covering

$$\nu_r : \widetilde{\Omega}^{[2]} \to \Omega^{[2]},$$

which is universal. Then $M^{[2]}$ is a free group, whose rank is equal to the number of fundamental cycles in $\Omega^{[2]} \cong K_{3,3}$ which is

$$|E(\Omega^{[2]})| - |V(\Omega^{[2]})| + 1 = 4.$$

Also because of the universality of ν_r the group $M^{[2]}$ acts regularly on $\nu_r^{-1}(w)$ for every vertex w of $\Omega^{[2]}$. Let $\widehat{\Omega}^{[2]}$ be the graph on the set of orbits of $M^{[2]}$ on $V(\widetilde{\Omega}^{[2]})$ in which two orbits are adjacent if there is at least one edge which joins them. Then by the above discussion we have the following.

Lemma 2.4.2 *The graphs $\widehat{\Omega}^{[2]}$ and $\Omega^{[2]}$ are isomorphic.*

Thus the graph $\Omega^{[2]}$ is already visible in the universal completion of $\mathcal{H}$. Notice also that the automorphism group of $\Omega^{[2]}$ is the subgroup in $\mathrm{Out}\, K^{[2]}$ generated by the natural images of $H^{[02]}$ and $H^{[12]}$.

2.5 Characterization

Let $\xi : \mathcal{H} \to X$ be an arbitrary faithful generating completion of $\mathcal{H}$, $\Xi = \Lambda(\mathcal{H}, \xi, X)$ be the corresponding coset graph. Let

$$\lambda : \widetilde{H} \to X$$

be the completion homomorphism and

$$\omega : \widetilde{\Omega} \to \Xi$$

be the graph covering as in (2.3.2). Let Y be the kernel of λ and put $Y^{[2]} = Y \cap \widetilde{H}^{[2]}$. Since $Y^{[2]}$ intersects $\rho(K^{[2]})$ trivially, $Y^{[2]} \leq M^{[2]} = C_{\widetilde{H}^{[2]}}(\rho(K^{[2]}))$.

Let $\Xi^{[2]} = \omega(\widetilde{\Omega}^{[2]})$ be the geometric cubic subgraph in Ξ which is the image of $\widetilde{\Omega}^{[2]}$. Then the orbits of $M^{[2]}$ on $\Xi^{[2]}$ have length

$$d = [M^{[2]} : Y^{[2]}]$$

Then $\Xi^{[2]}$ is a d-fold covering of the graph $\widehat{\Xi}^{[2]} \cong \Omega^{[2]} \cong K_{3,3}$ defined as in the paragraph before (2.4.2). We find it appropriate to call d the *defect* of the completion ξ. Thus the geometric cubic subgraph is the smallest possible if and only if the defect of the completion is 1.

We are going to show that H is the only completion group with defect 1. Thus we assume that $Y^{[2]} = M^{[2]} = C_{\widetilde{H}^{[2]}}(\rho(K^{[2]}))$ or, equivalently that $\Xi^{[2]} \cong \Omega^{[2]} \cong K_{3,3}$. We further assume that X is the largest completion with defect 1, which means that Y is the smallest normal subgroup in $\widetilde{H}$ which intersects $\widetilde{H}^{[2]}$ in $M^{[2]}$. Therefore Y is the normal closure in $\widetilde{H}$ of $M^{[2]} = C_{\widetilde{H}^{[2]}}(\rho(K^{[2]}))$. Under this assumption there is a completion homomorphism

$$\alpha : X \to H$$

and a covering

$$\beta : \Xi \to \Omega$$

of the corresponding coset graphs, such that μ is the composition of λ and α, while ν is the composition of ω and β.

In order to establish the stated uniqueness of (H, Ω) it is sufficient to show that at least one (and hence both) of α and β is bijective. We prefer to deal with β. By the assumption the restriction of β to any geometric cubic subgraph in Ξ is an isomorphism onto a geometric cubic subgraph in Ω. On the other hand, every cycle of length 4 in Ω is contained in some geometric cubic subgraph. Therefore every cycle of length 4 in Ω is *contractible* with respect to β (this means that for a 4-cycle δ in Ω the preimage $\beta^{-1}(\delta)$ is a disjoint union of 4-cycles in Ξ). Now in order to conclude that β is an isomorphism it is sufficient to establish the following.

Lemma 2.5.1 *The fundamental group of Ω is generated by the cycles of length 4.*

Proof We have to show that an arbitrary cycle in Ω can be decomposed into quadrangles. Let $\sigma = (x_0, x_1, \ldots, x_n)$ be such a cycle of length n, say. Since Ω is bipartite, n is even, say $n = 2m$. If $m = 2$, σ is itself a 4-cycle, so we assume that $m \geq 3$. One can observe that it is sufficient to decompose the cycles such that

$$d_{\Omega}(x_0, x_m) = m.$$

The 2-arc $\pi = (x_{m-1}, x_m, x_{m+1})$ satisfies $d_{\Omega}(x_0, x_{m-1}) = d_{\Omega}(x_0, x_{m-1}) = m - 1$. Let Σ be the unique geometric cubic subgraph in Ω containing π. Then by (2.1.1 (vi)) there is a unique vertex u in Σ nearest to x_0. Since Σ is of diameter 2 and it contains a vertex x_m at distance m from x_0 and at least two vertices x_{m-1} and x_{m+1} at distance $m - 1$ from x_0, we conclude that $d_{\Omega}(x_0, u) = m - 2$. Let

$(x_0 = y_0, y_1, \ldots, y_{m-2} = u)$ be a shortest arc which joins x_0 and u. Then σ decomposes into a 4-cycle δ and two $(2m - 2)$-cycles

$$(x_0, x_1, \ldots, x_{m-1}, u = y_{m-2}, y_{m-3}, \ldots, y_0 = x_0)$$

and

$$(x_0, x_{2m-1}, \ldots, x_{m+1}, u = y_{m-2}, y_{m-3}, \ldots, y_0 = x_0).$$

Now it only remains to apply the induction on m. ∎

By (2.5.1) whenever all 4-cycles are contractible with respect to a covering of Ω, this covering is necessarily an isomorphism and we are ready to formulate the main result of the chapter originally established in Cameron and Praeger (1982).

Proposition 2.5.2 *Let* $\mathcal{H} = \{H^{[0]}, H^{[1]}\}$ *be the amalgam formed by the vertex and edge stabilizers of* $H \cong O_{10}^+(2)$ *acting on the dual polar graph* $\Omega = D^+(10, 2)$. *Let* $\xi : \mathcal{H} \to X$ *be a faithful generating completion of* $\mathcal{H}$, *let* $\Xi = \Lambda(\mathcal{H}, \xi, X)$ *be the corresponding coset graph, let* $\Xi^{[2]}$ *be the geometric cubic subgraph in* Ξ *and* $X^{[2]}$ *be the stabilizer in* X *of* $\Xi^{[2]}$ *as a whole. Suppose that the following equivalent conditions holds where* $K^{[2]} \cong 2^{3+12} : L_3(2)$ *is the vertexwise stabilizer of* $\Xi^{[2]}$ *in* X:

(i) $\Xi^{[2]} \cong K_{3,3}$;
(ii) $C_{X^{[2]}}(K^{[2]}) = Z(K^{[2]}) = 1$.

Then $(X, \Xi) = (H, \Omega)$.

Exercises

1. The natural action of the orthogonal group $F \cong O_{2n}^+(2)$ on the dual polar graph $\Phi = D^+(2n, 2)$ is locally projective of type $(n, 2)$. Let $\{u, v\}$ be an edge of Φ and $\mathcal{F} = \{F(u), F\{u, v\}\}$. Show that F is the only completion of $\mathcal{F}$ subject to the condition that $C_{F^{[2]}}(K^{[2]}) = 1$.
2. Show that the dual polar space $\mathcal{O}^+(2n, 2)$ is simply connected for $n \geq 3$.
3. Let $\mathcal{F} = \{F^{[0]}, F^{[1]}\}$ be an amalgam of rank 2 such that $[F^{[1]}, F^{[01]}] = 2$ and $[F^{[0]} : F^{[01]}] \geq 2$. Let $\alpha : \mathcal{F} \to F$ be a faithful generating completion of $\mathcal{F}$. Suppose that (a) $F^{[01]}$ is a maximal subgroup in $F^{[0]}$ (equivalently $F^{[0]}$ acts primitively on the cosets of $F^{[01]}$); (b) the automorphism of $F^{[01]}$ induced by $f \in F^{[1]} \setminus F^{[01]}$ is not a restriction of an automorphism of $F^{[0]}$. Then $\Lambda(\mathcal{F}, \alpha, F)$ has no multiple edges.
4. The conclusion of exercise 3 holds if the condition (b) is substituted by the following two conditions: (c) no non-trivial normal subgroup in $F^{[01]}$ is normal in both $F^{[0]}$ and $F^{[1]}$; (d) $F^{[01]}$ contains a non-trivial normal subgroup of $F^{[0]}$.

3

MODIFYING THE RANK 2 AMALGAM

In this section we modify the classical amalgam $\mathcal{H}$ to obtain the amalgam $\mathcal{G}$ which will eventually lead us to J_4. The core of the modification is the triple isomorphism

$$\Omega_6^+(2) \cong \mathrm{Alt}_8 \cong L_4(2),$$

which explains the non-vanishing cohomology of the natural 6-dimensional orthogonal module of the triform group. Although the modifications looks like an innocent move, almost immediately exceptional structures like Mathieu groups appear.

3.1 Complements and first cohomology

Let V be an elementary abelian 2-group, let C be a group, let $\sigma : C \to \mathrm{Aut}\, V$ be a homomorphism (which turns V into a $GF(2)$-module for C) and let $G = V : C$ be the semidirect product of V and C (with respect to σ). We identify V and C with their images under the natural injections into G.

In this section we recall some standard tools for calculating the automorphism group A of $G = V : C$. The inner automorphisms are well understood. Some further automorphisms are those of C normalizing σ. The tricky part is

$$D(C, V) = C_A(V) \cap C_A(G/V).$$

The elements of $D(C, V)$ will be called the *deck automorphisms* of the semidirect product $G = V : C$. It is known that the deck automorphisms are controlled by the first cohomology group $H^1(C, V)$. We are going to discuss outer deck automorphisms and relate them to the classes of complements to V in G and also to indecomposable extensions of V by trivial C-modules. We follow section 17 in Aschbacher (1986).

If $d \in D(C, V)$ is a deck automorphism, then d centralizes V. For every $c \in C$ there is $v \in V$ (depending on d and c) such that $d(c) = cv$ and $d(C)$ is a complement to V in G. Any complement to V in G is a transversal of the left cosets of V in G. Considering C as a canonical transversal, any transversal can be represented as

$$C_\gamma = \{c\gamma(c) \mid c \in C\} \text{ for a function } \gamma : C \to V.$$

Lemma 3.1.1 *The transversal C_γ is a complement to V in G if and only if γ is a* cocycle *in the sense that*

$$\gamma(ab) = \gamma(a)^b \gamma(b)$$

for all $a, b \in C$.

Proof The transversal C_γ is a complement to V in G if and only if it is closed under multiplication. The multiplication law in the semidirect product gives the cocycle condition. ∎

Let $\Gamma(C,V)$ denote the set of V-valued functions on C which satisfy the cocycle condition in (3.1.1). By the definition every cocycle $\gamma \in \Gamma(C,V)$ maps the identity element of C onto the identity element of V.

The image $d(C)$ of C under a deck automorphism $d \in D(C,V)$ is a complement to V in C and by (3.1.1) $d(C) = C_\gamma$ for a cocycle $\gamma \in \Gamma(C,V)$. The correspondence works in the other direction as well.

Lemma 3.1.2 *Let* $\gamma \in \Gamma(C,V)$ *be a cocycle, so that* C_γ *is a complement to* V *in* C. *Then*

$$d_\gamma : cv \mapsto c\gamma(c)v$$

is a deck automorphism which maps C *onto* C_γ.

Proof It is straightforward to use the multiplication law in the semidirect product $G = V : C$ to check that d_γ is in fact an automorphism. It is also clear that d_γ centralizes both V and G/V and maps C onto C_γ. ∎

Thus we can define a mapping $\varphi : \Gamma(C,V) \to D(C,V)$ by setting $\varphi(\gamma) = d_\gamma$. Since $D(C,V)$ is a group of automorphisms of G, it carries a group structure under composition. We define a group structure on $\Gamma(C,V)$ via pointwise multiplication (for $\gamma, \delta \in \Gamma(C,V)$ we define $\gamma\delta(c) = \gamma(c)\delta(c)$ for all $c \in C$). It is easy to check that $\Gamma(C,V)$ is a group with respect to this multiplication.

Lemma 3.1.3 *The mapping* $\varphi : \Gamma(C,V) \to D(C,V)$ *is an isomorphism of groups.*

Proof Observe that φ is surjective by (3.1.1) and (3.1.2). It is immediate to check that φ respects the above defined group structures. Finally, d_γ is the identity automorphism if and only if $C_\gamma = C$, that is, if γ is the trivial cocycle which maps every element of C onto the identity of V. Thus φ is an isomorphism. ∎

Since V is an elementary abelian 2-group, so is the group of V-valued functions on C (with respect to the pointwise multiplication).

Lemma 3.1.4 *The group* $D(C,V)$ *of deck automorphisms is an elementary abelian 2-group. It acts regularly on the set of complements to* V *in* G.

Proof By the remark before the lemma $\Gamma(C,V)$ is an elementary abelian 2-group and by (3.1.3) so is $D(C,V)$. By (3.1.1) every complement to V in G is of the form C_γ for some $\gamma \in \Gamma(C,V)$ and by (3.1.3) $\varphi(\gamma)$ is the unique deck automorphism which maps C onto C_γ. ∎

Let ι_g denote the inner automorphism of G induced by $g \in G$ acting via conjugation, so that

$$\iota : g \mapsto \iota_g$$

is the natural homomorphism of G into its automorphism group A.

Lemma 3.1.5 *If $u \in V$ then*

(i) ι_u *is a deck automorphism;*
(ii) *if $\gamma_u = \varphi^{-1}(\iota_u)$ then $\gamma_u : c \mapsto u^c u$ for every $c \in C$.*

Proof Since V is an abelian normal subgroup in G we have that ι_u centralizes both V and G/V, which gives (i). Furthermore, $\iota_u(c) = ucu = cu^c u$ (since u is of exponent 2 we suppress the inverses) and (ii) follows. ∎

In terms introduced in the proof of (3.1.5) let

$$B(C,V) = \{\gamma_u \mid u \in V\},$$

so that $B(C,V)$ is the image of $\iota(V)$ under φ^{-1}.

The *first cohomology group* of the representation σ of C on V is defined as

$$H^1(C,V) \cong \Gamma(C,V)/B(C,V) \cong D(C,V)/\iota(V).$$

Lemma 3.1.6 *The following assertions hold:*

(i) $H^1(C,V)$ *acts regularly on the set of conjugacy classes of the complements to V in C;*
(ii) $H^1(C,V)$ *is isomorphic to the image of $D(C,V)$ in the outer automorphism group of $G = V : C$.*

Proof Since V and C factorize G, $\iota(V)$ acts regularly on set of complements conjugate to C in G, (i) is by (3.1.4) and the definition of $H^1(C,V)$. (ii) follows directly from the definition. ∎

Since both $\iota(G)$ and $D(C,V)$ are normal in $N_A(V)$, the elements ι_c for $c \in C$ act on $D(C,V)$ via conjugation. This action turns $D(C,V)$ into a $GF(2)$-module for C. Pushing this action through the isomorphism $\varphi^{-1} : D(C,V) \to \Gamma(C,V)$ we also turn $\Gamma(C,V)$ into a $GF(2)$-module for C. With respect to this definition, $B(C,V)$ is a submodule in $\Gamma(C,V)$ isomorphic to $\iota(V)$ (notice that the latter is isomorphic to $V/C_V(G)$).

Lemma 3.1.7 *With respect to the $GF(2)$-module structure of $\Gamma(C,V)$ for C defined above the following equivalent assertions hold:*

(i) $[\Gamma(C,V), C] \le B(C,V)$;
(ii) C *centralizes $\Gamma(C,V)/B(C,V)$.*

Proof We calculate in $A = \mathrm{Aut}\, G$. Since $\iota(V)$ centralizes $D(C,V)$, we have $[D(C,V), \iota(C)] = [D(C,V), \iota(G)]$ and since $D(V,C) \cap \iota(G) = \iota(V)$, (i) follows.

By (3.1.6) the elements of $\Gamma(C,V)/B(C,V) \cong H^1(C,V)$ are indexed by the conjugacy classes in G of the complements to V in G. Clearly $\iota(G)$ acts trivially on these classes, which gives (ii). ∎

Suppose now that $C_C(V) = 1$, so that $\iota(V) \cong B(C,V) \cong V$. In this case by (3.1.7) $\Gamma(C,V)$ is an indecomposable extension of V by trivial C-modules. The next lemma shows that this extension is the largest one.

Lemma 3.1.8 *Suppose that $C_V(C) = 1$ and let W be a $GF(2)$-module for C which contains V as a submodule. Suppose further that $[W,C] \leq V$ (this means that W is an indecomposable extension of V by a trivial module). Then there is an injection $\psi : W \to \Gamma(C,V)$ which sends V onto $B(C,V)$.*

Proof Consider the semidirect product $H = W : C$ of W and C with respect to the natural action. Then H contains $G = V : C$ as a normal subgroup and we can consider $\lambda : H \to A = \text{Aut } G$ via conjugation in H. Clearly the restriction of λ to G coincides with ι. Since $C_V(C) = 1$ and $[W,C] \leq V$ the restriction of λ to W is an isomorphism. Furthermore, since V is an abelian normal subgroup in W, $\lambda(W) \leq C_A(V)$ and since W/V is a trivial C-module, $\lambda(W) \leq C_A(G/V)$. Hence $\lambda(W) \leq D(C,V)$ and composing the restriction of λ to W with φ^{-1} we obtain the required injection ψ. ∎

We will also need the following straightforward generalization of (3.1.2).

Lemma 3.1.9 *Let $G = Q : C$ be a semidirect product of Q and C, let $\gamma : C \to Q$ be a function satisfying the cocycle condition in (3.1.1), so that $C_\gamma = \{c\gamma(c) \mid c \in C\}$ is a complement to Q in G. Suppose further that $\gamma(c) \in Z(Q)$ for every $c \in C$. Then*

$$d_\gamma : cq \mapsto c\gamma(c)q$$

is a deck automorphism of G which centralizes Q and maps C onto C_γ.

We conclude this section a special case of Schur–Zassenhaus Theorem (cf. (17.10) in Aschbacher (1986)) which we present without a proof.

Lemma 3.1.10 *Suppose that C is a group of odd order and V is a $GF(2)$-module for C. Then $H^1(C,V)$ is trivial, so that all complements to V in $G = V : C$ are conjugate in G.*

3.2 Permutation modules

We have seen in the previous section that the first cohomology group $H^1(C,V)$ can be understood through indecomposable extensions of V by trivial C-modules. An important source of such extensions are the $GF(2)$-permutation modules.

Let P_n be a set of n elements and $\mathcal{P}_n$ be the space of $GF(2)$-valued functions on P_n. Then $\mathcal{P}_n$ carries a $GF(2)$-vector space structure with respect to the pointwise addition. Identifying a function with its support, we treat $\mathcal{P}_n$ as the *power set* of P_n. In these terms the addition is performed by the symmetric difference operator $\triangle$ defined by

$$A \triangle B = (A \cup B) \backslash (A \cap B)$$

for $A, B \subseteq P_n$. For any group R of permutations of P_n the space $\mathcal{P}_n$ carries the $GF(2)$-module structure called the *permutation module* over $GF(2)$.

Let $h : \mathcal{P}_n \times \mathcal{P}_n \to GF(2)$ be defined by

$$h(A, B) = |A \cap B| \bmod 2$$

and put

$$\mathcal{P}_n^c = \{\emptyset, P_n\}, \quad \mathcal{P}_n^e = \{A \mid A \subseteq P_n, |A| = 0 \bmod 2\}.$$

The following result is standard.

Lemma 3.2.1 *Let R be the symmetric or alternating group of P_n. Then*

(i) *h is bilinear, R-invariant and non-singular;*

(ii) *$\mathcal{P}_n^c$ and $\mathcal{P}_n^e$ are the only proper R-submodules in $\mathcal{P}_n$.*

By (3.2.1) $\mathcal{P}_n$ is self-dual, $(\mathcal{P}_n^c)^\perp = \mathcal{P}_n^e$ and $(\mathcal{P}_n^e)^\perp = \mathcal{P}_n^c$. Furthermore, when n is odd $\mathcal{P}_n = \mathcal{P}_n^c \oplus \mathcal{P}_n^e$ while

$$0 < \mathcal{P}_n^c < \mathcal{P}_n^e < \mathcal{P}_n$$

is the only R-invariant composition series of $\mathcal{P}_n$ when n is even, in particular the *heart*

$$\mathcal{H}_n = \langle \mathcal{P}_n^e, \mathcal{P}_n^c \rangle / \mathcal{P}_n^c$$

of the permutation module is always irreducible for Alt_n. If n is even then $\mathcal{P}_n / \mathcal{P}_n^c$ is an indecomposable extension of $\mathcal{H}_n$ by the trivial 1-dimensional module.

Lemma 3.2.2 *Let $C^{(1)} = \Omega_6^+(2)$, $V^{(1)} = V_6^+$; $C^{(2)} = L_4(2)$, $V^{(2)}$ be the exterior square of the natural module of $C^{(2)}$; $C^{(3)} = \mathrm{Alt}_8$, $V^{(3)} = \mathcal{H}_8$. Then*

(i) *the groups $V^{(i)} : C^{(i)}$ are pairwise isomorphic for $1 \leq i \leq 3$;*

(ii) *an element from $\mathcal{P}_8 / \mathcal{P}_8^c : \mathrm{Alt}_8$ acting on the subgroup $V^{(3)} : C^{(3)}$ by conjugation induces a deck automorphism of the latter;*

(iii) *$H^1(C^{(i)}, V^{(i)}) \cong 2$ for $1 \leq i \leq 3$.*

Proof (i) is directly by (1.9.5) and its proof while (ii) is by (3.1.8) in view of the remark before that lemma. By (ii) we know that $H^1(C^{(i)}, V^{(i)})$ is non-trivial. The exact order is known in the literature (cf. exercise 6.3 in Aschbacher (1986) in terms of Alt_8 and Bell (1978) in terms of $L_4(2)$). ∎

We can reformulate the discussions in Section 1.10 to observe the following. Let $L \cong L_2(7)$ act on P_8 as on the projective line over $GF(7)$. Then

$$\mathcal{H}_8 \cong I_1 \oplus I_2$$

where I_1 and I_2 are the natural and the dual natural $L_3(2)$-modules of L.

Lemma 3.2.3 *Let J_2 be the preimage of I_2 in $\mathcal{P}_8^e$ and put $\widehat{I}_1 = \mathcal{P}_8/J_2$. Then*

(i) $\widehat{I}_1$ *is an indecomposable extension of I_1 by a 1-dimensional trivial module;*

(ii) $H^1(L, I_1) \cong 2$.

Proof Let S be a Sylow 2-subgroup in L. Then $|S| = 8$, S intersects trivially the stabilizer of a point from P_8 and hence S acts transitively on P_8. Since S must stabilize a non-zero vector in each L-submodule, we conclude that $\mathcal{P}_8^c$ is the unique minimal L-submodule and dually $\mathcal{P}_8^c$ is the only maximal submodule which gives (i).

By (i) $H^1(L, I_1)$ is non-trivial in view of (3.1.8) and $I_1 : L$ contains at least two conjugacy classes of complements (compare (3.1.6)). Hence in order to establish (ii) it is sufficient to show that $I_1 : L$ contains at most two classes of complements. Let $L^{(1)}$ be a complement which is not in the class of L. Then by (3.1.10) and the Sylow theorem we may assume without loss that $L \cap L^{(1)}$ is the normalizer F of a Sylow 7-subgroup in L (which is the Frobenius group of order 21). Let T be a Sylow 3-subgroup in F. Then $N_L(T) \cong \mathrm{Sym}_3$ and since F is maximal in L we conclude that L is generated by F together with an element a of order 2 inverting T. Clearly $L^{(1)}$ is generated by F and a similar element $a^{(1)}$. We assume that $aI_1 = a^{(1)}I_1$, so that $aa^{(1)}$ is a non-identity element in I_1 commuting with T. Since $C_{I_1}(T) \cong 2$ this element is uniquely determined and the result follows. ∎

We will make use of the following result, which is a special case of proposition 3.3.5 in Ivanov and Shpectorov (2002).

Lemma 3.2.4 *Let $M \cong L_4(2)$ and let U be the natural 4-dimensional $GF(2)$-module for M. Let P_{15} be the set of non-zero vectors of U on which M acts in the natural way and let $\mathcal{P}_{15}$ be the corresponding $GF(2)$-permutation module. Then*

$$\mathcal{P}_{15} = \mathcal{P}_{15}^c \oplus \mathcal{P}_{15}^e$$

and $\mathcal{P}_{15}^e$ possesses the unique M-invariant composition series

$$0 < \mathcal{R}_1 < \mathcal{R}_2 < \mathcal{P}_{15}^e,$$

where

$$\mathcal{P}_{15}^e/\mathcal{R}_2 \cong U, \quad \mathcal{R}_2/\mathcal{R}_1 \cong \bigwedge{}^2 U \cong \bigwedge{}^2 U^*, \quad \mathcal{R}_1 \cong U^*.$$

We use the above lemma to prove yet another fact on first cohomology which we will need later.

Lemma 3.2.5 *Let M and U be as in (3.2.4). Then $H^1(M, U)$ is trivial.*

Proof If $H^1(M, U)$ is non-trivial, then by (3.1.7) there exists an indecomposable extension W of U by a 1-dimensional trivial module. Consider the dual W^* of W. Then $[M, W^*] = W^*$ (since the extension is indecomposable), $W^*/C_{W^*}(M) \cong U^*$ and $C_{W^*}(M)$ is 1-dimensional (trivial). The group $M \cong L_4(2)$ acts on the set of non-zero vectors in U^* as it does on the cosets of a subgroup $K \cong 2^3 : L_3(2)$. The preimage in W^* of a vector from U^*, stabilized by K is of size 2. Since K has no subgroups of index 2, it fixes every vector in the preimage and hence there is a 15-orbit of M on W^* and since the extension is indecomposable, this orbit generates the whole of W^*. Thus W^* is a quotient of the dual of $\mathcal{P}_{15}$. By (3.2.4) the latter involves only one 1-dimensional composition factor and this factor splits as a direct summand. Hence there are no indecomposable extensions W^* as required and the result follows. ∎

Let $S \cong S_4(2) \cong \text{Sym}_6$, let P_6 be a set of six elements on which S induces the symmetric group and let $\mathcal{P}_6$ be the corresponding $GF(2)$-permutation module. By the results in Section 1.8, the heart $\mathcal{H}_6$ of the permutation module is isomorphic to the 4-dimensional symplectic module of S (recall that S possesses two such modules which are permuted by the outer automorphism group). By (3.2.1) $\mathcal{P}_6^e$ is an indecomposable extension of a trivial 1-dimensional module by $\mathcal{H}_6$ while $\mathcal{P}_6/\mathcal{P}_6^c$ is an indecomposable extension of $\mathcal{H}_6$ by a trivial 1-dimensional module.

Lemma 3.2.6 _Let $S \cong S_4(2) \cong \text{Sym}_6$, let V_4 be the natural symplectic module for S and let V_5 be an indecomposable extension of a trivial 1-dimensional module V_1 by V_4. Then_

   (i) $H^1(S, V_4) \cong 2$;
   (ii) $V_5 \cong \mathcal{P}_6^e$;
   (iii) $H^1(S, V_5) \cong 2^2$;
   (iv) _if $T \cong \text{Alt}_6$ is the commutator subgroup of S, then $H^1(T, V_4) \cong H^1(T, V_5) \cong 2$._

Proof By the paragraph before the lemma and (3.1.8), the first cohomology of S on V_4 is non-trivial. For the exact order in (i) we refer to exercise 6.3 in Aschbacher (1986). Assertion (i) and (3.1.8) (in view of the duality) give the uniqueness of V_5 and hence (ii). Since $\mathcal{P}_6$ is an indecomposable extension of $\mathcal{P}_6^e$ by a trivial 1-dimensional module, $H^1(S, V_5)$ is non-trivial. If $S^{(1)}$ is a complement to V_5 in the semidirect product R of V_5 and S, then $N_R(S^{(1)}) = S^{(1)} \times V_1$, where $V_1 = C_R(S^{(1)})$, contains two subgroups isomorphic to $S^{(1)}$. This construction doubles the number of classes of $S_4(2)$-complements in R/V_1 and (iii) follows. Now (iv) must also be rather clear. ∎

Lemma 3.2.7 _Let V_5 be a 5-dimensional $GF(2)$-space and $L \cong L_5(2)$ be the general linear group of V_5. Let X be the stabilizer in L of a 1-dimensional subspace V_1 of V_5 together with a non-singular symplectic form f on V_5/V_1. Then_

   (i) $X \cong 2^4 : S_4(2)$ _is the semidirect product of $S_4(2)$ and its natural symplectic module;_
   (ii) _X contains two classes of $S_4(2)$-complements with representatives $S^{(1)}$ and $S^{(2)}$, say;_

(iii) *the representatives can be chosen so that the action of* $S^{(1)}$ *on* V_5 *is semisimple while that of* $S^{(2)}$ *is indecomposable;*

(iv) *there is a deck automorphism of* X *which transposes* $S^{(1)}$ *onto* $S^{(2)}$.

Proof Statement (i) is clear while (ii) is by (3.2.6 (i)). We know that there are two extensions of a trivial 1-dimensional module by the natural symplectic module: the semisimple and the indecomposable ones. Hence there are two $S_4(2)$-subgroups $S^{(1)}$ and $S^{(2)}$ in L acting on V_5 in these two ways, preserving a 1-dimensional subspace V_1 and a fixed non-singular symplectic form f on V_5/V_1. Then both $S^{(1)}$ and $S^{(2)}$ are contained in X and it is easy to check that they are not conjugate in X. Hence (iii) follows. Now (iv) is immediate from (3.1.6). ∎

3.3 Deck automorphisms of $2^6 : L_4(2)$

Let $Z^{[1]}L^{[1]} \cong 2^6 : L_4(2)$ be a subgroup of $H^{[1]}$ (compare (2.1.2)). In this section we analyse the deck automorphisms of $Z^{[1]}L^{[1]}$ (which are the automorphisms acting trivially on both $Z^{[1]}$ and $Z^{[1]}L^{[1]}/Z^{[1]}$).

By (1.9.5) and (3.2.2) $Z^{[1]}L^{[1]}$ can be treated in the following four different languages (notice that $Z^{[1]}$ is self-dual as a module for $L^{[1]}$).

Linear: $L^{[1]}$ is the general linear group of U_4 and $Z^{[1]} \cong \bigwedge^2 U_4$ is the exterior square of U_4.

Symplectic: $L^{[1]}$ is the general linear group of U_4 and $Z^{[1]} \cong \mathcal{S}(U_4)$ is the space of symplectic forms on U_4.

Orthogonal: $L^{[1]} \cong \Omega_6^+(2)$ is the commutator subgroup of the orthogonal group of plus type in dimension 6 over $GF(2)$ and $Z^{[1]} \cong V_6^+$ is the natural orthogonal module.

Permutation: $L^{[1]} \cong \mathrm{Alt}_8$ is the alternating group of a set P_8 of eight elements and $Z^{[1]} \cong \mathcal{H}_8$ is the heart of the $GF(2)$-permutation module $\mathcal{P}_8$ of $L^{[1]}$ on P_8.

The following lemma (which is a direct consequence of (1.9.5), (3.2.2) and (1.4.3) gives a classification of the non-zero vectors in $Z^{[1]}$ in the above four languages.

Lemma 3.3.1 *The group* $L^{[1]}$ *acts by conjugation on the set of non-identity elements of* $Z^{[1]}$ *with two orbits* Υ_0 *and* Υ_1 *of length 35 and 28, respectively. Furthermore,*

(i) *if* $z \in \Upsilon_0$ *then*

$$L^{[1]}(z) \cong 2^4 : (\mathrm{Sym}_3 \times \mathrm{Sym}_3) \cong (\mathrm{Sym}_4 \wr \mathrm{Sym}_2)^+$$

(where + indicates the intersection with the alternating group);

(ii) *if* $y \in \Upsilon_1$ *then*

$$L^{[1]}(y) \cong S_4(2) \cong \mathrm{Sym}_6;$$

(iii) $z \in \Upsilon_0$ *if and only if the following equivalent conditions hold:*

(a) $z - u \wedge v \in \bigwedge^2 U_4$ *for some linearly independent* $u, v \in U_4$;

(b) $z \in \mathcal{S}(U_4)$ *is a form with 2-dimensional radical;*

(c) z is singular in V_6^+;

(d) z is a partition of P_8 into two 4-element subsets;

(e) z is the Siegel transformation s_W for a 2-dimensional subspace W in U_4;

(iv) $y \in \Upsilon_1$ if and only if the following equivalent conditions hold:

(f) $y = u \wedge v + w \wedge r \in \bigwedge^2 U_4$ for some linearly independent $u, v, w, r \in U_4$;

(g) $y \in \mathcal{S}(U_4)$ is a non-singular symplectic form;

(h) y is non-singular in V_6^+;

(i) y is a partition of P_8 into a 2-element subset and its 6-element complement.

Let $h \in \mathcal{S}(U_4)$ be a symplectic form on U_4 considered as an element of $Z^{[1]}$. Let Z_h be the orthogonal perp of h with respect to the non-singular symplectic form f on $Z^{[1]}$ preserved by $L^{[1]}$ (so that $Z^{[1]} \cong V_6^+ \cong (V, f, q)$ in the orthogonal language). Since f is non-singular, every hyperplane in $Z^{[1]}$ arises in this way. In the permutation terms two even partitions

$$P_8 = A_1 \cup A_2 \text{ and } P_8 = B_1 \cup B_2$$

(considered as elements of $Z^{[1]}$) are perpendicular with respect to f if and only if $|A_i \cap B_j|$ is even for $1 \le i, j \le 2$. This gives the following.

Lemma 3.3.2 *The following assertions hold:*

(i) *if $h \in \Upsilon_0$ then Z_h contains 19 vectors from Υ_0 and 12 vectors Υ_1;*

(ii) *if $h \in \Upsilon_1$ then Z_h contains 15 vectors from Υ_0 and 16 vectors from Υ_1.*

The deck automorphisms of $Z^{[1]}L^{[1]}$ are best understood in the permutation language, but we also will deal with them in the other languages.

Let $Z^{[1]u}$ be the indecomposable extension of the module $Z^{[1]}$ by the trivial 1-dimensional submodule. Then by (3.2.2), $Z^{[1]u}L^{[1]} \cong (\mathcal{P}_8/\mathcal{P}_8^c) : \mathrm{Alt}_8$. An element $d \in \mathcal{P}_8/\mathcal{P}_8^c$ naturally corresponds to a partition of P_8 into two subsets. Furthermore, $d \in Z^{[1]}$ if and only if the partition is even. Whenever $d \notin Z^{[1]}$ the subgroup $L^{[1]}$ and its image $(L^{[1]})^d$ under d represent different classes of complements to $Z^{[1]}$ in $Z^{[1]}L^{[1]}$. For $i = 1$ and 3 let d_i correspond to a partition of P_8 into two subsets of size i and $(8 - i)$, respectively.

Lemma 3.3.3 *There are two classes of elements in $Z^{[1]u} \setminus Z^{[1]}$ with representatives d_1 and d_3. Furthermore*

$$C_{L^{[1]}}(d_1) \cong \mathrm{Alt}_7, \quad C_{L^{[1]}}(d_3) \cong (\mathrm{Sym}_3 \times \mathrm{Sym}_5)^+$$

where $+$ indicates the intersection with the alternating group Alt_8.

In what follows we will need some further information on the action of elements from $Z^{[1]u}$ on the preimages of maximal parabolic subgroups of $Z^{[1]}L^{[1]}/Z^{[1]} \cong L^{[1]} \cong L_4(2)$.

Let $u \in U_4^{\#}$ and let $L^{[1]}(u) \cong 2^3 : L_3(2)$ be the stabilizer of u in $L^{[1]}$.

Lemma 3.3.4 *Let*

  (i) $K^{(1)}$ *be the stabilizer in* $L^{[1]}(u)$ *of a hyperplane* V *in* U_4 *not containing* u;

 (ii) $K^{(2)}$ *be the stabilizer in* $L^{[1]}(u)$ *of an element from* P_8;

(iii) $K^{(3)}$ *be the stabilizer in* $L^{[1]}(u)$ *of a ordered pair of disjoint maximal totally singular subspaces in* V_6^+.

Then $K^{(1)} \cong K^{(2)} \cong K^{(3)} \cong L_3(2)$, *furthermore* $K^{(1)}$ *is conjugate in* $L^{[1]}(u)$ *to* $K^{(3)}$ *but not to* $K^{(2)}$.

Proof Since $\langle v \wedge w | v, w \in V \rangle$ and $\langle u \wedge v | v \in V \rangle$ are disjoint maximal totally singular subspaces in V_6^+ stabilized by $K^{(1)}$, the latter is a conjugate of $K^{(3)}$. On the other hand by (1.10.1) the subgroup $K^{(3)}$ acts doubly transitively on P_8 as $PSL_2(7)$ acts on the projective line over $GF(7)$. Hence $K^{(2)}$ and $K^{(3)}$ cannot possibly be conjugate in $L^{[1]}(u)$. ∎

Notice that the actions of $K^{(1)}$ on U_4 and V_6^+ are semisimple and the action of $K^{(1)}$ on P_8 is doubly transitive, while the actions of $K^{(2)}$ on U_4 and V_6^+ are indecomposable while on P_8 it is intransitive.

Lemma 3.3.5 *Assume that* $K^{(2)}$ *is the stabilizer in* $L^{[1]}(u)$ *of a point* $p \in P_8$ *and that* $d_1 \in Z^{[1]u}$ *corresponds to the partition of* P_8 *into* $\{p\}$ *and its complement. Then* d_1 *centralizes* $K^{(2)}$ *but not* $K^{(1)}$ *and it maps* $O_2(L^{[1]}(u))$ *onto the unique* $K^{(2)}$-*invariant complement to* $Z^{[1]}$ *in* $Z^{[1]}O_2(L^{[1]}(u))$ *distinct from* $O_2(L^{[1]}(u))$.

Proof By (3.3.1) $C_{L^{[1]}}(d_1) = L^{[1]}(p) \cong \mathrm{Alt}_7$ intersects $L^{[1]}(u)$ in $K^{(2)}$. Since (as a $K^{(2)}$-module) the quotient of $Z^{[1]}O_2(L^{[1]}(u))$ over its centre is the direct sum of two copies of $O_2(L^{[1]}(u))$ (one of them being in $Z^{[1]}$), we have the uniqueness stated in the lemma. ∎

Let W be a 2-dimensional subspace in U_4. Then

$$L^{[1]}(W) \cong 2^4 : (\mathrm{Sym}_3 \times \mathrm{Sym}_3)$$

and in terms of the permutation action on P_8 it is the stabilizer of a partition of P_8 into two equal parts, say A and B, and

$$L^{[1]}(W) \cong (\mathrm{Sym}_4 \otimes \mathrm{Sym}_2)^+.$$

Let T be a Sylow 3-subgroup in $L^{[1]}(W)$, and let $T^{(1)}$ and $T^{(2)}$ be the subgroups of order 3 in T normal in $N_{L^{[1]}(W)}(T) \cong \mathrm{Sym}_3 \times \mathrm{Sym}_3$. Then in the linear terms $N_{L^{[1]}(W)}(T)$ is a Levi complement in $L^{[1]}(W)$ which is the stabilizer of a direct sum decomposition

$$U_4 = W^{(1)} \oplus W^{(2)},$$

where $W^{(1)} = W$. We assume that $T^{(i)}$ is the kernel of T on $W^{(3-i)}$ for $i = 1, 2$. Let u and v be non-zero vectors from $W^{(1)}$ and $W^{(2)}$, respectively and let $p(u, v)$

denote the transvection with centre u and axis $\langle W^{(1)}, v \rangle$. Then the elements $p(u, v)$ taking for all the nine choices for the pair (u, v) generate $O_2(L^{[1]}(W))$.

In the permutation terms $N_{L^{[1]}(W)}(T)$ is the stabilizer of a partition

$$P_8 = \{a, b\} \cup (A \setminus \{a\}) \cup (B \cup \{b\})$$

for some $a \in A$ and $b \in B$. The subgroup $T^{(i)}$ is generated by an element of cycle type 3^2, a Sylow 2-subgroup R of $N_{L^{[1]}(W)}(T)$ contains three involutions, say $r^{(1)}$, $r^{(2)}$ and $r^{(12)}$, such that $r^{(1)}$ and $r^{(2)}$ are of cycle type 2^4 and $r^{(12)}$ is of cycle type $2^2 1^4$. We assume $r^{(i)}$ centralizes $T^{(3-i)}$ and inverts $T^{(i)}$ for $i = 1, 2$ so that $r^{(12)} = r^{(1)} r^{(2)}$ acts on T fixed-point freely. Let d be a deck automorphism of $Z^{[1]} L^{[1]}$ which centralizes T. Then there are four choices for d, but independently of the choice we have the following.

Lemma 3.3.6 *In the above terms the following assertions hold:*

    (i) $C_{L^{[1]}}(d) \cap O_2(L^{[1]}(W)) \cong 2^2$;

    (ii) $C_{L^{[1]}}(d) \cap R = \langle r^{(12)} \rangle$;

    (iii) $[r^{(1)}, d] = [r^{(2)}, d] = \delta$, *where δ is the element from $Z^{[1]}$ corresponding to the partition of P_8 into $\{a, b\}$ and its complement.*

3.4 Automorphism group of $H^{[01]}$

As above let $\mathcal{H} = \{H^{[0]}, H^{[1]}\}$ be the amalgam formed by the stabilizers in $H \cong O_{10}^+(2)$ of a vertex x and an edge $\{x, y\}$ of the dual polar graph $\Omega = D^+(10, 2)$. In terms of the orthogonal space V_{10}^+, the vertex x is a maximal totally singular subspace U_5, while $\{x, y\}$ is a hyperplane U_4 in U_5. The amalgam $\mathcal{H}$ can be looked at in the following way:

  (H0) $H^{[0]}$ is the semidirect product of $Q^{[0]} \cong \bigwedge^2 U_5$ and $L^{[0]} \cong L_5(2)$ acting naturally on $Q^{[0]}$;

 (H01) $H^{[01]}$ is the semidirect product of $Q^{[0]}$ and the maximal parabolic subgroup $L^{[0]}(U_4) \cong 2^4 : L_4(2)$ in $L^{[0]}$, which is the stabilizer of the hyperplane U_4 in U_5;

  (H1) $H^{[1]} = \langle H^{[01]}, t_0 \rangle$, where t_0 is an element of order 2 which induces an automorphism τ_0 of $H^{[01]}$ which does not normalize $Q^{[0]}$.

We are going to modify $\mathcal{H}$ by keeping (H0) with (H01) and changing the element t_0 in (H1) by an element t_1 inducing a different automorphism of $H^{[01]}$ still not normalizing $Q^{[0]}$. Towards this end we calculate the automorphism group of $H^{[01]}$.

First let us recall and refine the geometrical setting from Section 2.1. Let W_2 be a complement to U_4 in $U_4^\perp$ (so that $W_2 \cong V_2^+$) and let T_4 be a maximal totally singular subspace in $W_2^\perp \cong V_8^+$, disjoint from U_4. Let u and v denote the non-zero singular vectors in W_2. We assume without loss that $\langle U_4, u \rangle = U_5$, while $\langle U_4, v \rangle$ is the only other maximal totally singular subspace in $V \cong V_{10}^+$

which contains U_4 (and corresponds to the vertex y of Ω). Then

$$C^{[1]} = L^{[1]} \times R^{[1]} \cong L_4(2) \times 2$$

is the stabilizer of T_4 in $H^{[1]}$ ($C^{[1]}$ also stabilizes $W_2 = \langle U_4, T_4 \rangle^{\perp}$). The kernel $Q^{[1]}$ of the action of $H^{[1]}$ on $U_4^{\perp}$ is a special 2-group of order 2^{14}. The centre $Z^{[1]}$ of $Q^{[1]}$ is generated by the Siegel transformations s_T taken for all the 2-dimensional subspaces T in U_4 so that by (1.3.5) $Z^{[1]}$ and $\bigwedge^2 U_4$ are isomorphic as $L^{[1]}$-modules. Let $A^{[1]}$ and $B^{[1]}$ be the subgroups in $Q^{[1]}$ generated by the Siegel transformations $s_{\langle u,w \rangle}$ and $s_{\langle v,w \rangle}$ taken for all $w \in U_4^{\#}$. Then by the proof of (1.3.5) we have the following commutator relations

$$[s_{\langle u,w_1 \rangle}, s_{\langle v,w_2 \rangle}] = \begin{cases} 1 & \text{if } w_1 = w_2; \\ s_{\langle w_1,w_2 \rangle} & \text{otherwise.} \end{cases}$$

Therefore the following lemma holds.

Lemma 3.4.1 *Let* $\alpha : U_4 \rightarrow A^{[1]}$, $\beta : U_4 \rightarrow B^{[1]}$ *and* $\zeta : \bigwedge^2 U_4 \rightarrow Z^{[1]}$ *be the isomorphism commuting with the action of* $L^{[1]}$. *Then the following key commutator relation holds*

$$[\alpha(w_1), \beta(w_2)] = \zeta(w_1 \wedge w_2)$$

for all $w_1, w_2 \in U_4$.

By (2.1.2) $Q^{[1]} = Z^{[1]} A^{[1]} B^{[1]}$, $Q^{[0]} = Z^{[1]} A^{[1]}$, and $H^{[01]} = Q^{[1]} L^{[1]}$. Finally, $R^{[1]}$ is of order 2 generated by the element t_0 which is the orthogonal transvection with respect to the unique non-singular vector $u + v$ of W_2. Then t_0 centralizes both $Z^{[1]}$ and $L^{[1]}$, and conjugates $s_{\langle u,w \rangle}$ onto $s_{\langle v,w \rangle}$ for every $w \in U_4^{\#}$. Therefore t_0 conjugates $A^{[1]}$ onto $B^{[1]}$. In particular it does not normalize $Q^{[0]} = Z^{[1]} A^{[1]}$.

Put

$$F^{[1]} = \langle Z^{[1]}, L^{[1]} \rangle \cong 2^6 : L_4(2).$$

Then by (3.2.3 (iii)) $H^1(L^{[1]}, Z^{[1]})$ is of order 2, so that by (3.1.6) there exists a cocycle $\gamma : L^{[1]} \rightarrow Z^{[1]}$ such that

$$L_{\gamma}^{[1]} = \{ l\gamma(l) \mid l \in L^{[1]} \}$$

is a complement to $Z^{[1]}$ in $F^{[1]}$ not conjugate to $L^{[1]}$. By (3.1.2) the mapping

$$d_{\gamma} : lz \mapsto l\gamma(l)z$$

(where $l \in L^{[1]}$ and $z \in Z^{[1]}$) is an outer deck automorphism of $F^{[1]}$ which maps $L^{[1]}$ onto $L_{\gamma}^{[1]}$. Since $Z^{[1]}$ is the centre of $Q^{[1]}$, by (3.1.9) the mapping

$$\sigma_{\gamma} : lq \mapsto l\gamma(l)q$$

(where $l \in L^{[1]}$, $q \in Q^{[1]}$) is an outer deck automorphism of

$$H^{[01]} = Q^{[1]} : L^{[1]}.$$

Proposition 3.4.2 *The following assertions hold:*

  (i) $Q^{[1]} = O_2(H^{[01]})$;

 (ii) $Z^{[1]} = Z(O_2(H^{[01]}))$;

(iii) *if $F^{[1]}$ is a normal subgroup of $H^{[01]}$ of order 2^a for $a \geq 10$ and $F^{[1]} < Q^{[1]}$ then $a = 10$, $F^{[1]}$ is elementary abelian and coincides with one of the following $A^{[1]}Z^{[1]}$, $B^{[1]}Z^{[1]}$ and*

$$D^{[1]} = \langle s_{\langle u,w\rangle} s_{\langle v,w\rangle} \mid w \in U_4^{\#}\};$$

 (iv) *$H^{[01]}$ contains exactly two conjugacy classes of $L_4(2)$-complements to $Q^{[1]}$ with representatives $L^{[1]}$ and $L_\gamma^{[1]}$;*

  (v) *if L is a complement to $Q^{[1]}$ in $H^{[01]}$ then $A^{[1]}Z^{[1]}$ and $B^{[1]}Z^{[1]}$ are semisimple while $D^{[1]}$ is indecomposable (as L-modules);*

 (vi) *the outer automorphism group of $H^{[01]}$ is elementary abelian of order 4 generated by the image of the automorphism τ_0 induced by t_0 and by the image of the deck automorphism σ_γ.*

Proof Assertions (i) and (ii) are (2.1.2 (i), (ii)) restated. By (2.1.2 (vii)) $Q^{[1]}/Z^{[1]}$ as a module for $H^{[01]}/Q^{[1]} \cong L_4(2)$ is isomorphic to the direct sum of two copies of U_4; therefore there are exactly three proper submodules, which are $A^{[1]}Z^{[1]}/Z^{[1]}$, $B^{[1]}Z^{[1]}/Z^{[1]}$ and the diagonal one denoted by $D^{[1]}$. The commutator relation in (3.4.1) shows that $D^{[1]}$ is elementary abelian and contains $Z^{[1]}$, so that (iii) follows. By (3.2.5) every $L_4(2)$-complement in

$$H^{[01]}/Z^{[1]} \cong (2^4 \times 2^4) : L_4(2)$$

is a conjugate of $L^{[1]}Z^{[1]}/Z^{[1]}$. The preimage of the latter in $H^{[01]}$ is $Z^{[1]} : L^{[1]} \cong 2^6 : L_4(2)$. By (3.2.2 (iii)) this preimage contains exactly two classes of complements and (iv) follows. Therefore in (v) we can assume without loss that $L = L^{[1]}$; then the assertion is immediate by (3.4.1).

Let δ be an automorphism of $H^{[01]}$. In view of (iv), multiplying δ by inner automorphisms and/or by σ_γ, if necessary, we can assume that δ normalizes $L^{[1]}$. It is known that

$$\text{Aut } L^{[1]} \cong L_4(2) : \langle \tau \rangle,$$

where τ is the contragredient automorphism. Since $Q^{[1]}/Z^{[1]}$ involves two isomorphic copies of U_4 (and does not involve the dual of U_4), δ cannot induce on $L^{[1]}$ a conjugate of the contragredient automorphism. Therefore, adjusting δ by inner automorphisms of $L^{[1]}$, we can assume that δ centralizes $L^{[1]}$. By (v), δ either normalizes each of $A^{[1]}Z^{[1]}$ and $B^{[1]}Z^{[1]}$, or permutes them. Multiplying δ by the automorphism τ_0, if necessary, we assume that the former possibility holds. Since $A^{[1]}Z^{[1]}$ is the direct sum of two non-isomorphic absolutely irreducible $L^{[1]}$-modules $A^{[1]}$ and $Z^{[1]}$ (isomorphic to U_4 and $\bigwedge^2 U_4$, respectively) and δ commutes with the action of $L^{[1]}$ on these modules, we conclude that δ centralizes both $A^{[1]}$ and $Z^{[1]}$. Similarly δ centralizes $B^{[1]}$ and (vi) follows, thereby completing the proof.   ∎

Notice that in terms on (3.2.4), the module $D^{[1]}$ for $L^{[1]} \cong L_4(2)$ is isomorphic to the section

$$\mathcal{P}^e_{15}/\mathcal{R}_1 \cong 2^{6+4}$$

of the permutation module on the non-zero vectors of $U^\#_4$.

3.5   Amalgam $\mathcal{G} = \{G^{[0]}, G^{[1]}\}$

In this section we modify the amalgam $\mathcal{H} = \{H^{[0]}, H^{[1]}\}$ (taken from $O^+_{10}(2)$) to obtain an amalgam $\mathcal{G} = \{G^{[0]}, G^{[1]}\}$ which will eventually lead us to the group J_4.

We preserve $H^{[0]}$ and $H^{[01]}$, so that we assume that $G^{[0]} = H^{[0]}$ and $G^{[01]} = H^{[01]}$, while $G^{[1]}$ is the semidirect product of $G^{[01]}$ and a group of order 2 generated by an element t_1 such that the automorphism τ_1 of $G^{[01]}$ induced by t_1 is the product of τ_0 and the automorphism σ_γ defined before (3.4.2).

Lemma 3.5.1 *The following assertions hold:*

(i) *the automorphism τ_1 is of order 2, so that $\mathcal{G} = \{G^{[0]}, G^{[1]}\}$ is correctly defined;*

(ii) *the above definition specifies $\mathcal{G}$ up to isomorphism.*

Proof The automorphism τ_0 centralizes $Z^{[1]}L^{[1]}$ and induces a permutation of order 2 on $A^{[1]} \cup B^{[1]}$ while σ_γ centralizes $Q^{[1]}$ and induces an automorphism of order 2 of the subgroup $Z^{[1]}L^{[1]}$. Therefore τ_1 is of order 2 and (i) follows. The isomorphism classes of rank two amalgams are controlled by Goldschmidt's theorem (cf. Goldschmidt (1980) or proposition 8.3.2 in Ivanov and Schpectorov (2002)). The number of such classes is the number of double cosets in $O :=$ Out $G^{[01]}$ of the subgroups $O^{[0]}$ and $O^{[1]}$, where $O^{[i]}$ is the natural image in O of $N_{\text{Aut } G^{[i]}}(G^{[01]})$ for $i = 0$ and 1. By (3.4.2 (vi)) O is elementary abelian of order 4 and $O^{[1]}$ is the whole of O, which gives (ii).   ∎

The next result shows that in a certain sense $\mathcal{G}$ is the only possible modification of the amalgam $\mathcal{H}$.

Proposition 3.5.2 *Let $\mathcal{X} = \{X^{[0]}, X^{[1]}\}$ be an amalgam of rank 2 satisfying the following conditions (i) to (iii), where $X^{[01]} = X^{[0]} \cap X^{[1]}$:*

(i) *there is an isomorphism $\alpha : H^{[0]} \to X^{[0]}$ such that $X^{[01]} = \alpha(H^{[01]})$;*

(ii) *$[X^{[1]} : X^{[01]}] = 2$;*

(iii) *no proper subgroup from $X^{[01]}$ is normal in both $X^{[0]}$ and $X^{[1]}$.*

Then $\mathcal{X}$ is isomorphic to either $\mathcal{H}$ or $\mathcal{G}$.

Proof The restriction of α to $H^{[01]}$ induces an isomorphism of Aut $H^{[01]}$ onto Aut $X^{[01]}$. Abusing the notation, we denote this isomorphism by the same letter α. Let $t \in X^{[1]} \setminus X^{[01]}$. Since $X^{[01]}$, being a subgroup of index 2, is normal in $X^{[1]}$, t induces an automorphism τ of $X^{[01]}$. The only proper normal subgroup of

$H^{[0]}$ contained in $H^{[01]}$ is $O_2(H^{[0]})$. Therefore condition (iii) states that $\alpha^{-1}(\tau)$ does not normalize $O_2(H^{[0]}) = Q^{[0]} = Z^{[1]}A^{[1]}$, while by (3.4.2 (vi))

$$N_{\text{Aut } H^{[01]}}(Q^{[0]}) = \langle \text{Inn } H^{[01]}, \sigma_\gamma \rangle.$$

By (3.4.2 (vi)) we can adjust τ by inner automorphisms of $X^{[01]}$ so that either

$$\alpha^{-1}(\tau) = \tau_0, \text{ or } \alpha^{-1}(\tau) = \tau_1 = \tau_0 \sigma_\gamma.$$

We claim that whenever t is chosen in this way, it is an involution. In fact, $t^2 \in X^{[01]}$ and hence t^2 induces an inner automorphism of $X^{[01]}$. On the other hand we have seen in the proof of (3.5.1) that both τ_0 and τ_1 are involutory automorphisms. Therefor t^2 is in the centre of $X^{[01]}$ (compare (3.4.2) and its proof). Hence the claim follows. Thus $X^{[1]}$ is the semidirect product of $X^{[01]}$ and a group of order 2 generated by t and the latter induces the automorphism $\alpha(\tau_i)$ for $i = 0$ or 1. Therefore there are two possibilities for the isomorphism type of $X^{[1]}$ and arguing as in the proof of (3.5.1) (applying Goldschmidt's theorem in a rather trivial way), we conclude that the isomorphism of $\mathcal{X}$ is uniquely determined by making the choice of τ between $\alpha(\tau_0)$ and $\alpha(\tau_1)$). ∎

3.6 Vectors and hyperplanes in $Q^{[0]}$

From now on we require a better understanding of the structure of $Q^{[0]} = O_2(G^{[0]})$, which we know is the exterior square of the 5-dimensional $GF(2)$-space U_5. We describe the set $\mathcal{P}$ of hyperplanes (subgroups of index 2) in $Q^{[0]}$. By (1.4.4) $\mathcal{P}$ is in a natural bijection with the set of non-zero vectors in $\bigwedge^2 U_5^*$ and by (1.4.3) these vectors in turn correspond to the non-zero symplectic forms on U_5. The group

$$L^{[0]} \cong G^{[0]}/Q^{[0]} \cong L_5(2)$$

has two orbits on the set of non-zero symplectic forms on U_5, consisting of the forms of rank 2 and 4, respectively. There are 155 and 15 subspaces in U_5 of dimension 2 and 4 respectively. Since there is a unique non-singular form in a 2-dimensional space and exactly $28 = \binom{8}{2}$ non-singular forms in a space of dimension 4 (compare (3.3.1)) we obtain the following.

Lemma 3.6.1 *The group $L^{[0]} \cong L_5(2)$ acting on the set $\mathcal{P}$ of hyperplanes in $Q^{[0]}$ has two orbits $\mathcal{P}_1$ and $\mathcal{P}_2$ with lengths 155 and 868, respectively.*

The hyperplanes in $\mathcal{P}_1$ correspond to the rank 2 forms on U_5 which are uniquely determined by their 3-dimensional kernels. Let V_3 be a 3-subspace in U_5 and let $P(V_3)$ be the corresponding hyperplane from $\mathcal{P}_1$. Then the Siegel transformation s_T for a 2-subspace T in U_5 is contained in $P(V_3)$ if and only if $V_3 \cap T \neq 0$. The stabilizer of $P(V_3)$ in $G^{[0]}$ coincides with that of V_3 and if $V_3 = U_3$ the stabilizer coincides with $G^{[02]}$. Recall that

$$G^{[02]}/Q^{[0]} \cong 2^2 : (L_3(2) \times \text{Sym}_3).$$

The hyperplanes in $\mathcal{P}_2$ correspond to the rank 4 symplectic forms on U_5. By (1.4.1 (iii)) such a form f is determined by its radical V_1 (which is a 1-dimensional subspace in U_5) and by the non-singular symplectic form $\bar{f}$ on U_5/V_1 induced by f. The hyperplane determined by f will be denoted by $P(V_1, f)$ (although V_1 is determined by f, we prefer to keep the explicit reference to the radical). If $V_1 = U_1$ then the stabilizer S of $P(V_1, f)$ is contained in $G^{[04]}$ with index 28 (which is the number of non-singular symplectic forms on U_5/U_1). Furthermore, $S = Q^{[0]} : L^{[0]}(f)$ and by (1.5.1)

$$L^{[0]}(f) = S_f \cong 2^4 : S_4(2) \cong 2^4 : \mathrm{Sym}_6.$$

We also have the following 'contragredient' version of (3.6.1).

Lemma 3.6.2 *The group $L^{[0]} \cong L_5(2)$ acting on the set of non-zero vectors of $Q^{[0]}$ (considered as a $GF(2)$-module) has two orbits $\mathcal{Q}_1$ and $\mathcal{Q}_2$ with lengths 155 and 868, respectively. The vectors in $\mathcal{Q}_1$ are indexed by the 2-dimensional subspace in U_5 while the vectors in $\mathcal{Q}_2$ are indexed by the pairs (V_4, g), where V_4 is a hyperplane in U_5 and g is a non-singular symplectic form on V_4.*

Finally (3.6.1) and (3.6.2) are combined in the following.

Lemma 3.6.3 *The following assertions hold:*

(i) *a hyperplane from $\mathcal{P}_1$ contains 91 vectors from $\mathcal{Q}_1$ and 420 vectors from $\mathcal{Q}_2$;*

(ii) *a hyperplane from $\mathcal{P}_2$ contains 75 vectors from $\mathcal{Q}_1$ and 436 vectors from $\mathcal{Q}_2$;*

(iii) *a vector from $\mathcal{Q}_1$ corresponding to a 2-dimensional subspace V_2 of U_5 is contained in the hyperplane from $\mathcal{P}_1$ corresponding to a 3-dimensional subspace V_3 of U_5 if and only if $V_2 \cap V_3$ is non-zero;*

(iv) *a vector from $\mathcal{Q}_1$ corresponding to a 2-dimensional subspace V_2 is contained in the hyperplane from $\mathcal{P}_2$ corresponding to a pair (V_1, f) (where V_1 is a 1-dimensional subspace in U_5 and f is a non-singular symplectic form on U_5/V_1) if and only if either $V_1 \subseteq V_2$ or the image of V_2 in U_5/V_1 is a maximal totally isotropic subspace in $(U_5/V_1, f)$.*

3.7   Structure of $G^{[1]}$

By the definition from Section 3.5, $G^{[1]}$ is a semidirect product of $G^{[01]}$ and a group of order 2 generated by an element t_1. Furthermore, $G^{[01]}$ (through its identification with $H^{[01]}$) can be written as

$$G^{[01]} = Z^{[1]} A^{[1]} B^{[1]} L^{[1]},$$

where $L^{[1]} \cong L_4(2)$ with the natural module U_4, $Z^{[1]}$, $A^{[1]}$ and $B^{[1]}$ are elementary abelian 2-subgroups normalized by $L^{[1]}$ with

$$Z^{[1]} \cong \bigwedge{}^2 U_4, \quad A^{[1]} \cong B^{[1]} \cong U_4$$

(as $L^{[1]}$-modules). Therefore there are $L^{[1]}$-isomorphisms

$$\alpha : U_4 \to A^{[1]}, \quad \beta : U_4 \to B^{[1]} \quad \text{and} \quad \zeta : \bigwedge\nolimits^2 U_4 \to Z^{[1]}.$$

If W_2 is a complement to U_4 in $U_4^\perp$ with singular vectors u, v and the only non-singular vector $u + v$, then (with $G^{[01]}$ identified with $H^{[01]}$) the mappings α, β, and ζ are defined in terms of Siegel transformations as follows:

$$\alpha : w \mapsto s_{\langle w,u \rangle}, \quad \beta : w \mapsto s_{\langle w,v \rangle}, \quad \text{and} \quad \zeta : w_1 \wedge w_2 \mapsto s_{\langle w_1, w_2 \rangle},$$

where w, w_1, w_2 are distinct non-zero vectors in U_4. Then the *key commutator relation* is

$$[\alpha(w_1), \beta(w_2)] = \zeta(w_1 \wedge w_2).$$

Let $D^{[1]}$ be the subgroup in $G^{[01]}$ generated by the elements $\alpha(w)\beta(w)$ taken for all $w \in U_4^\#$. Then $D^{[1]}$ is elementary abelian and as a module for $L^{[1]}$ it is an indecomposable extension of $\bigwedge^2 U_4$ by U_4, isomorphic to a quotient of the $GF(2)$-permutation module of $L^{[1]}$ acting on the set of non-zero vectors in $U_4^\#$ (compare with the last paragraph of Section 3.4).

The element t_0 which extends $G^{[01]} = H^{[01]}$ to $H^{[1]}$ can be taken to be the orthogonal transvection associated with $u + v$. Then t_0 centralizes $Z^{[1]}L^{[1]}$ and permutes $\alpha(w)$ and $\beta(w)$ for every $w \in U_4^\#$, so that

$$C_{G^{[01]}}(t_0) = D^{[1]}L^{[1]} \cong 2^{6+4} : L_4(2).$$

The element t_1 which extends $G^{[01]}$ to $G^{[1]}$ can still be taken to be of order 2 and to induce the automorphism which is the product of the one induced by t_0 and the deck automorphism σ_γ as in (3.4.2 (vi)). This automorphism still normalizes $D^{[1]}L^{[1]}$ centralizing $D^{[1]}$ but no longer normalizes any one of the $L_4(2)$-complements. Put $Q^{[m]} = \langle D^{[1]}, t_1 \rangle$.

Lemma 3.7.1 *The subgroup*

$$Q^{[m]} := \langle t_1, \alpha(w)\beta(w) \mid w \in U_4 \rangle$$

is elementary abelian of order 2^{11}. Moreover, it is the only subgroup of order 2^{11} normal in $G^{[1]}$.

Proof Since t_1 is an involution commuting with $D^{[1]}$, the subgroup $Q^{[m]}$ is elementary abelian. Let $F^{[1]}$ be a normal subgroup of order 2^{11} in $G^{[1]}$. Then $F^{[1]} \cap G^{[01]}$ is a normal subgroup of order 2^a in $G^{[01]}$ for $a \in \{10, 11\}$. By (3.4.2 (iii)), $a = 10$ and $F^{[1]} \cap G^{[01]}$ is one of the following: $Z^{[1]}A^{[1]}$, $Z^{[1]}B^{[1]}$, or $D^{[1]}$. By the order consideration $F^{[1]}G^{[01]} = G^{[1]}$ and hence $F^{[1]} \cap G^{[01]}$ must be normal in $G^{[1]}$. Therefore $F^{[1]} = C_{G^{[1]}}(D^{[1]}) = Q^{[m]}$ and the result follows. ∎

By the above lemma we can treat $G^{[1]}$ as a semidirect product of $Q^{[m]} \cong 2^{11}$ and $A^{[1]}L^{[1]} \cong 2^4 : L_4(2)$.

3.8 A shade of a Mathieu group

Already at this stage we are in a position to relate $G^{[1]}$ and the semidirect product $\bar{\mathcal{C}}_{11} : M_{24}$ where M_{24} is the Mathieu group of degree 24 and $\bar{\mathcal{C}}_{11}$ is the irreducible Todd module (cf. Section 11).

Lemma 3.8.1 *Let X be the semidirect product of the Todd module $\bar{\mathcal{C}}_{12}$ and the stabilizer M_b of an octad in M_{24} (the semidirect product is with respect to the natural action of M_{24} restricted to M_b). Then*

(i) *the commutator subgroup X' of X is isomorphic to $H^{[01]} = G^{[01]}$;*
(ii) *$X \cong \operatorname{Aut} G^{[01]}$;*
(iii) *the subgroup $\bar{\mathcal{C}}_{11} : M_b$ of X is isomorphic to $G^{[1]}$.*

Proof Let $Q_b = O_2(M_b)$. By (11.3.1) X' is the semidirect product $\bar{\mathcal{E}}_{10} : M_b$, where $\bar{\mathcal{E}}_{10} = [\bar{\mathcal{C}}_{11}, Q_b]$ is the only hyperplane in $\bar{\mathcal{C}}_{11}$ stabilized by M_b. If B is the octad stabilized by M_b then $\bar{\mathcal{E}}_{10}$ consists of the 2-element subsets of P_{24} and the sextets which intersect B evenly (for a sextet this means that every tetrad intersects B in an even number of elements). By (11.3.1) the only proper M_b-submodule in $\bar{\mathcal{E}}_{10}$ is

$$C_{\bar{\mathcal{E}}_{10}}(Q_b) \cong \bigwedge{}^2 Q_b$$

and the corresponding quotient is isomorphic to Q_b.

We claim that $\bar{\mathcal{E}}_{10}$, as a module for $K_b \cong L_4(2)$, is an indecomposable 10-dimensional quotient of the permutation module on the non-identity elements of Q_b (compare the last paragraph from Section 3.4). Let $p \in P_{24} \backslash B$ be the element stabilized by K_b, put

$$S = \{\{p, r\} \mid r \in P_{24} \backslash (B \cup \{p\})\}$$

and define a mapping $\omega : Q_b \to S$ by $\omega : q \mapsto \{p, p^q\}$. Then clearly ω is a bijection which commutes with the action of K_b. Now it only remains to show that S (considered as a subset of $\bar{\mathcal{E}}_{10}$) generates the whole of $\bar{\mathcal{E}}_{10}$. But since every 2-element subset of $P_{24} \backslash B$ is either in S already or is the symmetric difference of a pair of subsets from S, this is quite clear. Now take a look at

$$G^{[01]} = Z^{[1]} A^{[1]} B^{[1]} L^{[1]}.$$

Let λ be an isomorphism of $L^{[1]}$ onto K_b which induces an isomorphism μ of U_4 onto Q_b. Then we can uniquely define isomorphisms of $A^{[1]}$ onto Q_b and of $D^{[1]}$ onto $\bar{\mathcal{E}}_{10}$ (the latter maps $\alpha(w)\beta(w)$ onto $\{p, p^{\mu(w)}\}$ for every $w \in U_4$) commuting with λ. It is straightforward to check using the uniqueness property of the key commutator relations in (3.4.1) that in this way we get an isomorphism of $G^{[01]}$ onto $\bar{\mathcal{E}}_{10} : M_b$ establishing (i). By (3.4.2 (vi)) Out $(G^{[01]})$ has order 4. Since $[X : X'] = 4$ and the centralizer of X' in X is trivial, (i) implies (ii).

By (3.4.2 (vi)) the outer automorphism group of $G^{[01]}$ is elementary abelian of order 4 and its non-identity elements are the images of the automorphisms τ_0, τ_1 and $\tau_0\tau_1 = \sigma_\gamma$. Furthermore, τ_0 commutes with $L^{[1]}$, $\tau_0\tau_1$ commutes with $A^{[1]}$ while τ_1 commutes with none of those two subgroups. On the other hand, let $r \in B$. Then $\{p\}$, $\{p, r\}$, and $\{r\}$ (considered as vectors in $\bar{C}_{12}$) generate a 2-dimensional subspace which complements $\bar{\mathcal{E}}_{10}$. The automorphism of X' induced by $\{p\}$ commutes with K_p while the automorphism induced by r commutes with Q_b. Since $\{p, r\} \in \bar{C}_{11}$, (iii) follows. ∎

Lemma 3.8.2 *The group* $A^{[1]}L^{[1]} \cong 2^4 : L_4(2)$ *acting on the non-identity elements of the subgroup* $Q^{[m]} \cong 2^{11}$ *as in (3.7.1) has exactly six orbits with lengths 28, 35, 120, 840, 128, and 896 with stabilizers isomorphic to* $2^4 : S_4(2)$, $2^4 : 2^4 : (\text{Sym}_3 \times \text{Sym}_3)$, $2^4 : L_3(2)$, $\text{Sym}_4 \times 2$, Alt_7, $(\text{Sym}_3 \times \text{Sym}_5)$. *The former two orbits are in* $Z^{[1]}$ *while the former four are in* $D^{[1]}$.

Proof Because of the isomorphism established in (3.8.1) in order to determine the orbits in question it is sufficient to calculate the orbits of M_b on the 2-element subsets of P_{24} and on the sextets. An orbit on 2-element subsets is determined by the size of the intersections with B. These sizes are 2, 1, or 0 while the corresponding orbit lengths are 28, 128, and 120. The orbits on the sextets can be read from the diagram $D_b(M_{24})$ in Section 11.5, their lengths are 35, 840, and 896. For the structure of the stabilizers we refer to Lemma 3.7.2 in Ivanov (1999). ∎

It is worth noticing that the orbit lengths of the actions of $A^{[1]}L^{[1]}$ on $D^{[1]}$ and $Z^{[1]}B^{[1]}$ are the same, although the actions are not isomorphic.

Exercises

1. Prove (3.1.10) and (3.6.3).
2. Calculate directly the orbit lengths of $A^{[1]}L^{[1]}$ on $Q^{[m]}$.
3. Define the Cayley graph Ξ on $Q^{[m]}$ with respect to the union of the $A^{[1]}L^{[1]}$-orbits of lengths 28, 120, and 128 (so that Ξ is regular of valency 276). Prove that the automorphism group of Ξ is the semidirect product $\bar{C}_{11} : M_{24}$.

4

PENTAD GROUP $2^{3+12} \cdot (L_3(2) \times \mathrm{Sym}_5)$

The geometric subgroup at level 2 of the modified amalgam $\mathcal{G}$ is known as the *pentad group* and it has a rather complicated structure. In a very beautiful way the structure can be described in terms of the non-singular 4-dimensional symplectic space over $GF(2)$ and the associated orthogonal spaces. This makes the pentad group an example of a class of so-called tri-extraspecial groups studied by S. V. Shpectorov and the author in Ivanov and Shpectorov (submitted).

4.1 Geometric subgroups and subgraphs

Let $\mathcal{G} = \{G^{[0]}, G^{[1]}\}$ be the amalgam defined in (3.5.1). We believe it is appropriate to recall here the abstract definition of $\mathcal{G}$ (we continue to assume that $G^{[0]} = H^{[0]}$ and $G^{[01]} = H^{[01]}$).

Let U_5 be a 5-dimensional $GF(2)$-vector space, let $L^{[0]} \cong L_5(2)$ be the general linear group of U_5, and let $Q^{[0]} = \bigwedge^2 U_5$ be the exterior square of U_5. Then $G^{[0]}$ is the semidirect product of $Q^{[0]}$ and $L^{[0]}$ with respect to the natural action. We assume that $G^{[0]}$ acts on U_5 so that $Q^{[0]}$ is the kernel and $G^{[0]}/Q^{[0]} \cong L_5(2)$ acts in the natural way. Let U_4 be a hyperplane in U_5. Then $G^{[01]} = G^{[0]} \cap G^{[1]}$ is the stabilizer of U_4 in $G^{[0]}$, so that $G^{[01]}$ is the semidirect product of $Q^{[0]}$ and $L^{[0]}(U_4) \cong 2^4 : L_4(2)$. The structure of $G^{[01]}$ can be described as follows. Let

$$d : \begin{bmatrix} U_5 \\ 2 \end{bmatrix} \to Q^{[0]}$$

be the injection which commutes with the action of $L^{[0]}$ (cf. (1.3.4 (i))). Let $Z^{[1]}$ be the subgroup in $Q^{[0]}$ generated by the images under d of the 2-dimensional subspaces contained in U_4. Then $Z^{[1]}$ is normal in $G^{[01]}$. Furthermore, restricting d to $\begin{bmatrix} U_4 \\ 2 \end{bmatrix}$ and applying (1.3.4) we obtain an isomorphism

$$\zeta : \bigwedge^2 U_4 \to Z^{[1]},$$

which commutes with the action of $G^{[01]}$. Let $u \in U_5 \backslash U_4$, so that

$$L^{[1]} := L^{[0]}(U_4, u) \cong L_4(2).$$

For $w \in U_4$ let $\alpha(w) = d(\langle w, u \rangle)$, let $\beta(w)$ be the transvection of U_5 with centre w and axis U_4 (this transvection is an element of $L^{[0]}$). Put

$$A^{[1]} = \langle \alpha(w) \mid w \in U_4 \rangle, \quad B^{[1]} = \langle \beta(w) \mid w \in U_4 \rangle.$$

Then $A^{[1]}$ is a complement to $Z^{[1]}$ in $Q^{[0]}$, $B^{[1]} = O_2(L^{[0]}(U_4))$, the above defined mappings $\alpha : U_4 \to A^{[1]}$ and $\beta : U_4 \to B^{[1]}$ are isomorphisms of $L^{[1]}$-modules and the *key commutator relation*

$$[\alpha(w_1), \beta(w_2)] = \zeta(w_1 \wedge w_2)$$

(which holds for all $w_1, w_2 \in U_4$) describes $Q^{[1]} := Z^{[1]} A^{[1]} B^{[1]} = O_2(G^{[01]})$ up to isomorphism. Then $G^{[1]}$ is a semidirect product of $G^{[01]}$ and a group of order 2 with generator t_1, such that t_1 (acting by conjugation) permutes $\alpha(w)$ and $\beta(w)$ for every $w \in U_4$ and t_1 induces a non-trivial deck automorphism of $Z^{[1]} L^{[1]} \cong 2^6 : L_4(2)$. This means that t_1 centralizes both $Z^{[1]}$ and $Z^{[1]} L^{[1]}/Z^{[1]} \cong L_4(2)$.

The subgroup $P := \langle Z^{[1]} L^{[1]}, t_1 \rangle$ is isomorphic to the semidirect product with respect to the natural action of $\mathcal{P}_8/\mathcal{P}_8^c \cong 2^7$ and $L^{[1]}$, where $\mathcal{P}_8$ is the $GF(2)$-permutation module of the natural action of $L^{[1]} \cong \mathrm{Alt}_8$ of degree 8. The group $G^{[1]}$ can be described as a partial semidirect product of $Q^{[1]}$ and P (for the definition of the partial semidirect product see p. 28 in Gorenstein (1968)). By (3.8.1) and (3.8.2) $G^{[1]}$ can also be described as a semidirect product of $Q^{[m]} = \langle t_1, \alpha(w)\beta(w) \mid w \in U_4 \rangle \cong 2^{11}$ and $A^{[1]} L^{[1]} \cong 2^4 : L_4(2)$.

Let

$$\varphi : \mathcal{G} \to G$$

be a faithful generating completion of $\mathcal{G}$ (so far we have not got any particular completion to think about, but we can always think about the universal completion).

As in Section 2.3 we associate with the triple $(\mathcal{G}, \varphi, G)$ a graph

$$\Gamma = \Lambda(\mathcal{G}, \varphi, G)$$

whose vertices and edges are the cosets of $\varphi(G^{[0]})$ and $\varphi(G^{[1]})$ in G and the incidence is via non-emptiness of intersections. Notice that Γ is the incidence graph of the cosets geometry $\mathcal{F}(G, \mathcal{G})$. The vertex and the edge of Γ corresponding to the cosets of $G^{[0]}$ and $G^{[1]}$ containing the identity will be denoted by x and $\{x, y\}$, respectively. Hopefully it will always be clear from the context which graph is under consideration.

Lemma 4.1.1 *The following assertions hold:*

(i) Γ *is an undirected connected graph of valency* 31 *without multiple edges;*

(ii) G *acts faithfully on* Γ *and this action is transitive on the set of incident vertex–edge pairs;*

(iii) $G(x) = G^{[0]}$, $G\{x, y\} = G^{[1]}$;

(iv) *there is a unique projective space structure* Π_x *on* $\Gamma(x)$ *preserved by* $G(x)$ *with points indexed by the hyperplanes in* U_5;

(v) $G(x)^{\Gamma(x)} \cong L_5(2)$ *is the full automorphism group of* Π_x;

(vi) $G_1(x) = Q^{[0]}$;

(vii) $G_1(y) = Z^{[1]} B^{[1]}$ *induces on* $\Gamma(x)$ *an elementary abelian group of order* 2^4 *whose non-identity elements are the transvections of* Π_x *with centre* y;

(viii) $G\{x, y\}$ *stabilizes a unique bijection between* $L_x(y)$ *and* $L_y(x)$ *(where* $L_u(v)$ *is the set of lines in* Π_u *passing through* $v \in \Gamma(u)$*).*

Proof Since the properties of $\mathcal{H}$ recorded in the third column of Table 1 in Section 2.3 are inherited by the amalgam $\mathcal{G}$, all the properties in the second column hold with (G, Γ) in place of (H, Ω) which gives (i) to (vi). The element t_1 maps x onto y and conjugates $G_1(x) = Z^{[1]}A^{[1]}$ onto $G_1(y) = Z^{[1]}B^{[1]}$. On the other hand, $B^{[1]}$ is of order 2^4 trivially intersecting $Q^{[0]}$, which gives (vii). Since t_1 commutes with $G(x, y)/(G_1(x)G_1(y)) \cong L_4(2)$, (viii) follows. ∎

We can define geometric subgroups in G in the following way. For $2 \leq i \leq 4$ set $G^{[0i]} = H^{[0i]}$, $G^{[01i]} = H^{[01i]}$, $G^{[1i]} = \langle G^{[01i]}, t_1 \rangle = N_{G^{[1]}}(G^{[01i]})$ and put

$$G^{[i]} = \langle \varphi(G^{[0i]}), \varphi(G^{[1i]}) \rangle.$$

The important property of t_1 we use here is that it centralizes $G^{[01]}/O_2(G^{[01]})$.

It is clear that there is a maximal flag

$$0 < U_1 < U_2 < U_3 < U_4 < U_5$$

in U_5 such that the geometric subgroup $H^{[i]}$ in $H \cong O_{10}^+(2)$ is the stabilizer of U_{5-i} in H. Since $G^{[0]}$ still acts on U_5 we can define $G^{[0i]}$ and $G^{[01i]}$ to be stabilizers of U_{5-i} in $G^{[0]}$ and $G^{[01]}$, respectively (here $2 \leq i \leq 5$), while the rest of the definition stays as it is.

Having the geometric subgroups we can define the geometric subgraphs $\Gamma^{[i]}$ for $2 \leq i \leq 4$ as follows. The vertices and edges of $\Gamma^{[i]}$ are the images under $G^{[i]}$ of the vertex x and the edge $\{x, y\}$, respectively. By the definition $(x, \{x, y\})$ is an incident pair and every image of an incident pair is incident as well.

4.2 Kernels and actions

We continue to use the notation introduced in the previous section. Put $\mathcal{G}^{[i]} = \{G^{[0i]}, G^{[1i]}\}$, denote by $\varphi^{[i]}$ the restriction of φ to $\mathcal{G}^{[i]}$ and consider the cosets graph

$$\Lambda^{[i]} = \Lambda(\mathcal{G}^{[i]}, \varphi^{[i]}, G^{[i]}).$$

Then $\Lambda^{[i]}$ is of valency $2^i - 1 = [G^{[0i]} : G^{[01i]}]$ and the natural action of $G^{[i]}$ on $\Lambda^{[i]}$ is locally projective of type $(i, 2)$. We can further define a mapping

$$\mu^{[i]} : \Lambda^{[i]} \to \Gamma$$

via $gG^{[0i]} \mapsto gG^{[0]}$. The mapping $\mu^{[i]}$ is an injection if and only if $G^{[i]}$ is a proper subgroup in G (cf. Lemma 9.6.4 in Ivanov (1999)) in which case the image of μ_i is the geometric subgraph $\Gamma^{[i]}$.

Let $N^{[i]}$ denote the largest subgroup from $G^{[01i]}$ which is normal in both $G^{[0i]}$ and $G^{[1i]}$ for $2 \leq i \leq 4$.

Lemma 4.2.1 *The following assertions hold:*

(i) *the subgroup $N^{[i]}$ is independent of the choice of the completion $\varphi : \mathcal{G} \to G$;*

(ii) $N^{[i]}$ *is the kernel of the action of $G^{[i]}$ on $\Lambda^{[i]}$;*
(iii) *if $G^{[i]}$ is proper in G, then $N^{[i]}$ is the vertexwise stabilizer of $\Gamma^{[i]}$ in $G^{[i]}$;*
(iv) *whenever $N^{[i]}$ is non-identity, $G^{[i]}$ is proper in G;*
(v) $N^{[4]} \leq N^{[3]} \leq N^{[2]}$.

Proof Since $N^{[i]}$ is defined solely in terms of $\mathcal{G}^{[i]}$, (i) is immediate. Since $N^{[i]}$ is the largest subgroup in $G^{[01i]}$ which is normal in both $G^{[0i]}$ and $G^{[1i]}$, (ii) follows and implies (iii). Since G acts faithfully on Γ, it does not contain normal subgroups inside $G^{[0]}$. Therefore $N^{[i]}$, in a non-identity group, cannot possibly be normal in G and (iv) follows. Suppose that G is the universal completion of $\mathcal{G}$. Then $N^{[i]}$ is the vertexwise stabilizer in G of the geometric subgraph $\Gamma^{[i]}$. Since

$$\Gamma^{[2]} \subset \Gamma^{[3]} \subset \Gamma^{[4]},$$

we obtain (v). ∎

Lemma 4.2.2 *Let $\widehat{G^{[i]}}$ be the image of $G^{[i]}$ in the outer automorphism group of $N^{[i]}$ for $2 \leq i \leq 4$. Then*

(i) $\widehat{G^{[i]}} = G^{[i]}/(N^{[i]}C_{G^{[i]}}(N^{[i]}))$;
(ii) $\widehat{G^{[i]}}$ *is generated by the images in $\mathrm{Out}\, N^{[i]}$ of $G^{[0i]}$ and $G^{[1i]}$ (equivalently of $G^{[0i]}$ and t_1);*
(iii) $\widehat{G^{[i]}}$ *is independent of the choice of the completion $\varphi : \mathcal{G} \to G$;*
(iv) $|G^{[i]}| \geq |N^{[i]}| \cdot |\widehat{G^{[i]}}|$.

Proof Statement (i) holds by the definition, while (ii) is clear since $G^{[i]} = \langle G^{[0i]}, G^{[1i]} \rangle$. Since, in (ii), the group $\widehat{G^{[i]}}$ is defined in terms of $\mathcal{G}$, (iii) follows. Statement (iv) also follows from (i) and the order formula for products. ∎

Recall that if the completion $\varphi : \mathcal{G} \to G$ is universal then G is the free product of $G^{[0]}$ and $G^{[1]}$ amalgamated over the common subgroup $G^{[01]}$. In this case $G^{[i]}$ is also a free amalgamated product and it is proper in G. Finally each of Γ, $\Gamma^{[i]}$, and $\Lambda^{[i]}$ is a tree of a suitable valency.

We will explore the relationship between the $N^{[i]}$ and the subgroups $K^{[i]}$ associated with the geometric subgroups in $H \cong O_{10}^+(2)$ as in (2.1.3 (vi)). The following data can be read from (2.1.2) and (2.1.3 (vi)). Because of our identification of $G^{[0]}$ and $H^{[0]}$ we consider the $K^{[i]}$'s as subgroups in $G^{[0]}$.

Lemma 4.2.3 *For $2 \leq i \leq 4$, let $K^{[i]}$ be the kernel of the action of the geometric subgroup $H^{[i]}$ in $H \cong O_{10}^+(2)$ on the geometric subgraph $\Omega^{[i]}$ in the dual polar graph $\Omega = D^+(10, 2)$. Then*

(i) $K^{[2]}$ *is a semidirect product of a special group $Q^{[2]}$ (of order 2^{15} with centre $Z^{[2]}$ of order 2^3) and $L^{[2]} \cong L_3(2)$, every chief factor of $K^{[2]}$ inside $Q^{[2]}$ is either the natural or the dual natural module of $K^{[2]}/Q^{[2]} \cong L_3(2)$;*

(ii) $K^{[3]}$ *is a semidirect product of an extraspecial group* $Q^{[3]} \cong 2^{1+12}_+$ *and a group* $L^{[3]} \cong L_2(2) \cong \mathrm{Sym}_3$, *a Sylow 3-subgroup of* $L^{[3]}$ *acts fixed-point freely on* $Q^{[3]}/Z^{[3]} \cong 2^{12}$ *where* $Z^{[3]}$ *is the centre of* $Q^{[3]}$;

(iii) $K^{[4]} = Q^{[4]}$ *is elementary abelian of order* 2^8.

In order to get a lower bound for the order of the $N^{[i]}$ we present a restriction on the possible composition factors of $G^{[0i]}/N^{[i]}$ which comes from a well-known property of graphs with a locally projective action.

Lemma 4.2.4 *Let* Ψ *be a connected graph of valency* $2^k - 1$ *for* $k \geq 2$ *and* F *be a group of automorphisms of* Ψ *such that*

(i) $F(a)$ *is finite for every vertex* a *of* Ψ;

(ii) F *acts transitively on the set of incident vertex–edge pairs in* Ψ;

(iii) *for every vertex* a *in* Ψ *the group* $F(a)$ *induces on the set* $\Psi(a)$ *of neighbours of* a *in* Ψ *the natural doubly transitive action of* $L_k(2)$;

(iv) *if* $\{a,b\}$ *is an edge of* Ψ *then there is a bijection* ψ_{ab} *of* $L_a(b)$ *onto* $L_b(a)$ *commuting with the action of the stabilizer of* $\{a,b\}$ *in* F *(where* $L_a(b)$ *is the set of lines passing through* b *in the unique projective space* Π_a *on* $\Psi(a)$ *preserved by* $F(a)$ *and similarly for* $L_b(a)$*).*

Then $F_1(a)$ *(which is the vertexwise stabilizer of* $\Psi(a)$ *in* $F(a)$*) is a 2-group, so that* $F(a)/O_2(F(a)) \cong L_k(2)$.

Proof We claim that $F_1(a)$ induces a 2-group on $\Psi(b)$ for every $b \in \Psi(a)$. In fact $F_1(a)$ fixes every line $L_a(b)$ pointwise and by (iv) it stabilizes every line in $l \in L_b(a)$ as a set. Since $l = \{a,c,d\}$ for some $c,d \in \Psi(b)\backslash\{a\}$ and a is fixed by $F_1(a)$, the latter can only permute the vertices c and d thus inducing on l an action of order at most 2. Since every vertex from $\Psi(b)$ is on a line from $L_b(a)$, the claim follows. Thus $O^2(F_1(a))$ acts trivially on $\Psi(b)$ and hence $O^2(F_1(a)) \leq F_1(b)$. Now it is easy to apply induction and to use the connectivity of Ψ to conclude that $O^2(F_1(a))$ fixes all the vertices of Ψ. This gives the result since by definition the action of F on Ψ is faithful. ∎

Notice that the existence and uniqueness of the projective space structure Π_a in (4.2.4 (iv)) can be established as in the proof of (2.2.1).

Lemma 4.2.5 *The following assertions hold:*

(i) $N^{[2]}$ *contains* $K^{[2]}$;

(ii) $N^{[3]}$ *contains the commutator subgroup of* $K^{[3]}$, *which is of index 2 in* $K^{[3]}$.

Proof By (4.2.4) we know that the only chief factor of $G^{[0i]}/N^{[i]}$ of order greater than 2 is isomorphic to $L_i(2)$ if $i = 3$ or 4 and to the cyclic group of order 3 if $i = 2$. On the other hand it is immediate from the structure of the maximal parabolic subgroups in $G^{[0]}/O_2(G^{[0]}) \cong L_5(2)$ that $G^{[0i]}/O_2(G^{[0i]})$ is isomorphic to

$$\mathrm{Sym}_3 \times L_3(2), \quad L_3(2) \times \mathrm{Sym}_3, \quad \text{and} \quad L_4(2)$$

for $i = 2$, 3, and 4, respectively. Therefore $L^{[2]} \cong L_3(2)$ is contained in $N^{[2]}$. By (4.2.3 (i)) $[L^{[2]}, G^{[02]}] = K^{[2]}$ which gives (i). Similarly, if T is a Sylow 3-subgroup of $L^{[3]} \cong \mathrm{Sym}_3$ then T is in $N^{[3]}$ while by (4.2.3 (ii)) $[T, G^{[03]}] = Q^{[3]}T$. By noticing that $Q^{[3]}T$ is the only index 2 subgroup in $K^{[3]}$ we obtain (ii). ∎

The above discussion says nothing about $N^{[4]}$. We will see in due course that in fact this group is trivial.

Lemma 4.2.6 $N^{[2]} = K^{[2]}$.

Proof By (4.2.5 (i)) and (4.2.1 (iv)) we know that $G^{[2]}$ is a proper subgroup in G and hence the geometric subgraph $\Gamma^{[2]}$ exists. Since $G^{[02]}/K^{[2]} \cong H^{[02]}/K^{[2]} \cong \mathrm{Sym}_3 \times 2$, it is sufficient to show that the action induced by $G^{[02]}$ on $\Gamma^{[2]}$ has order 12 at least. Clearly $G^{[02]}$ induces Sym_3 on $\Gamma^{[2]}(x)$. By (4.1.1 (vii)) it is easy to see that $G_1(x)$ induces on $\Gamma^{[2]}(y)$ an action of order 2. Hence the result. ∎

Lemma 4.2.7 *The following assertions hold:*

(i) $O_2(G^{[02]}/N^{[2]}) = O_2(G^{[02]})N^{[2]}/N^{[2]}$;
(ii) $G^{[12]}/N^{[2]} \cong D_8$;
(iii) $A^{[1]}B^{[1]}N^{[2]}/N^{[2]}$ *and* $Q^{[m]}N^{[2]}/N^{[2]}$ *are the two elementary abelian subgroups of order 4 in* $G^{[12]}/N^{[2]}$.

Proof A comparison of the proofs (4.2.5 (i)) and (4.2.6) gives (i). Since $G^{[012]}/N^{[2]} \cong 2^2$ and since $G^{[12]}\backslash G^{[012]}$ contains the involution t_1, (ii) follows. Since D_8 contains only two elementary abelian subgroups of order 4, we also obtain (iii). ∎

Lemma 4.2.8 $[K^{[3]} : N^{[3]}] = 2$.

Proof We follow the notation in the proof of (4.2.5). So that T is a Sylow 3-subgroup of $L^{[3]} \cong \mathrm{Sym}_3$ which is also a Sylow 3-subgroup of $O_{2,3}(K^{[3]})$. By (4.2.3) $Q^{[3]}T$ is normal in both $G^{[03]}$ and $G^{[13]}$ and we have to show that $K^{[3]}$ is not normal in at least one of these groups. It is clear that $K^{[3]}$ is normal in both $G^{[03]} = H^{[03]}$ and in $H^{[13]}$. In addition

$$H^{[13]} = \langle H^{[013]}, t_0 \rangle, \quad G^{[13]} = \langle H^{[013]}, t_1 \rangle,$$

where the automorphism of $G^{[01]} = H^{[01]}$ induced by t_1 is that induced by t_0 multiplied by the deck automorphism σ_γ (compare Section 3.5). Therefore in order to show that $K^{[3]}$ is not normal in $G^{[13]}$, it is sufficient to take an involution $s \in L^{[3]}$ and show that $[s, \sigma_\gamma] \notin Q^{[3]}T$.

By the definition $L^{[3]} \cong L_2(2) \cong \mathrm{Sym}_3$ is the vectorwise stabilizer in $G^{[03]} = H^{[03]}$ of W_6 and also stabilizes T_2. Here as in Section 2.1 W_6 is a complement to U_2 in $U_2^\perp$ and T_2 is a maximal totally singular subspace in $W_6^\perp \cong V_4^+$ disjoint from U_2. By our notation convention in Section 2.1, $U_2 < U_4$ and $U_4 \cap W_6$ is a complement to U_2 in U_4. Therefore $L^{[3]} < L^{[1]}$ and $[L^{[3]}, \sigma_\gamma] \leq Z^{[1]} L^{[1]}$ and we only have to deal with the restriction of σ_γ to $Z^{[1]} L^{[1]}$ which is a deck automorphism d as in Section 3.3. Now it only remains to relate our current notation with that in the paragraph after the proof of (3.3.5). We have $U = U_4$, $W = W^{(1)} = U_2$ and $W^{(2)} = U_4 \cap W_6$, $T^{(2)} \langle r^{(1)} \rangle = L^{[3]} = T \langle s \rangle$. Then by (3.3.6 (iii)) $[s, d] = \delta$, where δ corresponds to a partition of P_8 into 2- and 6- element subsets (when $Z^{[1]}$ is treated as the heart of the $GF(2)$-permutation module). Suppose that $\delta \in Q^{[3]} T$. Since δ is of order 2, we must have $\delta \in Q^{[3]}$. Since T acts fixed-point freely on $Q^{[3]}/Z^{[3]}$, while T centralizes δ, we further get $\delta \in Z^{[3]}$. On the other hand $Z^{[3]}$ is of order 2, generated by the Siegel transformation s_{U_2} corresponding to U_2. This is a contradiction since s_{U_2} is a singular vector while δ is a non-singular vector with respect to the quadratic form on $Z^{[1]} \cong V_6^+$ preserved by $L^{[1]} \cong \Omega_6^+(2)$. In other terms s_{U_2} belongs to the 35-orbit while δ belongs to the 28-orbit of $L^{[1]}$ on the set of non-identity elements of $Z^{[1]}$. ∎

4.3 Inspecting $N^{[2]}$

The ultimate goal of this chapter is to show that $\widehat{G^{[2]}}$ is isomorphic to Sym_5. For this we first analyse the structure of $N^{[2]}$ and calculate its automorphism group. Since $N^{[2]} = K^{[2]}$ is contained in $G^{[0]}$, now identified with $H^{[0]}$, we can carry out calculations inside H, being careful not to miss the difference between the abstract properties of $N^{[2]}$ and those associated with its embedding into H.

Let us recall the geometric setting from Section 2.1. Let U_3 be the 3-dimensional totally singular subspace in

$$(V, f, q) \cong V_{10}^+$$

such that $H^{[2]}$ is the stabilizer of U_3 in $H \cong O_{10}^+(2)$. Let W_4 be a complement to U_3 in $U_3^\perp$, so that $W_4 \cong V_4^+$ and let T_3 be a maximal totally singular subspace in $W_4^\perp \cong V_6^+$ disjoint from U_3. The following lemma makes a modest refinement of (2.1.2) and (2.1.3 (vi)).

Lemma 4.3.1 *The group $N^{[2]} = K^{[2]}$ is the semidirect product of $Q^{[2]}$ and $L^{[2]}$, where*

(i) *$L^{[2]} \cong L_3(2)$ is generated by the Siegel transformations $s_{\langle u,t \rangle}$ taken for all the orthogonal pairs (u,t), $u \in U_3^\#$, $t \in T_3^\#$;*

(ii) *$Q^{[2]}$ is a special group of order 2^{15} whose centre $Z^{[2]}$ is elementary abelian of order 2^3 generated by the Siegel transformations $s_{\langle u,v \rangle}$ taken for all distinct pairs $u, v \in U_3^\#$;*

(iii) *if w is a singular vector in W_4, then the Siegel transformations $s_{\langle v,w \rangle}$ taken for all $v \in U_3^\#$ generate in $Q^{[2]}$ an elementary abelian subgroup $E(w)$ of order 2^3 disjoint from $Z^{[2]}$ and normalized by $L^{[2]}$.*

By (2.1.2 (vii)) $Q^{[2]}/Z^{[2]}$, as a module for $H^{[2]}/Q^{[2]} \cong L_3(2) \times O_4^+(2)$, is isomorphic to the tensor product of U_3 and W_4. Thus, as a module for $L^{[2]}$ it is isomorphic to the direct sum of four copies of U_3. Therefore $Q^{[2]}$ contains exactly $15 = 2^4 - 1$ subgroups of order 2^6 containing $Z^{[2]}$ and normal in $N^{[2]}$. For a singular vector $w \in W_4$ put $D(w) = Z^{[2]} E(w)$, where $E(w)$ is as in (4.3.1 (iii)). For a non-singular vector $t \in W_4$ write $t = a + b$ where a and b are singular vectors in W_4 and put

$$D(t) = \langle s_{\langle a,u \rangle} s_{\langle b,u \rangle} \mid u \in U_3^{\#} \rangle.$$

Lemma 4.3.2 *In the above notation the following assertions hold:*

(i) *for every $v \in W_4^{\#}$ the subgroup $D(v)$ is elementary abelian of order 2^6, contains $Z^{[2]}$, and is normalized by $L^{[2]}$;*

(ii) *the subgroup $D(v)$ for non-singular v is independent of the choice of a and b.*

Proof If v is singular then $D(v)$ contains $Z^{[2]}$ by definition and, by (4.3.1 (iii)), it is elementary abelian normalized by $L^{[2]}$. If v is non-singular the result follows from the commutator relations in (3.4.1). Since there are exactly 15 subgroups of order 2^6 in $Q^{[2]}$ containing $Z^{[2]}$ and normal in $N^{[2]}$ they all are of the form $D(v)$ for some $v \in W_4^{\#}$. In particular (ii) holds. ∎

By (4.3.2),

$$\mathcal{E} = \{D(v) \mid v \in W_4^{\#}\}$$

is the set of all subgroups of order 2^6 in $Q^{[2]}$ containing $Z^{[2]}$ and normal in $N^{[2]}$. The subgroups from $\mathcal{E}$ will be called *dents*. Define a binary operation on $\mathcal{E}$ in the following way. For distinct $D_1, D_2 \in \mathcal{E}$ put $F = \langle D_1, D_2 \rangle$. Then $F/Z^{[2]}$ is the direct sum of $D_1/Z^{[2]}$ and $D_2/Z^{[2]}$, and therefore it contains just one further 3-dimensional $N^{[2]}$-submodule whose preimage is a dent from $\mathcal{E}$ which we denote by $D_1 + D_2$.

Lemma 4.3.3 *In the above terms $D(v) + D(w) = D(v + w)$.*

Proof The equality is immediate from the fact that $Q^{[2]}/Z^{[2]}$ carries the tensor product structure $U_3 \otimes W_4$. ∎

By (4.3.3) we can recover the vector space structure of W_4 by adjoining zero to the set $\mathcal{E}$ of dents and defining the addition as in the paragraph before (4.3.3) together with the obvious rule for the zero element. Next we show that the symplectic space structure can also be recovered. For $D_1, D_2 \in \mathcal{E}$ put

$$h(D_1, D_2) = \begin{cases} 0 & \text{if } \langle D_1, D_2 \rangle \text{ is abelian;} \\ 1 & \text{otherwise.} \end{cases}$$

Lemma 4.3.4 $h(D(v), D(w)) = 0$ *if and only if v and w are perpendicular in W_4.*

Proof By (4.3.3) $\langle D(v), D(w)\rangle = \langle D(v), D(v+w)\rangle$. Hence without loss we can assume that v and w are either both singular or both non-singular. In the former case, if v and w are perpendicular, then $\langle U_3, v, w\rangle$ is totally singular and $[E(w), E(v)] = 1$ by (1.3.5). If v and w are singular but non-perpendicular, then $v + w$ is non-singular and the commutator relations in (3.4.1) imply that $[E(v), E(w)] = Z^{[2]}$.

Suppose now that v and w are non-singular and perpendicular. Then we can find perpendicular hyperbolic pairs (a_1, a_2) and (b_1, b_2) such that $v = a_1 + a_2$ and $w = b_1 + b_2$. Since $\langle U_3, a_i, b_j\rangle$ is totally singular for $1 \le i, j \le 2$, every generator of $D(w)$ (as defined before (4.3.2)) commutes with every generator of $D(w)$ and hence $[D(v), D(w)] = 1$. Finally, if v and w are non-singular and non-perpendicular, the result is easy to deduce from the commutator relations in (3.4.1).   ∎

By (4.3.4) the mapping $v \mapsto D(v)$ extends to an isomorphism of the symplectic spaces $(W_4, f|_{W_4})$ and $(\mathcal{E} \cup \{0\}, h)$.

Define $q^{L^{[2]}}$ to be a $GF(2)$-valued function on $\mathcal{E}$ such that $q^{L^{[2]}}(D) = 0$ if D is a semisimple module for $L^{[2]}$ and $q^{L^{[2]}}(D) = 1$ if D is indecomposable.

Lemma 4.3.5 $q^{L^{[2]}}(D(v)) = 0$ *if and only if v is singular.*

Proof If v is singular then $D(v)$ is the direct sum of $L^{[2]}$-submodules $Z^{[2]}$ and $E(v)$ (where the latter is defined in (4.3.1 (iii))). If v is non-singular then the generators of $D(v)$ (as defined before (4.3.2)) form the only $L^{[2]}$-orbit of length 7 on $D(v)\backslash Z^{[2]}$. The commutator relations in (3.4.1) now show that $D(v)$ is indecomposable.   ∎

By (4.3.5) the mapping $v \mapsto D(v)$ extends to an isomorphism

$$(W_4, f|_{W_4}, q|_{W_4}) \cong (\mathcal{E} \cup \{0\}, h, q^{L^{[2]}})$$

of quadratic spaces. Thus it appears that the quadratic space structure of W_4 can be recovered on $\mathcal{E} \cup \{0\}$. But actually the recovery of $q^{L^{[2]}}$ has not been done in purely abstract terms, but rather in terms of a particular complement $L^{[2]}$ to $Q^{[2]}$ in $N^{[2]}$. This complement is distinguished in $H \cong O_{10}^+(2)$ since it is a so-called *Levi complement*, but abstractly it is just one of the complements. In fact we will show that $N^{[2]}$ contains quite a few conjugacy classes of $L_3(2)$-complements which lead to different quadratic forms on $\mathcal{E} \cup \{0\}$.

4.4   Cohomology of $L_3(2)$

Let $L \cong L_3(2)$, let U_3 be the natural module of L and let U_3^* be the dual of U_3, which is also isomorphic to the exterior square of U_3. By (1.10.1) we know that $L_3(2) \cong L_2(7)$ and $\text{Aut } L_3(2) \cong PGL_2(7) \cong L_3(2) : \langle \tau\rangle$, where τ is a contragredient automorphism which permutes U_3 and U_3^*. By (3.2.3 (ii)) and (3.1.8) we have the following.

Lemma 4.4.1 *Let W be U_3 or U_3^*. Then*

   (i) *$H^1(L, W)$ is of order 2;*
   (ii) *the semidirect product $W : L$ with respect to the natural action contains exactly two classes of complements;*
  (iii) *if $L^{(1)}$ and $L^{(2)}$ are representatives of the classes of $L_3(2)$-complements in $W : L$, then $L^{(1)} \cap L^{(2)} \cong F_7^3$ (the Frobenius group of order 21);*
  (iv) *there exists a unique indecomposable extension W^u of W by the trivial 1-dimensional module and a unique extension W^d of the trivial 1-dimensional module by W (here 'u' is for* up *and 'd' is for* down*);*
   (v) *$(U_3^u)^* \cong (U_3^*)^d$;*
  (vi) *$W^u : L$ is isomorphic to the automorphism group of $W : L$.*

Let P_7 be the set of non-zero vectors in U_3. Then L acts doubly transitively on P_7 with stabilizer $L(p) \cong \mathrm{Sym}_4$ for $p \in P_7$. Let $A \cong \mathrm{Alt}_4$ be the only index 2 subgroup in $L(p)$ and let P_{14} be the set of cosets of A in L on which L acts in the natural way. Then there is a unique surjective map

$$\pi : P_{14} \to P_7$$

which commutes with the action of L. Let $\mathcal{P}_7$ and $\mathcal{P}_{14}$ be the $GF(2)$-permutation modules of L on P_7 and P_{14}, respectively.

Lemma 4.4.2 (i) *$\mathcal{P}_7 = \mathcal{P}_7^c \oplus \mathcal{P}_7^e$ and $\mathcal{P}_7^e$ is an indecomposable extension of U_3^* by U_3;*
  (ii) *U_3^d is a quotient of $\mathcal{P}_{14}$ but not a quotient of $\mathcal{P}_7$;*
  (iii) *if $\mathcal{C}$ is the submodule in $\mathcal{P}_{14}$ formed by the functions constant on $\pi^{-1}(p)$ for every $p \in P_7$, then $\mathcal{C} \cong \mathcal{P}_{14}/\mathcal{C} \cong \mathcal{P}_7$;*
  (iv) *if $\mathcal{C}^e$ is the unique codimension 1 submodule in $\mathcal{C}$, then $\mathcal{P}_{14}^e/\mathcal{C}^e$ is an indecomposable extension of the trivial 1-dimensional module by $\mathcal{P}_7^e$.*

Proof Statement (i) is well known and easy to check (cf. proposition 3.3.5 in Ivanov and Shpectorov (2002)). Notice that if $\mathcal{P}_7$ is identified with the points of the Fano plane, then U_3^* is generated by the vectors whose support is the complement of a line. Let $\chi : U_3^d \to U_3$ be the natural homomorphism. We claim that $L(p)$ permutes transitively the two vectors in $\chi^{-1}(p)$. In fact, otherwise $L(p)$ would fix these vectors and U_3^d would be a quotient of $\mathcal{P}_7$ which contradicts (i). Thus $L(p)$ acts on the vectors in $\chi^{-1}(p)$ as on the cosets of A and (ii) follows. (iii) is easy and of a rather general nature. Let $\alpha : \mathcal{P}_{14} \to U_3^d$ be the homomorphism whose existence follows from (ii) and let $\mathcal{K}$ be the kernel of α. If $\mathcal{K}$ would contain $\mathcal{C}$, U_3^d would be a quotient of $\mathcal{P}_7$, which contradicts (i). On the other hand, since U_3^d possesses the unique proper quotient which is isomorphic to U_3, it is easy to see that $\mathcal{K}$ contains the unique codimension 1 submodule in $\mathcal{C}$ and that $\mathcal{K}\mathcal{P}_{14}^e = \mathcal{P}_{14}$. Thus (iv) follows. ∎

Lemma 4.4.3 $H^1(L, \mathcal{P}_7^e)$ *is of order* 2.

Proof Since $\mathcal{P}_7^e$ is self-dual, $H^1(L, \mathcal{P}_7^e)$ is of order at least 2 by (3.1.8). Suppose the order of $H^1(L, \mathcal{P}_7^e)$ is greater than 2. Then there exists an indecomposable extension Y of the trivial 1-dimensional module by $\mathcal{P}_{14}^e/\mathcal{C}^e$. The latter module is a quotient of $\mathcal{P}_{14}$, since it contains a generating set of 14 vectors on which L acts as on the cosets of A. Since A has no subgroups of index two, this orbit lifts to a similar orbit in Y. Hence Y also must be a quotient of $\mathcal{P}_{14}$, which contradicts (4.4.2 (i), (iii)). ∎

In order to simplify the notation in what follows we denote $\mathcal{P}_7^e$ by D (where 'D' is for *dent*). Let Z be the unique L-submodule in D. Then $Z \cong U_3^*$ and $D/Z \cong U_3$. In fact D is the only indecomposable extension of U_3^* by U_3 (cf. Bell (1978)). Let $Y = D : L$ be the semidirect product with respect to the natural action and let $Q = ZL$. Then $Q \cong 2^3 : L_3(2)$ is as in (4.4.1 (ii)) and it contains two classes of $L_3(2)$-complements with representatives L and L_Z, where L_Z is the image of L under an outer automorphism of Q which acts trivially both on Z and Q/Z. It is easy to see that L and L_Z are not conjugate in Y and by (4.4.3) any complement to D in Y is conjugate either to L or to L_Z. On the other hand, $Y/Z \cong 2^3 : L_3(2)$ is also as in (4.4.1 (i)) and $Q/Z \cong L_3(2)$ is a complement to D/Z in Y/Z. Let Q^D be the preimage in Y of an $L_3(2)$-complement to D/Z in Y/Z which is not conjugate to Q/Z in Y/Z. Then Q^D is an extension of $Z \cong U_3^*$ by $L_3(2)$ and the extension cannot split, since we have already accounted all the complements to D in Y. Therefore we have the following.

Lemma 4.4.4 *The subgroup* Q^D *is a non-split extension* $2^3 \cdot L_3(2)$ *of* U_3^* *by* $L_3(2)$.

It is well-known (Bell 1978) that the second cohomology group $H^2(L, U_3^*)$ is of order 2 and therefore there is a unique non-split extension as in (4.4.4).

Lemma 4.4.5 *The outer automorphism group of* $Q^D \cong 2^3 \cdot L_3(2)$ *is of order* 2.

Proof First we show that Q^D possess an outer automorphism. Let σ be a non-trivial automorphism of $Y = D : L$ which acts trivially both on D and on Y/D (such an automorphism exists by (3.1.6) and (4.4.3)). We claim that σ normalizes Q^D and induces on it an outer automorphism. Consider $Y/Z \cong 2^3 : L_3(2)$. Since σ is trivial on D/Z and on Y/D, it either centralizes Y/Z or permutes two classes of $L_3(2)$-complements in Y/D. We know that Q/Z and Q^D/Z are representatives of the classes of complements. Since Q splits over Z while Q^D does not split, these complements cannot be permuted by σ. Hence σ normalizes Q^D. On the other hand σ cannot centralize Q^D since it already centralizes D and since $Y = DQ^D$.

Next we claim that the order of Out Q^D is at most 2. Let δ be an automorphism of Q^D and let $F \cong F_7^3$ be the normalizer of a Sylow 7-subgroup in Q^D. By a

Frattini argument (working modulo inner automorphisms) we can assume that δ normalizes F. By 3-subgroup lemma in this case δ centralizes F. Then δ normalizes a Sylow 2-subgroup S in the normalizer in Q^D of a Sylow 3-subgroup in F. It is easy to see that S contains three involutions, say z, x, and y, where $z \in Z$. Since Q^D does not split

$$\langle F, x \rangle = \langle F, y \rangle = Q^D$$

and the automorphism δ is uniquely determined by its action on $\{x, y\}$. If the action is trivial, it is the identity automorphism. If δ permutes x and y then it is an automorphism from the previous paragraph. ∎

Let S_3 be a Sylow 3-subgroup of L (which is also a Sylow 3-subgroup of Y) and let $I = I(Y)$ be a Sylow 2-subgroup in $N_Y(S_3)$. Since $N_L(S_3) \cong \mathrm{Sym}_3$ and the centralizer in U_3 of S_3 and $N_{U_3}(S_3)$ coincide and have dimension 1, we conclude that I is elementary abelian of order 2^3. By a Frattini argument we may assume that the non-identity automorphism σ of Y which centralizes both D and Y/D also normalizes I. Then it is easy to see that σ acts on I as the transvection whose axis is $I \cap D$ and whose centre is the non-identity element $z \in Z \cap I$.

4.5 Trident group

In this section U is a 4-dimensional $GF(2)$-space, $M \cong L_4(2)$ is the linear group of U and W is the exterior square of U (which is also the exterior square of the dual of U). If we treat M as $\Omega_6^+(2)$, then $W \cong V_6^+$ and if we treat M as Alt_8, then W is the heart $\mathcal{H}_8$ of the $GF(2)$-permutation module of the natural action on a set P_8 of size 8.

Let $u \in U^\#$, $M(u) \cong 2^3 : L_3(2)$ be the stabilizer of u in M and let $T = W : M(u)$ be the semidirect product with respect to the natural action. We call T the *trident group*. In this section we analyse the structure of T and calculate its automorphism group.

First observe that $M(u)$ is the semidirect product of $B \cong 2^3$ and $L \cong L_3(2)$, where the non-identity elements of B are the transvections with centre u and L is the stabilizer in $M(u)$ of a hyperplane which does not contain u. We identify this hyperplane with the dual natural module of L and denote it by U_3^*. The subgroup $M(u)$ stabilizes in W a unique proper submodule Z which is 3-dimensional, generated by the elements $u \wedge v$ taken for all $v \in U_3 \backslash \{u\}$; this submodule is centralized by B, while L acts on it as on U_3^*. Although W is indecomposable as an $M(u)$-module, L stabilizes a complement A to Z, which is generated by the elements $w \wedge v$ taken for all $w, v \in U_3^*$. This gives us the following description of T.

Lemma 4.5.1 *We have $T = ZABL$, where $L \cong L_3(2)$, while Z, A and B are elementary abelian of order 2^3. There are L-module isomorphisms $\alpha : U_3 \to A$, $\beta : U_3 \to B$ and $\zeta : U_3^* \to Z$ and the only non-trivial commutator relations* *among the elements of ZAB is*

$$[\alpha(v), \beta(w)] = \zeta(\langle v, w \rangle),$$

where v and w are distinct non-zero elements of U_3.

Proof Almost everything follows from the paragraph before the lemma. It is clear that ZAB is non-abelian with centre Z and the unique non-trivial bilinear L-invariant mapping from $A \times B$ into Z is the one described by the above commutator relation. ∎

For an $L_3(2)$-complement K to ZAB in T let K_Z denote the image of L under an outer deck automorphism of $ZK \cong 2^3 : L_3(2)$.

Suppose that E is a subgroup of order 2^3 in ZAB disjoint from Z and normalized by K. Then K^E denotes the image of K under an outer deck automorphism of $EK \cong 2^3 : L_3(2)$. Put

$$C = \langle \alpha(v)\beta(v) \mid v \in U_3 \rangle.$$

Lemma 4.5.2 *The following assertions hold:*

(i) *ZAB contains exactly three subgroups of order 2^6 normal in T which are AZ, BZ and C;*

(ii) *as L-modules AZ and BZ are semisimple while C is indecomposable;*

(iii) *T contains six classes of $L_3(2)$-complements and*

$$\mathcal{L} = \{L, \ L_Z, \ L^A, \ L_Z^A, \ L^B, \ L_Z^B\}$$

is a set of representatives for these classes.

Proof (i) and (ii) are similar to (3.4.2 (v)). Since ZAB/Z is the direct product of two copies of U_3, T/Z contains exactly four classes of complements with representatives

$$LZ/Z, \ L^A Z/Z, \ L^B Z/Z, \ (LZ)^C/Z.$$

Here $(LZ)^C$ can be defined as follows. The group $LC/Z \cong 2^3 : L_3(2)$ contains $LZ/Z \cong L_3(2)$ and possesses a non-trivial deck automorphism d. Then $(LZ)^C$ is the preimage in T of the image of LZ/Z under d. By (ii) and (4.4.4) the preimage of $(LZ)^C/Z$ is the non-split extensions $2^3 \cdot L_3(2)$, while the preimages of the former three complements split over Z. Since ZL has two classes of complements with representatives L and L_Z (iii) follows. ∎

A normal subgroup of order 2^6 in T will be called a *dent*. By (4.5.2)

$$\mathcal{E}^T = \{AZ, BZ, C\}$$

is the set of dents. The name *trident* that we gave to T is explained by the fact that it contains exactly three dents. A subgroup Q in T will be called a *quasi-complement* if $Q \cap ZAB = Z$ and $QAB = T$, so that Q is an extension of Z by

TABLE 2. Dents under various
quasi-complements

	LZ	$L^A Z$	$L^B Z$	$(LZ)^C$
AZ	0	0	1	1
BZ	0	1	0	1
C	1	0	0	1

$L_3(2)$ (which might or might not split). By the proof of (4.5.2)

$$\mathcal{Q} = \{LZ, L^A Z, L^B Z, (LZ)^C\}$$

is the set of quasi-complements. The following table shows which dents are semisimple '0' and which are indecomposable '1' under various quasi-complements.

We are going to introduce some automorphisms of T. Let λ be the automorphism which centralizes ZAB, and maps L onto L_Z (its existence is guaranteed by (3.1.9)). Thus λ is a deck automorphism of T, when T is regarded as a semi-direct product of ZAB and L. Let μ be the automorphism of T which centralizes Z and L and which transposes $\alpha(v)$ and $\beta(v)$ for every $v \in U_3$. It is immediate from (4.5.2) that μ is indeed an automorphism of T. Finally, let ν be the restriction to T of an outer deck automorphism of $W : M \cong 2^6 : L_4(2)$. We will assume without loss that the complements in $\mathcal{L}$ contain a common normalizer of a Sylow 7-subgroup and that this normalizer is centralized by the automorphisms λ, μ, and ν.

Lemma 4.5.3 *The automorphisms λ, μ and ν induce on $\mathcal{L}$ the following permutations:*

$$\lambda : \ (L, L_Z)(L^A, L_Z^A)(L^B, L_Z^B);$$

$$\mu : \ (L)(L_Z)(L^A, L^B)(L_Z^A, L_Z^B);$$

$$\nu : \ (L, L^A)(L_Z, L_Z^A)(L^B)(L_Z^B).$$

In particular Aut T *permutes transitively the classes of $L_3(2)$-complements.*

Proof The actions of λ and μ are immediate, while the action of ν follows from (3.3.5) and Table 2. ∎

Lemma 4.5.4 *The following assertions hold:*

(i) *the automorphism group of T permutes transitively the classes of $L_3(2)$-complements;*

(ii) *μ is the only non-identity automorphism of T which centralizes L;*

(iii) Out $T \cong \mathrm{Sym}_3 \times 2$;

(iv) Out T *induces* Sym_3 *on the set of quasi-complements, the kernel of the action is generated by the image of* λ.

Proof (i) follows directly from (4.5.3) while (ii) is easy to deduce from (4.5.2) and (4.5.3). Considering the actions of λ, μ, and ν on $\mathcal{L}$ we obtain (iii) and (iv).   ∎

Notice that if we treat $\mathcal{E}^T$ as the set of non-zero vectors of a non-singular symplectic 2-dimensional $GF(2)$-space $(\mathcal{E}^T \cup \{0\}, f)$ and consider columns of Table 2 as $GF(2)$-valued functions, then (assuming that zero always maps to zero) we obtain all the orthogonal forms associated with f. This can be formulated as follows.

Lemma 4.5.5 *For a quasi-complement* Q *define* $p^Q : (\mathcal{E}^T \cup \{0\}) \to GF(2)$ *as follows:* $p^Q(0) = 0$ *and for* $D \in \mathcal{E}$ *we have* $p^Q(D) = 0$ *if and only if* Q *acts semisimply on* D. *Then*

   (i) $(\mathcal{E}^T \cup \{0\}, f, p^Q)$ *is an orthogonal space for every* $Q \in \mathcal{Q}$;
   (ii) $Q \mapsto (\mathcal{E}^T \cup \{0\}, f, p^Q)$ *is a bijection between* $\mathcal{Q}$ *and the set of orthogonal spaces associated with* (V, f);
   (iii) p^Q *is of plus type if* Q *splits over* Z *and of minus type if* Q *does not split.*

We conclude this section with the following lemma which describes how to switch a quasi-complement over a dent.

Lemma 4.5.6 *Let* $Q \in \mathcal{Q}$ *and* $D \in \mathcal{E}^T$. *Then* QD *contains exactly two classes of quasi-complements with representatives* Q *and* Q^D. *Furthermore, in terms of* (4.5.5) *we have* $p^Q(D) = p^{Q^D}(D)$ *while* $p^Q(D_1) \neq p^{Q^D}(D_1)$ *for every* $D_1 \in \mathcal{E}^T \setminus D$.

Proof Since $QD/Z \cong 2^3 : L_3(2)$ and Q/Z is an $L_3(2)$-complement in this group, by (4.4.1) we have two classes of quasi-complements in QD. The second claim is immediate from Table 2.   ∎

Let S_3 be a Sylow 3-subgroup of L (which is also a Sylow 3-subgroup of T) and let $I = I(T)$ be a Sylow 2-subgroup of $N_T(S_3)$. Let z, a, b and l be the non-identity elements in the intersection of I with Z, A, B and L, respectively. These four elements are pairwise commuting involutions which generate I. Since they are also independent, $I \cong 2^4$. Without loss of generality we can assume that the automorphisms λ, μ and ν in (4.5.3) normalize I. It is obvious for instance that λ induces the transvection with axis $\langle z, a, b \rangle$ and centre z. The action of μ to be used later is recorded in the following lemma which is immediate from Table 2 and the action of μ on the classes of $L_3(2)$-complements.

Lemma 4.5.7 *The automorphism* μ *(which is the restriction to* T *of a deck automorphism of* $2^6 : L_4(2)$) *acts on* $I = I(T)$ *as the transvection with axis* $\langle z, a, lb \rangle$ *and centre* a. *In particular* μ *centralizes the non-split quasi-complement* $(LZ)^C$.

4.6   Automorphism group of $N^{[2]}$

In this section we apply the isomorphism

$$(W_4, f|_{W_4}, q|_{W_4}) \cong (\mathcal{E} \cup \{0\}, h, q^{L^{[2]}}),$$

established in Section 4.3 to obtain an abstract description of $N^{[2]}$. Next we apply results from Section 4.4, to classify all the $L_3(2)$-complements and all the quasi-complements in $N^{[2]}$. Then, using the properties of the trident group from Section 4.5 we observe that all the $L_3(2)$-complements lead to equivalent abstract descriptions of $N^{[2]}$. This will allow us to conclude that Aut $N^{[2]}$ acts transitively on the classes of $L_3(2)$-complements.

Thus $N^{[2]}$ is a semidirect product of $L^{[2]}$ and $Q^{[2]}$, where $L^{[2]} \cong L_3(2)$ and $Q^{[2]}$ is special of order 2^{15} with centre $Z^{[2]}$ of order 2^3. Let U_3 and U_3^* be the natural and dual natural modules of $L^{[2]}$. Then there is an L-isomorphism

$$\zeta : U_3^* \to Z^{[2]}.$$

The group $Q^{[2]}$ contains a set $\mathcal{E}$ of 15 dents which are the elementary abelian subgroups of order 2^6 containing $Z^{[2]}$ and normal in $N^{[2]}$. The dents generate the whole $Q^{[2]}$ while every dent is normalized by $L^{[2]}$. Furthermore a dent D is semisimple (as an $L^{[2]}$-module) if it is singular with respect to the form $q^{L^{[2]}}$, and it is an indecomposable extension of U_3^* by U_3 if D is nonsingular. We have seen in Section 4.4 that the indecomposable extension is the even half $\mathcal{P}_7^e$ of the permutation module of $L^{[2]}$ on the set P_7 of non-zero vectors of U_3.

Lemma 4.6.1 *Let D be a dent from $\mathcal{E}$. Then there is a unique $L^{[2]}$-invariant injection*

$$\delta_D : U_3^\# \to D.$$

The image of δ_D generates a complement to $Z^{[2]}$ if D is singular with respect to $q^{L^{[2]}}$ and the whole of D if it is non-singular. If $\{u, v, w\}$ is the set of non-zero vectors of a 2-subspace V in U_3, then

$$\delta_D(u)\delta_D(v)\delta_D(w) = \begin{cases} 1 & \text{if } q^{L^{[2]}}(D) = 0; \\ \zeta(V) & \text{otherwise.} \end{cases}$$

Proof The uniqueness of δ_D can be seen as follows. If there were two distinct such injections δ_1 and δ_2, then

$$u \mapsto \delta_1(u)\delta_2(u)$$

would be a non-trivial L-invariant mapping of U_3 into U_3^*, which is impossible. The existence of δ follows from the paragraph before the lemma. The 2-subspace condition is quite clear.   ∎

The vector space structure on $\mathcal{E} \cup \{0\}$ is reflected in the following.

Lemma 4.6.2 *Let D_1 and D_2 be distinct dents. Then the mapping of $U_3^{\#}$ into $D_1 + D_2$ as in (4.6.1) is the following:*

$$u \mapsto \delta_{D_1}(u)\delta_{D_2}(u)$$

for $u \in U_3^{\#}$.

Finally the symplectic form h on $\mathcal{E} \cup \{0\}$ is reflected in the following (recall that D_1 and D_2 commute whenever $h(D_1, D_2) = 0$.)

Lemma 4.6.3 *Suppose that $h(D_1, D_2) = 1$ and $u, v \in U_3^{\#}$. Then*

$$[\delta_{D_1}(u), \delta_{D_2}(v)] = \begin{cases} 1 & \text{if } u = v; \\ \zeta(\langle u, v \rangle) & \text{otherwise.} \end{cases}$$

Proof The subgroup $L^{[2]}(u)$ does not stabilize non-zero vectors in U_3^* while $L^{[2]}(u) \cap L^{[2]}(v)$ stabilizes a unique such vector, namely the one corresponding to $\langle u, v \rangle$. ∎

Let us turn to the classes of $L_3(2)$-complements in $N^{[2]}$. First of all

$$Z^{[2]} L^{[2]} \cong 2^3 : L_3(2)$$

contains two classes of complements whose representatives are $L^{[2]}$ and the image $L_Z^{[2]}$ of $L^{[2]}$ under an outer deck automorphism of $Z^{[2]} L^{[2]}$. Since $Z^{[2]}$ is the centre of $Q^{[2]}$, the complements $L^{[2]}$ and $L_Z^{[2]}$ have the same image in Aut $Q^{[2]}$.

Lemma 4.6.4 *The form defined with respect to $L_Z^{[2]}$ as in the paragraph before (4.3.5) coincides with $q^{L^{[2]}}$. There is a deck automorphism ξ of $N^{[2]}$ which centralizes $Q^{[2]}$ and maps $L^{[2]}$ onto $L_Z^{[2]}$.*

Proof The existence of ξ follows from (3.1.9). ∎

An $L_3(2)$-complement in $N^{[2]}$ and its image under ξ act on $Q^{[2]}$ in exactly the same way. Therefore it is convenient for us to deal with *quasi-complements* which are subgroups R in $N^{[2]}$ such that $R \cap Q^{[2]} = Z^{[2]}$ and $Q^{[2]} R = N^{[2]}$. For instance

$$Q := L^{[2]} Z^{[2]} = L_Z^{[2]} Z^{[2]} = L^{[2]} L_Z^{[2]}$$

is a quasi-complement.

The quasi-complements are bijective with the complements to $Q^{[2]}/Z^{[2]}$ in $N^{[2]}/Z^{[2]}$. Since the former (as a module for $N^{[2]}/Q^{[2]} \cong L_3(2)$) is isomorphic to the direct sum of four copies of U_3, it follows from (4.4.1) that there are exactly 2^4 classes of quasi-complements.

Lemma 4.6.5 *Let R be a quasi-complement in $N^{[2]}$. Then after conjugation of R by a suitable element of $N^{[2]}$ one of the following holds:*

(i) $R = Q = L^{[2]}Z^{[2]}$;
(ii) *there is a unique dent $D \in \mathcal{E}$ such that Q and R are representatives of two classes of quasi-complements in DQ.*

Proof Since both $Q/Z^{[2]}$ and $R/Z^{[2]}$ are $L_3(2)$-complements in $N^{[2]}/Z^{[2]}$ and R is chosen up to conjugation in $N^{[2]}$, we can assume that there is an invariant submodule $\bar{D}$ in $Q^{[2]}/Z^{[2]}$ such that

$$(Q/Z^{[2]})\bar{D} = (R/Z^{[2]})\bar{D}.$$

Since $\bar{D} = D/Z^{[2]}$ for a dent D, the result follows. ∎

The above lemma implies that once we have a quasi-complement R, we can obtain the representatives of the remaining 15 classes of quasi-complements by switching R over the 15 dents in $\mathcal{E}$. Therefore

$$\mathcal{Q} = \{Q \cup \{Q^D \mid D \in \mathcal{E}\}\}$$

is a set of representatives of the classes of quasi-complements in $N^{[2]}$. Some of these quasi-complements split, some do not. We know that Q splits. In order to specify the choice of representatives we will assume that the quasi-complements in $\mathcal{Q}$ share the normalizer of a Sylow 7-subgroup (which is the Frobenius group of order 21).

Lemma 4.6.6 *The quasi-complement Q^D splits if and only if D is singular with respect to $q^{L^{[2]}}$.*

Proof If D is singular then there is an $L^{[2]}$-invariant complement E to $Z^{[2]}$ in D. If L^E is the image of $L^{[2]}$ under an outer deck automorphism of $EL^{[2]} \cong 2^3 : L_3(2)$ then $Q^D = L^E Z^{[2]}$. If D is non-singular, then it is indecomposable with respect to $L^{[2]}$ and the result follows from (4.4.4). ∎

For a quasi-complement $R \in \mathcal{Q}$ define, q^R to be the $GF(2)$-valued function on $\mathcal{E} \cup \{0\}$ such that $q^R(0) = 0$ and

$$q^R(D) = \begin{cases} 0 & \text{if } R \text{ acts on } D \text{ semisimply;} \\ 1 & \text{if } R \text{ acts on } D \text{ indecomposably.} \end{cases}$$

Comparing this definition with the paragraph before (4.3.5), we observe that $q^{L^{[2]}} = q^Q$ is a quadratic form of plus type on $\mathcal{E} \cup \{0\}$ associated with the symplectic form h (recall that $Q = L^{[2]}Z^{[2]}$). The following crucial lemma shows how the function q^R changes when we switch the quasi-complement R over a dent (recall that the quasi-complements in $\mathcal{Q}$ share a Sylow 7-normalizer).

Lemma 4.6.7 *Let* $R \in \mathcal{Q}$ *be a quasi-complement, let* $D \in \mathcal{E}$ *be a dent and let* $R_1 = R^D \in \mathcal{Q}$ *be* R *switched over* D. *Then for a dent* $D_1 \in \mathcal{E}$ *we have* $q^{R_1}(D_1) = q^R(D_1)$ *if and only if* $h(D, D_1) = 0$.

Proof Suppose that $h(D, D_1) = 0$. Then $\langle D, D_1 \rangle$ is abelian which means that the actions of R and R^D on D_1 coincide. If $h(D, D_1) = 1$ and R splits over $Z^{[2]}$ then DD_1R is isomorphic to the trident group from the previous section and the result follows from (4.5.6). But in fact one can observe that whether or not R splits is irrelevant, since this does not effect the action of R on DD_1. ∎

In terms of Section 1.7 the above lemma can be reformulated as follows.

Corollary 4.6.8 *For* $D \in \mathcal{E}$ *let* l_D *be the linear function on* $\mathcal{E} \cup \{0\}$ *defined by* $l_D : D_1 \mapsto h(D, D_1)$. *Then* $q^{R^D} = q^R + l_D$.

Since q^Q is a quadratic form of plus type associated with h, combining (1.7.2), (4.6.6), and (4.6.8) we obtain the following.

Proposition 4.6.9 *The set* $\{q^R \mid R \in \mathcal{Q}\}$ *consists of all the quadratic forms on* $\mathcal{E} \cup \{0\}$ *associated with the symplectic form* h. *Furthermore, for* $R = Q^D$, *we have the following:*

(i) *if* D *is semisimple under* Q *then* q^R *is of plus type and* R *splits over* $Z^{[2]}$;
(ii) *if* D *is indecomposable under* Q *then* q^R *is of minus type and* R *does not split over* $Z^{[2]}$.

By (1.2.4) we observe that $N^{[2]}$ contains 10 classes of split and 6 classes of non-split quasi-complements and, particularly there are exactly 20 classes of $L_3(2)$-complements.

Lemma 4.6.10 *The automorphism group of* $N^{[2]}$ *permutes transitively the classes of* $L_3(2)$-*complements in* $N^{[2]}$.

Proof Let $L \cong L_3(2)$ be a complement and let $R = Z^{[2]}L$ be the corresponding quasi-complement. Let α be an isomorphism of L onto $L^{[2]}$ such that $l^{-1}\alpha(l) \in Q^{[2]}$ for every $l \in L$. Let β be an isomorphism

$$\beta : (\mathcal{E} \cup \{0\}, h, q^R) \to (\mathcal{E} \cup \{0\}, h, q^{L^{[2]}})$$

of orthogonal spaces whose existence follows from (4.6.9) and (1.1.8). We claim that the pair (α, β) can be extended to an automorphism of $N^{[2]}$. In fact, using (4.6.1), (4.6.2), and (4.6.3) we can obtain a description of $N^{[2]}$ using L instead of $L^{[2]}$. In particular for every $D \in \mathcal{E}$ there is a unique L-invariant bijection

$$\sigma_D : U_3^\# \to D.$$

We refine β to

$$\gamma : \sigma_D(u) \mapsto \delta_{\beta(D)}(u),$$

where $u \in U_3^{\#}$. Since $q^R(D) = q^Q(\beta(D))$ for every $D \in \mathcal{E}$, the relations for $\sigma_D(u)$ and $\delta_{\beta(D)}(u)$ coming from (4.6.2) and (4.6.3) are identical. Finally, by the choice of α, the actions of l and $\alpha(l)$ on U_3 are also identical. Therefore the mapping $l\sigma_D(u) \mapsto \alpha(l)\delta_{\beta(D)}(u)$ extends to an automorphism of $N^{[2]}$ which maps L onto $L^{[2]}$.   ∎

Proposition 4.6.11 *The following assertions hold;*

(i)   Aut $N^{[2]}$ *induces* $S_4(2) \cong \mathrm{Sym}_6$ *on the dent space* $\mathcal{E} \cup \{0\}$ *as well as on the set* $\mathcal{Q}$ *of quasi-complements;*

(ii) *the kernel of the action of* Out $N^{[2]}$ *on the set of quasi-complements is of order 2 and is generated by the image of the deck automorphism* ξ *as in* (4.6.4);

(iii) Out $N^{[2]} \cong \mathrm{Sym}_6 \times 2$.

Proof Since the symplectic space structure $(\mathcal{E} \cup \{0\}, h)$ is described in Section 4.3 in abstract terms of $N^{[2]}$, the action of Aut $N^{[2]}$ on the set of dents is a subgroup of $S_4(2) \cong \mathrm{Sym}_6$. On the other hand, $H^{[2]}$ stabilizes a class of $L_3(2)$-complements and induces $O_4^+(2)$ on the set of dents. By (4.6.9) and (4.6.10), the action of Aut $N^{[2]}$ on the set of quadratic forms of plus type associated with h is transitive. This implies (i). By (4.6.5) an automorphism α of $N^{[2]}$ which normalizes every quasi-complement also normalizes every dent. Adjusting α by an outer automorphism of Q if necessary we can assume that α induces a deck automorphism of Q transposing $L^{[2]}$ and $L_Z^{[2]}$. Since the actions of these two complements are identical on every dent D, α must commute with this action, forcing α to centralize D and hence to centralize every dent. Therefore α is a non-trivial deck automorphism of $N^{[2]}$, and hence coincides with ξ, which gives (ii). Thus $O = $ Out $N^{[2]}$ is an extension of a group Y of order 2 (generated by the image of ξ as in (4.6.4)) by $O/Y \cong \mathrm{Sym}_6$ and to establish (iii) we have to show that O splits over Y. Notice that the image of $H^{[2]}$ in O is isomorphic to $O_4^+(2)$ and intersects Y trivially (since $H^{[2]}$ stabilizes the class of $L_3(2)$-complements containing $L^{[2]}$). Let A be the preimage of the subgroup Alt_6 in O/Y. We claim that A splits over Y. If not, Y would be the unique non-split extension $SL_2(9)$ of a group of order 2 by Alt_6. But the group $SL_2(9)$ contains no involutions except the one in the centre while, on the other hand

$$\Omega_4^+(2) = O_4^+(2) \cap \mathrm{Alt}_6$$

(where $O_4^+(2)$ is the image of $H^{[2]}$ in O) contains more than one involution. Hence A splits over Y. Now it only remains to show that O contains involutions outside A. Let D_1 and D_2 be non-commuting dents which are semisimple under $L^{[2]}$ (equivalently singular with respect to $q^{L^{[2]}}$). Then (D_1, D_2) is a hyperbolic pair and hence

$$\langle D_1, D_2 \rangle \cong V_2^+.$$

The elementwise stabilizer S of the perp of $\langle D_1, D_2 \rangle$ in the symplectic group of the dent space is $S_2(2) \cong \mathrm{Sym}_3$. Using the model of the dent space in terms of a 6-element set, or otherwise, we observe that S contains involutions which map onto odd elements of $O/Y \cong \mathrm{Sym}_6$. On the other hand $T = D_1 D_2 L^{[2]}$ is the trident group from Section 4.5. By (4.5.4 (iii)) Out $T \cong \mathrm{Sym}_3 \times 2$ (which is a split extension of Y by S) supplies us with the required involutions. ∎

By (4.6.10) and the remark before it the group $N^{[2]}$ contains exactly 20 classes of $L_3(2)$-complements transitively permuted by Aut $N^{[2]} \cong \mathrm{Sym}_6 \times 2$. On the other hand, $H^{[2]}$ stabilizes the class of such complements containing $L^{[2]}$. Since the centralizer of $N^{[2]}$ in $H^{[2]}$ is trivial, the latter maps isomorphically onto its natural image in Aut $N^{[2]}$. Therefore, by a consideration of group orders we obtain the following.

Lemma 4.6.12 *The image of* $H^{[2]}$ *in* Aut $N^{[2]}$ *coincides with the stabilizer of the conjugacy class of* $L_3(2)$-*complements containing* $L^{[2]}$.

4.7 $\{D_{12}, D_8\}$-amalgams in Sym_6

Recall that the goal of this chapter is to show that

$$\widehat{G^{[2]}} \cong \mathrm{Sym}_5,$$

where $\widehat{G^{[2]}}$ is the image of $G^{[2]}$ in $O = $ Out $N^{[2]}$. By (4.2.6), $N^{[2]} = K^{[2]}$. We know that the image $\widehat{H^{[2]}}$ of $H^{[2]}$ in O is $O_4^+(2)$. On the other hand, in view of (4.2.7) both $\widehat{H^{[2]}}$ and $\widehat{G^{[2]}}$ are generated by amalgams consisting of

$$H^{[02]}/N^{[2]} = G^{[02]}/N^{[2]} \cong \mathrm{Sym}_3 \times 2 \cong D_{12}; \text{ and } H^{[12]}/N^{[2]} \cong G^{[12]}/N^{[2]} \cong D_8$$

intersecting in a subgroup of order 4 (recall that we have identified $G^{[0]}$ and $H^{[0]}$). By (4.6.11 (iii)) we know that Out $N^{[2]} \cong \mathrm{Sym}_6 \times 2$ and that it induces Sym_6 on the dent space. In this section we are making a further step towards the identification of $\widehat{G^{[2]}}$ by classifying the subgroups in Sym_6 generated by $\{D_{12}, D_8\}$-subamalgams.

We start with the following easy lemma whose proof we leave as an exercise.

Lemma 4.7.1 *Up to isomorphism there exists a unique amalgam* $\mathcal{E} = \{D^{[0]}, D^{[1]}\}$ *such that* $D^{[0]} \cong D_{12} \cong \mathrm{Sym}_3 \times \mathrm{Sym}_2$, $D^{[1]} \cong D_8$, $D^{[01]} := D^{[0]} \cap D^{[1]} \cong 2^2$, *no non-identity subgroup of* $D^{[01]}$ *is normal in both* $D^{[0]}$ *and* $D^{[1]}$.

Let (V, f) be a non-singular 4-dimensional symplectic space and let $(V, f, q_\varepsilon) \cong V_4^\varepsilon$ be an associated orthogonal space, where as usual ε is $+$ or $-$. By (1.2.4 (iii)) the number of non-singular vectors in V_4^+ is 6 while in V_4^- there are 10 such vectors.

Define Σ_ε to be the graph on the set of non-singular vectors in V_4^ε in which two such vectors are adjacent if they are perpendicular (with respect to f). Then using the combinatorial model $V(P_6)$ or otherwise, we observe that Σ_+ is the complete bipartite graph $K_{3,3}$ while Σ_- is the Petersen graph (cf. Lemma C in Preface for the pictures of these graphs). Consider the action on Σ_ε of the orthogonal group $X_\varepsilon = O_4^\varepsilon(2)$ which is the stabilizer of q_ε in the symplectic group associated with (V, f). Let $X_\varepsilon^{[0]}$ and $X_\varepsilon^{[1]}$ be the stabilizers in X_ε of a vertex in Σ_ε and an edge containing that vertex. The following result is immediate.

Lemma 4.7.2 *Independently on whether ε is $+$ or $-$ the amalgam $\{X_\varepsilon^{[0]}, X_\varepsilon^{[1]}\}$ is isomorphic to the amalgam $\mathcal{E}$ from (4.7.1).*

We need a description of the subamalgams in Sym_6 isomorphic to the amalgam $\mathcal{E}$ from (4.7.1).

Lemma 4.7.3 *Up to conjugation the group Sym_6 contains exactly four subamalgams $\mathcal{E}$ as in (4.7.1). The representatives*

$$\{\mathcal{D}_j = \{D_j^{[0]}, D_j^{[1]}\} \mid 1 \le j \le 4\}$$

of these subamalgams can be chosen in such a way that

  (i) *$\langle \mathcal{D}_1 \rangle \cong \langle \mathcal{D}_4 \rangle \cong O_4^-(2) \cong \mathrm{Sym}_5$ and there is an outer automorphism of Sym_6 which permutes $\mathcal{D}_1$ and $\mathcal{D}_4$;*
 (ii) *$\langle \mathcal{D}_2 \rangle = \langle \mathcal{D}_3 \rangle \cong O_4^+(2) \cong \mathrm{Sym}_3 \wr \mathrm{Sym}_2$ and there is an outer automorphism of Sym_6 which permutes $\mathcal{D}_2$ and $\mathcal{D}_3$ (and hence normalizes the subgroup generated by the subamalgam);*
(iii) *$D_i^{[0]} = D_j^{[0]}$ if and only if $\{i, j\} = \{1, 2\}$ or $\{i, j\} = \{3, 4\}$.*

Proof Let $\mathcal{D} = \{D^{[0]}, D^{[1]}\}$ be a subamalgam in $X \cong \mathrm{Sym}_6$ and let P_6 be a set of size 6 on which X induces the symmetric group. Then (taking into account the outer automorphisms of X) we assume without loss of generality that $T = O_3(D^{[0]})$ is generated by a 3-cycle on P_6. Since $N := N_X(T)/T \cong \mathrm{Sym}_3 \times \mathrm{Sym}_2$, $D^{[0]}/T$ is a Sylow 2-subgroup in N and hence is uniquely determined up to conjugation. Furthermore $N_X(D^{[01]})/D^{[01]} \cong 2^2$ while $C_X(D^{[01]})/D^{[01]} \cong 2$ and hence there are exactly two ways to extend $D^{[01]}$ to $D^{[1]}$, and the result follows. ∎

Consider $X \cong \mathrm{Sym}_6$ as an abstract group and let us follow the notation from Section 1.8. Then P_6 and R_6 are two sets of size 6 on which X acts in two inequivalent ways (permuted by an outer automorphism). Let $s(P_6)$ and $s(R_6)$ be the symplectic spaces associated with these two actions. Let $F_a = X(a)$ for $a \in P_6$ and $F_b = X(b)$ for $b \in R_6$, so that $F_a \cong F_b \cong \mathrm{Sym}_5$. Let $\{A, C\}$ be a partition of P_6 and let $\{B, D\}$ be a partition of R_6 such that $|A| = |B| = |C| = |D| = 3$, $a \in A$, $b \in B$. Finally let $E \cong \mathrm{Sym}_3 \wr \mathrm{Sym}_2$ be the stabilizer of both the partitions (compare the paragraph before (1.8.5)). Then in terms of (4.7.2), we can assume that $\langle \mathcal{D}_1 \rangle = F_a$, $\langle \mathcal{D}_2 \rangle = \langle \mathcal{D}_3 \rangle = E$ and that $\langle \mathcal{D}_4 \rangle = F_b$. The subgroup $D_1^{[0]} = D_2^{[0]} = E(a)$ is the stabilizer in E of a non-singular vector $\{A \setminus \{a\}, C \cup \{a\}\}$ of the orthogonal space $Q(P_6, \{A, C\})$ and also it is the stabilizer in E of a maximal totally singular subspace in the orthogonal space $Q(R_6, \{B, D\})$ (recall that $U_5 \cap W_4$ is assumed to be 2-dimensional).

Let us now relate this to $N^{[2]}$. Without loss we identify the dent space $\mathcal{D} \cup \{0\}$ with $s(R_6)$ via the isomorphism $O/Y \cong X$ where $O = \text{Out } N^{[2]}$ and Y is the kernel of the action of O on the dent space. Then $H^{[02]}/N^{[2]} = G^{[02]}/N^{[2]}$ is the stabilizer in $H^{[2]}/N^{[2]} \cong O_4^+(2)$ of a maximal totally singular subspace in $(\mathcal{E} \cup \{0\}, h, q^{L^{[2]}})$. Therefore by (4.7.3) we assume without loss that

$$\{H^{[02]}/N^{[2]}, H^{[12]}/N^{[2]}\} = \mathcal{D}_2,$$

$$\{G^{[02]}/N^{[2]}, G^{[12]}/N^{[2]}\} = \mathcal{D}_1.$$

Thus by (4.7.3) and in view of (1.8.5) we obtain the following.

Proposition 4.7.4 *The group $G^{[2]}$ induces Sym_5 on the dent space and permutes the dents transitively.*

In terms of another remarkable isomorphism

$$\text{Sym}_5 \cong P\Sigma L_2(4)$$

(where the latter is the extension of $L_2(4)$ by the non-identity automorphism of $GF(4)$), the dent space is the natural (2-dimensional $GF(4)$-) module for this group.

Now in order to get the precise structure of $\widehat{G^{[2]}}$, we have to decide about the action of $G^{[2]}$ on the 20 classes of complements to $Q^{[2]}$ in $N^{[2]}$. At this stage we can say that the action is either $\text{Sym}_5 \times 2$ or Sym_5 and in the latter case there are still two possibilities. The action on the 20 classes is either transitive or with two orbits of length 10 each (notice that $\text{Sym}_6 \times 2$ contains exactly 4 conjugacy classes of subgroups isomorphic to Sym_5).

4.8 The last inch

In order to get the precise structure of $\widehat{G^{[2]}}$ consider a Sylow 2-subgroup $I = I(N^{[2]})$ of the normalizer in $N^{[2]}$ of a Sylow 3-subgroup S_3 of $L^{[2]}$ (of course S_3 is also a Sylow 3-subgroup of the whole of $N^{[2]}$).

Lemma 4.8.1 $I = I(N^{[2]})$ *is elementary abelian of order 2^6 and $C_I(S_3)$ is a hyperplane in I.*

Proof Let (D_1, D_2) and (D_3, D_4) be perpendicular hyperbolic pairs in the orthogonal space $(\mathcal{E} \cup \{0\}, h, q^{L^{[2]}})$. Then both $T_1 = D_1 D_2 L^{[2]}$ and $T_2 = D_3 D_4 L^{[2]}$ are trident groups and $I \cap T_i = I(T_i)$ is elementary abelian of order 2^4 for $i = 1, 2$ (compare the paragraph before (4.5.7)). Since

$$I(T_1) \cap I(T_2) = I \cap Z^{[2]} L^{[2]}$$

and $\langle T_1, T_2 \rangle = N^{[2]}$, the result follows. ∎

Let $F \cong F_7^3$ be normalizer of a Sylow 7-subgroup of $L^{[2]}$. Then F is also the normalizer of a Sylow 7-subgroup in $N^{[2]}$. Let C be the centralizer of F in Aut $N^{[2]}$. The subgroup F has trivial centralizer in $N^{[2]}$ and

$$C \, \mathrm{Inn} \, N^{[2]} = \mathrm{Aut} \, N^{[2]},$$

by a Frattini argument. Therefore we conclude that $C \cong \mathrm{Out} \, N^{[2]} \cong \mathrm{Sym}_6 \times 2$ (compare (4.6.11)). Without loss of generality we assume that $S_3 \leq F$ and since there are three Sylow 2-subgroups in $N_{N^{[2]}}(S_3)$, we can further assume that C normalizes I. Our next goal is to describe I as a module for C.

As in the paragraph before (4.7.4) we consider C as the direct product of the symmetric group $X \cong \mathrm{Sym}_6$ of R_6 of size 6 and a group Y of order 2, so that $s(R_6)$ is canonically isomorphic to the dent space and the quadratic forms of minus type associated with $(\mathcal{E} \cup \{0\}, h)$ are indexed by the elements of R_6. Let R_6 be the $GF(2)$-permutation module of X acting on R_6. Then the subsets of size 1 form a basis $\mathcal{B}_1$ of R_6 and the subsets of size 5 form another basis $\mathcal{B}_2$. Let γ be a linear transformation of R_6 which sends $\mathcal{B}_1$ onto $\mathcal{B}_2$ by mapping a 1-element subset onto its 5-element complement in R_6. Then γ is of order 2 and it normalizes the natural action of X on R_6. Now consider R_6 as a module for $C \cong X \times Y$ where X acts naturally and the generator of Y acts as γ (we will denote this generator also by γ).

Lemma 4.8.2 *The action of C on I is isomorphic to the above defined action on R_6.*

Proof By (4.6.9) and (4.6.11), there are exactly 6 classes of non-split quasi-complements in $N^{[2]}$. The group C acts on these classes as on R_6 with Y being the kernel. If R is such a quasi-complement containing S_3, then $I \cap R$ is of order 4, and contains three involutions, one of them, say z, is inside $Z^{[2]}$ while two others, which are transposed by Y, invert S_3. Recall that Y is generated by the image of the deck automorphism ξ of $N^{[2]}$ and ξ acts on I as the transvection with centre z and axis $C_I(S_3)$. Thus C has an orbit Ω on I of length 12 on which it acts with imprimitivity classes of size 2 indexed by the elements of R_6. Exactly one of the two subgroups in C isomorphic to Sym_6 acts on Ω intransitively with two orbits of length 6, and without loss of generality we assume that X is chosen to be this subgroup. It is easy to see that each of the two X-orbits of length 6 span the whole I. Therefore I, as a module for X is isomorphic to R_6. The generator of Y fuses the orbits of length 6 and commutes with the action of X, so its action is also uniquely determined. ∎

Let $\mathcal{J} = \{H^{[02]} = G^{[02]}, H^{[12]}, H^{[2]}, G^{[12]}\}$. Since $C_J(N^{[2]}) = 1$ for every $J \in \mathcal{J}$, each such J can and will be identified with its natural image in Aut $N^{[2]}$. Therefore, if we put

$$J_c = J \cap C$$

then our goal is just to identify the subgroup in C generated by $G_c^{[02]}$ and $G_c^{[12]}$.

First we locate in C the subgroup $H_c^{[2]}$ and the subamalgam $\{H_c^{[02]}, H_c^{[12]}\}$. By (4.6.12) $H^{[2]}$ is the stabilizer in Aut $N^{[2]}$ of the conjugacy class of $L_3(2)$-complements containing $L^{[2]}$. Therefore $H_c^{[2]}$ is the centralizer of l in C, where $\{l\} = I \cap L^{[2]}$. The element l is contained in the (unique) orbit of length 20 of C on I (this orbit corresponds to the classes of $L_3(2)$-complements in $N^{[2]}$). In terms of $\mathcal{R}_6$ this orbit corresponds to the 3-element subsets of $\mathcal{R}_6$. Let A be the 3-element subset corresponding to l and set $B = \mathcal{R}_6 \setminus A$. The stabilizer in X of the partition $\{A, B\}$ is $\mathrm{Sym}_3 \wr \mathrm{Sym}_2 \cong O_4^+(2)$ and the generator γ of Y permutes A and B (compare the definition of γ in the paragraph before (4.6.8)). Therefore we have the following lemma.

Lemma 4.8.3 *The subgroup $H_c^{[2]}$ consists of the elements from X stabilizing each of the subsets A and B and the elements from X permuting A and B, multiplied by γ.*

At this final stage we require some explicit description of the automorphism σ_γ defined in the paragraph before (3.4.2) and of the restriction of σ_γ to $N^{[2]}$. We assume without loss that $\sigma \in C$ and as above identify $G^{[0]}$ with $H^{[0]}$ and $K^{[2]}$ with $N^{[2]}$.

Lemma 4.8.4 *Let σ be the restriction to $N^{[2]}$ of the automorphism σ_γ defined before (3.4.2). Then*

(i) *$N^{[2]}$ is the commutator subgroup (of index 4) of the normalizer in $H^{[01]}$ of $Z^{[2]}$;*

(ii) *$Z^{[1]}$ is a dent in $N^{[2]}$;*

(iii) *$T := Z^{[1]}L^{[1]} \cap N^{[2]}$ is isomorphic to the trident group;*

(iv) *σ centralizes a non-split quasi-complement in T;*

(v) *σ acts on the dent space $(\mathcal{E} \cup \{0\}, h)$ as the transvection whose centre is $Z^{[1]}$ and whose axis is the perp of $Z^{[1]}$;*

(vi) *σ acts on $\mathcal{R}_6$ as a transposition and $\sigma \in X$.*

Proof In view of the discussion at the beginning of Section 4.3, $H^{[012]} = N_{H^{[01]}}(U_3)$ and $H^{[012]}/N^{[2]} \cong 2^2$, since $H^{[012]}$ is the vertexwise stabilizer of an edge in the action of $H^{[2]}$ on the geometric subgraph $\Omega^{[2]}$ of valency 3 in Ω. Since $N^{[2]}$ coincides with its commutator subgroup, (i) follows. Since $Z^{[1]}$ is elementary abelian of order 2^6 and is normal in $N^{[2]}$, we obtain (ii). Since $Z^{[1]}L^{[1]} \cong 2^6 : L_4(2)$ and $L^{[1]} \cap N^{[2]} \cong 2^3 : L_3(2)$ is the stabilizer of a hyperplane in the natural module of $L^{[1]}$, (iii) follows directly from the definition of the trident group at the beginning of Section 4.5. The assertion (iv) now follows from (4.5.7). Since σ_γ acts trivially on $Q^{[1]}$ and the latter contains the centralizer of $Z^{[1]}$ in $N^{[2]}$, σ centralizes every dent perpendicular to $Z^{[1]}$. Since σ must also preserve the symplectic form h, we obtain (v). Since the dent space is isomorphic to $s(\mathcal{R}_6)$, symplectic transvections of the dent space act as transpositions on $\mathcal{R}_6$. By (v) σ centralizes a non-split quasi-complement in the trident group T. Therefore it centralizes at least two involutions from I contained in the orbit of length

12 of C on I. Therefore σ projects trivially on the direct factor Y of $C = X \times Y$ and (vi) follows. ■

Proposition 4.8.5 *Let $G_c^{[2]}$ be the subgroup in C generated by $G_c^{[02]}$ and $G_c^{[12]}$. Then*

(i) $G_c^{[2]} \cong \mathrm{Sym}_5$;

(ii) $G_c^{[2]}$ *acts transitively on the set of* 12 *involutions in I contained in non-split quasi-complements;*

(iii) $\widehat{G^{[2]}} \cong \mathrm{Sym}_5$.

Proof We turn (4.8.3) into explicit permutations. Put $R_6 = \{1, 2, \ldots, 6\}$, $A = \{1, 2, 3\}$, $B = \{4, 5, 6\}$. Since $H^{[02]}$ does not stabilize quadratic forms of minus type associated with $(\mathcal{E} \cup \{0\}, h) \cong s(R_6)$, we conclude that $O_3(H_c^{[02]})$ acts transitively on R_6 and without loss we assume that $H^{[02]} = \langle y, d_1, d_2 \rangle$, where

$$y = (1, 2, 3)(4, 5, 6), \ d_1 = (1, 4)(2, 5)(3, 6)\gamma, \ d_2 = (1, 4)(2, 6)(3, 5)\gamma.$$

The permutation of R_6 induced by y stabilizes each of A and B, while those induced by d_1 and d_2 permute A and B. Therefore these two permutations are multiplied by γ in accordance with (4.8.3). It is easy to deduce from (4.7.3) that the element x which extends $H_c^{[012]} = \langle d_1, d_2 \rangle$ to $H_c^{[12]}$ can be taken to be

$$x = (5, 6).$$

In order to obtain an element t which extends $H_c^{[012]}$ to $G_c^{[12]}$ we have to multiply x by an element $s \in C$ acting on R_6 as a transposition. On the other hand s is the action of σ on I. The crucial consequence of (4.8.4 (vi)) is that s is a pure permutation of R_6 (i.e. one which does non-project non-trivially on Y), so that

$$t = (1, 4)(5, 6).$$

The action of $G_c^{[2]} = \langle y, d_1, d_2, t \rangle$ on R_6 is generated by a $\{D_{12}, D_8\}$-amalgam and by (4.7.3) the action is isomorphic to Sym_5. Now we observe that the generators d_1, d_2 which induce odd permutations on R_6 are multiplied by γ while the even generators y, t are not. Therefore $G_c^{[2]}$ is a diagonal subgroup in

$$\mathrm{Sym}_5 \times Y \leq X \times Y,$$

which gives (i). Since $G_c^{[2]}$ projects onto Y, (ii) follows. Finally (iii) is from the general setting after the proof of (4.8.1). ■

4.9 Some properties of the pentad group

Since the centre of $N^{[2]}$ is trivial, $N^{[2]}$ maps isomorphically onto its inner automorphism group. Let $\overline{G^{[2]}}$ be the image of $G^{[2]}$ in the automorphism group of

$N^{[2]}$. Then

$$\overline{G^{[2]}} \cong G^{[2]}/C_{G^{[2]}}(N^{[2]})$$

by the definition and

$$\overline{G^{[2]}} \cong 2^{3+12} \cdot (L_3(2) \times \mathrm{Sym}_5)$$

by (4.8.5 (iii)). In what follows we call $\overline{G^{[2]}}$ the *pentad* group and identify the subgroups of $N^{[2]}$ (including $N^{[2]}$ itself) with their images in the pentad group.

Lemma 4.9.1 *Let S_7 and S_3 be subgroups of order 7 and 3 from $L^{[2]}$, respectively such that S_3 normalizes S_7 and let S_5 be a Sylow 5-subgroup in $\overline{G^{[2]}}$. For $p = 3, 5, 7$ let N_p be the normalizer of S_p in the pentad group $\overline{G^{[2]}}$. Then*

(i) $Q^{[2]}/Z^{[2]} \cong 2^{12}$, *as a module for $\overline{G^{[2]}}/Q^{[2]} \cong L_3(2) \times \mathrm{Sym}_5$ is isomorphic to the tensor product of the natural module of $L_3(2)$ and the natural module of $P\Sigma L_2(4) \cong \mathrm{Sym}_5$;*

(ii) $N_7 \cong F_7^3 \times \mathrm{Sym}_5$;

(iii) $N_3/S_3 \cong 2^6 : \mathrm{Sym}_5$, $O_2(N_3/S_3)$ *is a uniserial module for $N_3/(S_3 O_2(N_3)) \cong \mathrm{Sym}_5$ with orbit lengths on the non-zero vectors 1, 12, 15, 15, and 20;*

(iv) $N_5 \cong (2^3 \cdot L_3(2)) : F_5^4$ *is the semidirect product of a subgroup X isomorphic to the non-split extension of 2^3 by $L_3(2)$ and the Frobenius group Y of order 20, $C_Y(X) \cong D_{10}$ and N_5 maps onto $\mathrm{Aut}\, X$.*

Proof (i) is immediate from (4.7.4). Since $Z^{[2]}$ is the dual natural module for $L^{[2]}$, we have $C_{N^{[2]}}(S_7) = 1$ and (ii) follows from (i) and a Frattini argument. By (4.8.1) N_3 is an extension of $2^6 = N_3 \cap N^{[2]}$ by Sym_5 and by (i) the extension splits. The module structure follows from (4.8.5 (ii)). By (i) and (4.8.5 (ii)) $C_{N^{[2]}}(S_5)$ is a non-split quasi-complement on which an F_5^4-subgroup of N_5 induces an outer automorphism, hence (iv) follows (notice that by (4.4.5) $\mathrm{Out}\, X$ is of order 2). ∎

Exercises

1. Let $L \cong L_3(2)$ and U be the natural module for L. Show that up to isomorphism there exists a unique non-split extension $2^3 \cdot L_3(2)$ of U by L.
2. Prove the isomorphism $\mathrm{Sym}_5 \cong P\Sigma L_2(4)$.
3. Performing direct calculations in the amalgam $\mathcal{G}$ show that $N^{[4]} = 1$.

5

TOWARDS $2^{1+12}_+ \cdot 3 \cdot \text{AUT} \ (M_{22})$

In the chapter we show that $N^{[3]}$ is the extraspecial group $Q^{[3]} \cong 2^{1+12}_+$ extended by a group R of order 3 acting fixed-point freely on $Q^{[3]}/Z^{[3]}$ (where $Z^{[3]}$ is the centre of $Q^{[3]}$) and that $\widehat{G^{[3]}}$ is the automorphism group Aut (M_{22}) of the sporadic Mathieu group M_{22}. We use Shpectorov's geometric characterization of M_{22} in terms of a Petersen type geometry. Notice that the sporadic Mathieu group M_{22} appears here before we have even started considering any specific completions of $\mathcal{G}$.

5.1 Automorphism group of $N^{[3]}$

The ultimate goal of this chapter is to calculate the image of $G^{[3]}$ in the (outer) automorphism group of $N^{[3]}$. It is natural first to ask what the automorphism group of $N^{[3]}$ is.

Lemma 5.1.1 *The group $N^{[3]}$ is the extraspecial group $Q^{[3]} \cong 2^{1+12}_+$ with centre $Z^{[3]}$ extended by the subgroup R of order 3 which acts fixed-point freely on $Q^{[3]}/Z^{[3]}$.*

Proof Everything is immediate from (2.1.2) except possibly the type of $Q^{[3]}$ as an extraspecial group. By the proof (4.2.5) $Q^{[0]} \cap Q^{[3]}$ is of order 2^7 and the intersection is elementary abelian since $Q^{[0]}$ is already elementary abelian. By (1.6.7) $Q^{[3]}$ is of plus type. ∎

Lemma 5.1.2 *The automorphism group of $N^{[3]}$ is the semidirect product of $\bar{Q}^{[3]} = Q^{[3]}/Z^{[3]} \cong 2^{12}$ and $\Gamma U_6(2)$, where the latter group acts on $\bar{Q}^{[3]}$ as on the natural unitary module.*

Proof Since the orders of R and $Q^{[3]}$ are coprime, Aut $(N^{[3]})$ acts faithfully on $Q^{[3]}$ and by a Frattini argument Aut $(N^{[3]})$ is a split extension of $\bar{Q}^{[3]}$ (which is the image of $Q^{[3]}$ in Inn $(N^{[3]})$) and $M := N_{\text{Aut} \ (N^{[3]})}(R)$. Furthermore, since $\bar{Q}^{[3]}$ is the Frattini factor of $Q^{[3]}$, the action of M on $\bar{Q}^{[3]}$ is also faithful. Since $Q^{[3]} \cong 2^{1+12}_+$, by (1.6.1) we have $O := \text{Out } Q^{[3]} \cong O^+_{12}(2)$ and hence $M = N_O(R)$, where R is identified with its image in O. Let f and q be the symplectic and associated quadratic forms on $\bar{Q}^{[3]}$ defined as in (1.6.1). Then O is the automorphism group of the orthogonal space $(\bar{Q}^{[3]}, f, q) \cong V^+_{12}$. The action of R on $\bar{Q}^{[3]}$ defines on the latter two $GF(4)$-space structures preserved by $C_O(R)$ as follows. Let α be one of the two isomorphisms

$$\alpha : GF(4)^* \to R,$$

where $GF(4)^*$ is the multiplicative group of the field of four elements. Then for $\bar{v} \in \bar{Q}^{[3]}$ and $\omega \in GF(4)^*$ we define the scalar multiplication by $\omega\bar{v} = \bar{v}^{\alpha(\omega)}$, where on the right-hand side we have the image of $\bar{v}$ under the element $\alpha(\omega) \in R$ acting by conjugation. Then q becomes a non-singular Hermitian form on $\bar{Q}^{[3]}$ where the latter is now treated as a 6-dimensional $GF(4)$-space. The stabilizer of this form in $GL_6(4)$ is the general unitary group $GU_6(2)$. By the definition the unitary group preserves q and the latter in turn determines the homology class of 2-cocycles responsible for the extension of $Z^{[3]}$ by $\bar{Q}^{[3]}$. Therefore $GU_6(2)$ acts on $Q^{[3]}$ as an automorphism group. The centre of $GU_6(2)$ is of order 3 acting fixed-point freely on $\bar{Q}^{[3]}$ and hence this centre can be identified with R. Thus $C_O(R) \cong GU_6(2)$. Finally the non-identity automorphism of $GF(4)$ inverts R and also acts on $\bar{Q}^{[3]}$ permuting the two $GF(4)$-space structures preserved by R. This automorphism extends $C_O(R) \cong GU_6(2)$ to $\Gamma U_6(2)$. ∎

5.2 A Petersen-type amalgam in $\widehat{G}^{[3]}$

We would like to start the section with the following remark. We know exactly the structure of the amalgam $\mathcal{G} = \{G^{[0]}, G^{[1]}\}$ and the location of $N^{[3]}$ inside this amalgam. In view of (5.1.2) this enables us to write down a collection of generators for $\widehat{G}^{[3]}$ as semilinear transformations of the 6-dimensional $GF(4)$-space $\bar{Q}^{[3]}$. Then one way or another we should be able to identify the subgroup in $\Gamma U_6(2)$ generated by these transformations. In the last resort we could apply computer calculations. But we prefer to operate less explicitly and to apply Shpectorov's characterization (11.4.2) of the Mathieu group M_{22} and its automorphism group in order to identify $\widehat{G}^{[3]}$ as a completion of the amalgam formed by the images in $\widehat{G}^{[3]}$ of the subgroups $G^{[i3]}$ taken for $i = 0, 1$, and 2.

Up to some extend the above remark applies to $\widehat{G}^{[2]}$ as well in the sense that the calculations could be performed more explicitly in the automorphism group of $N^{[2]}$, but this way we would probably learn less about the structure of $\widehat{G}^{[2]}$.

As above let

$$\varphi : \mathcal{G} \to G$$

be a faithful generating completion of the amalgam $\mathcal{G} = \{G^{[0]}, G^{[1]}\}$. The structure of $\widehat{G}^{[3]}$ depends only on the amalgam, so there is no harm of taking G to be the universal completion. For $i = 0, 1$, and 2 let $\widehat{G}^{[i3]}$ be the image of the restriction to $G^{[i3]}$ of the natural homomorphism of $G^{[3]}$ onto $\widehat{G}^{[3]}$. Let

$$\widehat{\mathcal{A}}^{[3]} = \{\widehat{G^{[i3]}} \mid 0 \le i \le 2\}$$

be the subamalgam in $\widehat{G^{[3]}}$ formed by these subgroups. Since $N^{[3]}$ is contained in each of the $G^{[i3]}$'s, it is by the definition that

$$\widehat{G^{[i3]}} \cong G^{[i3]}/(N^{[3]}C_{G^{[i3]}}(N^{[3]})).$$

Proposition 5.2.1 *Let $\widehat{A^{[3]}}$ be the above defined subamalgam in $\widehat{G^{[3]}}$. Then*

(i) $\widehat{G^{[03]}} \cong G^{[03]}/N^{[3]} \cong 2^3 : L_3(2) \times 2$;

(ii) $\widehat{G^{[13]}} \cong G^{[13]}/N^{[3]} \cong (2^{1+4}_{+} : \mathrm{Sym}_3 \times 2) : 2$;

(iii) $\widehat{G^{[23]}} \cong 2^5 : \mathrm{Sym}_5$;

(iv) $\widehat{A^{[3]}}$ *corresponds to the diagram*

$$P_3 : \quad \underset{1}{\circ} \overset{\mathrm{P}}{\underline{\qquad}} \underset{2}{\circ} \underline{\qquad} \underset{2}{\circ}.$$

(v) $\widehat{G^{[3]}}$ *is isomorphic to either* Aut (M_{22}) *or* $3 \cdot$ Aut (M_{22}).

Proof Since $G^{[0]}$ is identified with $H^{[0]}$, while

$$H^{[03]} = Q^{[3]}L^{[3]}(R^{[3]} \cap H^{[0]}),$$

where $R^{[3]} \cap H^{[0]} \cong 2^3 : L_3(2)$ and since $N^{[3]} = O^2(K^{[3]}) = Q^{[3]}(O^2(L^{[3]}))$, (i) follows. From (i) we immediately obtain the structure of $G^{[013]}$ having index 2 in $G^{[13]}$, so (ii) follows. The subgroup $N^{[3]}$ is contained in $N^{[2]}$ (this inclusion is easily seen in terms of the locally projective graph $\Lambda(\mathcal{G}, \varphi, G)$ and the geometric subgraphs in that graph). Furthermore we can assume without loss of generality that the subgroup R of order 3 from $N^{[3]}$ is contained in the complement $L^{[2]} \cong L_3(2)$ to $Q^{[2]}$ in $N^{[2]}$, so that R is a Sylow 3-subgroup S_3 in $L^{[2]}$ as in Section 4.8. Now (iii) follows from (4.9.1 (iii)), since $C_{Q^{[3]}}(R) = Z^{[3]}$. Recall that $G^{[3]}$ is the stabilizer in G of a geometric subgraph $\Gamma^{[3]}$ in $\Lambda(\mathcal{G}, \varphi, G)$ of valency 7. Furthermore, $G^{[3]}$ induces on $\Gamma^{[3]}$ a locally projective action of type $(3, 2)$ with kernel $N^{[3]}$. Finally $G^{[03]}$, $G^{[13]}$, and $G^{[23]}$ are stabilizers in $G^{[3]}$ of incident vertex, edge, and cubic geometric subgraph in $\Gamma^{[3]}$, respectively. By (4.8.1) and the above proof of (iii) we observe that $O_2(\widehat{G^{[23]}})$ is the image of $G^{[23]} \cap G^{[01]}$ in $\widehat{G^{[3]}}$, (iv) follows from (iii). Finally (v) is what (11.4.2) is for. ∎

Applying (4.9.1 (ii)) or otherwise, it is easy to see that the following holds.

Lemma 5.2.2 *A Sylow 7-subgroup S_7 of $G^{[03]}$ acts fixed-point freely on $\bar{Q}^{[3]} = Q^{[3]}/Z^{[3]}$.*

Lemma 5.2.3 $\widehat{G^{[3]}} \cong$ Aut (M_{22}).

Proof By (5.2.1 (v)) either the assertion holds or the image X of $G^{[3]}$ in Aut $(N^{[3]})$ is of the form $2^{12} : 3^2 \cdot$ Aut (M_{22}). Suppose the latter and let

$Q = O_2(X)$ and Y be a Sylow 3-subgroup of $O_{2,3}(X)$. Since Q is the image of $Q^{[3]}$ in the automorphism group of $N^{[3]}$, it is elementary abelian of order 2^{12}. We claim that Y is elementary abelian of order 9. In fact, otherwise it would be the cyclic group of order 9. Since Y contains the image of a conjugate of R, it acts fixed-point freely on Q and therefore endows it with a $GF(64)$-vector space structure of dimension 2. Since clearly M_{22} does not possess 2-dimensional projective representations, this is impossible. Thus Y is elementary abelian and hence it contains a subgroup P of order 3 whose action on Q is not fixed-point free. In other terms both $C_Q(P)$ and $[Q, P]$ are non-trivial. By a Frattini argument the commutator subgroup C of $C_X(P)$ is either M_{22} or a non-split central extension of M_{22}. By (5.2.2) C acts non-trivially both on $C_Q(P)$ and on $[Q, P]$. On the other hand 11 divides the order of M_{22} and the minimal faithful representation over $GF(2)$ of a cyclic group of order 11 has dimension 10. Since

$$\dim C_Q(P) + \dim [Q, P] = \dim Q = 12,$$

this leads to a contradiction.   ■

Let $\widetilde{G^{[3]}}$ denote the image of $G^{[3]}$ in the automorphism group of $N^{[3]}$. Then (unlike in the $G^{[2]}$-case) the natural homomorphism of $N^{[3]}$ into $\widetilde{G^{[3]}}$ has non-trivial kernel, which is $Z^{[3]} = Z(N^{[3]})$.

Lemma 5.2.4 *The group* $\widetilde{G^{[3]}}$ *is a semidirect product of* $\widetilde{Q^{[3]}} \cong 2^{12}$ *which is the image of* $Q^{[3]}$ *and a subgroup*

$$\widetilde{K} = N_{\widetilde{G^{[3]}}}(\widetilde{R}) \cong 3 \cdot \mathrm{Aut}\,(M_{22}),$$

where $\widetilde{R}$ *is the image of* R. *The action of* $\widetilde{K}$ *on* $\widetilde{Q^{[3]}}$ *is irreducible.*

Proof We follow notation of the proof of (5.2.3) only that here Y is of order 3 identified with the image of R. All we have to show is that $N_X(Y)$ does not split over Y. This is the case, since otherwise the commutator subgroup of $N_Y(X)$ would be isomorphic to M_{22}. By (5.2.2) the irreducible constituents of the commutator subgroup in X are all non-trivial. This contradicts the list of 2-modular representations of M_{22} calculated in James (1973): the minimal dimension of a faithful representation is 10 and the second minimal is 34. Also by James (1973) 12 is the minimal dimension of faithful GF(2)-module for $3 \cdot M_{22}$. Hence the irreducibility of $\widetilde{K}$ on $\widetilde{Q^{[3]}}$ follows.   ■

Lemma 5.2.5 *For* $i = 0$, 1, *and* 2 $\widetilde{G^{[i3]}}$ *is the image in* $\widetilde{G^{[3]}}$ *of* $N_{G^{[i]}}(R)$.

Proof For $i = 0$ and 1 $C_{G^{[i]}}(N^{[3]}) = Z^{[3]}$ and $Q^{[3]} = [Q^{[3]}, R]$ so the assertion holds for $i = 0$ and 1. Since $\widetilde{G^{[23]}}$ is generated by the images of $G^{[023]}$ and $G^{[123]}$, for $i = 2$ the assertion also holds.   ■

Let $\widetilde{N^{[i]}}$ be the image of $N_{G^{[i]}}(R)$ in $\widetilde{K}$ for $i = 0$, 1, and 2. By (5.2.5) the amalgam

$$\mathcal{N} = \{\widetilde{N^{[i]}}/\widetilde{R} \mid 0 \le i \le 2\}$$

(where the intersections are chosen in the obvious way) is isomorphic to the amalgam $\mathcal{A}^{[3]}$ and hence also to the amalgam $\mathcal{Z}$ from (11.4.2). We would like to find a 'complement' $\mathcal{K}$ to $\widetilde{R}$ in the amalgam $\{\widetilde{N^{[i]}} \mid 0 \le i \le 2\}$. Since each of the three $\widetilde{N^{[i]}}$'s split over $\widetilde{R}$ in $\widetilde{N^{[i]}}$, the only problem is that there are three complements $\widetilde{K^{[i]}}$ for each i. The complements have to be chosen consistently so that a Sylow 2-subgroup of $\widetilde{K^{[0]}}$ is contained in both $\widetilde{K^{[1]}}$ and $\widetilde{K^{[2]}}$. This can be achieved in the following geometric way.

Assume for a moment that G is the universal completion of $\mathcal{G}$, so that $\Lambda(\mathcal{G}, \varphi, G)$ is a tree containing the complete family of geometric subgraphs. Then the geometric subgraph $\Gamma^{[3]}$ of valency 7 stabilized by $G^{[3]}$ is contained in three geometric subgraphs of valency 15 and the subgraph $\Gamma^{[4]}$ stabilized by $G^{[4]}$ is one of them. These three subgraphs are transitively permuted by R. For every $i = 0$, 1, and 2 the subgroup $N_{G^{[i]}}(R)$ induces the symmetric group Sym$_3$ on the set of geometric subgraphs properly containing $\Gamma^{[3]}$. This suggests a way to maintain the consistency.

Lemma 5.2.6 *Let* $\widetilde{K^{[i]}}$ *be the image in* $\widetilde{K}$ *of* $N_{G^{[i]}}(R) \cap G^{[4]}$. *Then the amalgam* $\widetilde{\mathcal{K}} = \{\widetilde{K^{[i]}} \mid 0 \le i \le 2\}$ *is isomorphic to the amalgam* $\mathcal{Z}$ *in* (11.4.2) *and* $\widetilde{K}$ *is the universal completion of* $\widetilde{\mathcal{K}}$.

5.3   The 12-dimensional module

In this section we discuss some properties of the $GF(2)$-module $\bar{Q}^{[3]}$ of the group

$$\widetilde{K} \cong 3 \cdot \text{Aut} \ (M_{22}).$$

According to James (1973) this is the unique 12-dimensional irreducible $GF(2)$-module of $\widetilde{K}$. In James (1973) the existence of this module is justified by the embedding of M_{22} into $U_6(2)$.

A possible way to see this embedding is in terms of the Leech lattice (cf. Conway (1969), Wilson (1989), and chapter 4 in Ivanov (1999) for details). Let Λ be the Leech lattice, $\bar{\Lambda} = \Lambda/2\Lambda$ and $C \cong Co_1$ be the first Conway sporadic simple group, acting naturally on $\bar{\Lambda} \cong 2^{24}$. We use the standard notation for vectors in the Leech lattice, assuming their coordinates to be in a basis of pairwise orthogonal vectors of squared length 1/8 (cf. section 4.4 in Ivanov (1999)). Let

$$\begin{aligned} \mu_0 &= (\ \ 4, \ \ 4, 0, 0^{21}), \\ \mu_1 &= (\ \ 0, \ \ 4, 4, 0^{21}), \\ \nu &= (-3, -3, 5, 1^{21}). \end{aligned}$$

Let $\bar{X} \cong 2^2$ be the subspace in $\bar{\Lambda}$ spanned by the images of μ_0 and μ_1, and let $\bar{Y} \cong 2^3$ be the subspace spanned by the images of μ_0, μ_1, and ν. Then by lemma 4.10.6 in Ivanov (1999)

$$C_C(\bar{X}) \cong U_6(2).$$

According to the paragraph after lemma 4.10.10 in Ivanov (1999)

$$C_C(\bar{Y}) \cong M_{22}$$

which gives the required embedding.

The 12-dimensional module of $\widetilde{K}$ was studied intensively mostly in the context of J_4.

Lemma 5.3.1 *Let $\widetilde{G^{[3]}} \cong 2^{12} : 3 \cdot \mathrm{Aut}\,(M_{22})$, let $\widetilde{Q^{[3]}} \cong Q^{[3]}/Z^{[3]} \cong 2^{12}$ be the image of $Q^{[3]} \cong 2^{1+12}_+$ and let*

$$\widetilde{K} = N_{\widetilde{G^{[3]}}}(\widetilde{R}) \cong 3 \cdot \mathrm{Aut}\,(M_{22}).$$

Let f and q be the symplectic and associated quadratic forms on $\widetilde{Q^{[3]}}$ defined as in (1.6.1). Then $\widetilde{K}$ acting on the non-zero vectors of $\widetilde{Q^{[3]}}$ has exactly three orbits Ψ_1, Ψ_2, and Ψ_3, such that the following assertions hold (where the singularity/non-singularity of vectors in $\widetilde{Q^{[3]}}$ is defined with respect to q):

(i) *Ψ_1 is of length 693, it consists of singular vectors and $\widetilde{K^{[2]}} \cong 2^5 : \mathrm{Sym}_5$ is the stabilizer of a vector from Ψ_1;*

(ii) *Ψ_2 is of length 1386, it consists of singular vectors with stabilizers of the form $2^4 : PGL_2(5)$;*

(iii) *Ψ_3 is of length 2016, it consists of the non-singular vectors with stabilizers isomorphic to $PGL_2(11)$.*

Proof The orbits on the non-zero vectors and the stabilizers were first determined in Janko (1976). The image W of $Z^{[2]}$ in $\widetilde{Q^{[3]}} \cong Q^{[3]}/Z^{[3]}$ is a 2-subspace normalized by the image of $N_{G^{[2]}}(R)$ in $\widetilde{K}$. Therefore $\widetilde{K^{[2]}}$ stabilizes a vector from W and this vector is in Ψ_1. ∎

By (5.2.6) $\widetilde{K^{[2]}}$ is the stabilizer of a point in the rank 3 Petersen type geometry $\mathcal{G}(3 \cdot M_{22})$ of which $\widetilde{K}$ is the automorphism group. By (5.3.1 (i)) there is a bijection β of the point-set of $\mathcal{G}(3 \cdot M_{22})$ onto Ψ_1 which commutes with the action of $\widetilde{K}$. This bijection induces what is called a *representation* of the geometry $\mathcal{G}(3 \cdot M_{22})$. In Proposition 4.4.6 in Ivanov and Shpectorov (2002) the following intersection diagram of the collinearity graph of $\mathcal{G}(3 \cdot M_{22})$ was given.

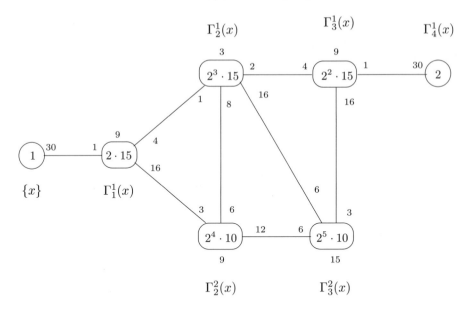

Lemma 5.3.2 *In the above terms if x and y are points of $\mathcal{G}(3 \cdot M_{22})$, then*

$$f(\beta(x), \beta(y)) = 1$$

if and only if $y \in \Gamma_3^2(x)$. In particular for every $v \in \Psi_1$ there are exactly 320 vectors $u \in \Psi_1$ such that $v \notin u^\perp$ (where perp is with respect to f).

Proof This is Corollary 4.4.7 in Ivanov and Shpectorov (2002). ∎

5.4   The triviality of $N^{[4]}$

Recall that $N^{[4]}$ is the largest subgroup in $G^{[014]}$ which is normal in both $G^{[04]}$ and in $G^{[14]}$. By (4.2.1 (v)) $N^{[4]} \leq N^{[3]} \cong 2_+^{1+12} : 3$. Since $H^{[4]} \cong 2^8 : O_8^+(2)$, it is easy to see that

$$G^{[04]} = H^{[04]} \cong 2^{4+4} : 2^6 : L_4(2)$$

so that the chief factors of $G^{[04]}$ inside $O_2(G^{[04]})$ are 4- and 6-dimensional. Therefore $N^{[4]}$ is a 2-group of order 2^m, where $m \in \{0, 4, 6, 8\}$ and $N^{[4]}$ is contained in $Q^{[3]}$.

On the other hand, clearly

$$N_{G^{[0]}}(R) \cap G^{[4]} \leq G^{[04]}$$

and

$$N_{G^{[1]}}(R) \cap G^{[4]} \leq G^{[14]}$$

and since $\mathcal{G}(3 \cdot M_{22})$ is connected, $\widetilde{K}$ is generated by $\widetilde{K^{[0]}}$ and $\widetilde{K^{[1]}}$ (compare (5.2.6)) while $\widetilde{K^{[i]}} = \widetilde{K^{[i]} \cap G^{[4]}}$ for $i = 0$ and 1. By (5.2.4) $\widetilde{K}$ acts on $Q^{[3]}/Z^{[3]}$ irreducibly. So from this point of view $N^{[4]}$, if non-trivial, must be $Q^{[3]}$ or $Z^{[3]}$. This, compared with the above spectrum of possible orders of $N^{[4]}$ gives the required result.

Lemma 5.4.1 *The subgroup $N^{[4]}$ is trivial.*

Since $N^{[4]}$ is trivial one can not guarantee the existence of the geometric subgraph $\Gamma^{[4]}$ for an arbitrary completion of $\mathcal{G}$ (compare (4.2.1 (iv))). In fact this subgraph is missing in the graph $\Lambda(\mathcal{G}, \varphi, J_4)$. Notice also that because of the triviality of $N^{[4]}$ the action of $G^{[4]}$ on the graph $\Lambda^{[4]} = \Lambda(\mathcal{G}^{[4]}, \varphi^{[4]}, G^{[4]})$ is faithful. This action is locally projective of type $(4, 2)$ corresponding to the amalgam $\mathcal{G}^{[4]} = \{G^{[04]}, G^{[14]}\}$. This is the amalgam $\mathcal{A}_4^{(4)}$ in Table 4 in Section 9.11.

Exercises

1. Use the embedding of M_{22} into $U_6(2)$ in terms of the Leech lattice to show that the orbits of $3 \cdot \text{Aut}(M_{22})$ on the non-zero vectors of its irreducible 12-dimensional module are as in (5.3.1).
2. Establish the isomorphism $\widehat{G}_3 \cong \text{Aut}(M_{22})$ performing direct calculations in the amalgam $\mathcal{G}$.

6

THE 1333-DIMENSIONAL REPRESENTATION

In this chapter we prove the existence of a completion of the amalgam $\mathcal{G}$ which is constrained at level 2. We achieve this by studying the non-trivial representations of $\mathcal{G}$ (the representations of the universal completion group of $\mathcal{G}$) of minimal degree. This degree turns out to be 1333 which is the famous minimal character degree of J_4.

6.1 Representations of rank 2 amalgams

Let $\varphi : \mathcal{G} \to A$ be a faithful completion of the amalgam $\mathcal{G} = \{G^{[0]}, G^{[1]}\}$ where $A = GL_m(\mathbf{C})$. Such a completion will be called a faithful representation of $\mathcal{G}$ in dimension m over the complex numbers or simply a *representation* of $\mathcal{G}$. The complex-valued function on the element set of $\mathcal{G}$ defined by

$$\chi(g) = \mathrm{tr}(\varphi(g))$$

will be called the *character* of φ.

Let $\varphi^{[0]}$, $\varphi^{[1]}$, and $\varphi^{[01]}$ be the restrictions of φ to $G^{[0]}$, $G^{[1]}$, and $G^{[01]}$, respectively. Then $\varphi^{[x]}$ is a group representation for $x = 0$, 1, and 01 and both $\varphi^{[0]}$ and $\varphi^{[1]}$, restricted to $G^{[01]}$ equal to $\varphi^{[01]}$. On the other hand, having a pair of group representations

$$\varphi^{[0]} : G^{[0]} \to GL_m(\mathbf{C}) \ \text{and} \ \varphi^{[1]} : G^{[1]} \to GL_m(\mathbf{C})$$

whose restrictions to $G^{[01]}$ coincide, we can reconstruct the whole completion map φ.

Lemma 6.1.1 *Let $\chi^{[0]}$ and $\chi^{[1]}$ be characters of $G^{[0]}$ and $G^{[1]}$, respectively. Then for the existence of a representation $(GL_m(\mathbf{C}), \varphi)$ such that the restrictions of φ to $G^{[0]}$ and $G^{[1]}$ afford $\chi^{[0]}$ and $\chi^{[1]}$, respectively, it is necessary and sufficient for the restrictions to $G^{[01]}$ of $\chi^{[0]}$ and $\chi^{[1]}$ to coincide.*

Proof The necessity is clear by considering the characters of the representations $\varphi^{[0]}$ and $\varphi^{[1]}$ in the paragraph before the lemma. To establish the sufficiency suppose that $\chi^{[0]}$ and $\chi^{[1]}$ satisfy the hypothesis. Let $\psi^{[0]}$ and $\psi^{[1]}$ be representations affording these characters. Since the restrictions of the characters to $G^{[01]}$ coincide, there is $l \in GL_m(\mathbf{C})$, such that $\psi^{[0]}(a) = l^{-1}\psi^{[1]}(a)l$ for every $a \in G^{[01]}$. Define $\varphi : \mathcal{G} \to GL_m(\mathbf{C})$ by putting $\varphi(a) = \psi^{[0]}(a)$ for $a \in G^{[0]}$ and $\varphi(a) = l^{-1}\psi^{[1]}(a)l$ for $a \in G^{[1]}$. Then it only remains to check that φ is injective. Since both $\chi^{[0]}$ and $\chi^{[1]}$ are faithful, the only obstacle is that the whole of $l^{-1}\psi^{[1]}(G^{[1]})l$ might be in $\psi^{[0]}(G^{[0]})$. But this is impossible since $G^{[01]}$ is normal in $G^{[1]}$ and self-normalized in $G^{[0]}$. ∎

The next lemma, which is the main result of (Thompson 1981) supplies the number $k(\chi^{[0]}, \chi^{[1]})$ of equivalence classes of representations corresponding to a pair of characters satisfying the hypothesis of (6.1.1).

Lemma 6.1.2 *Let $(GL_m(\mathbf{C}), \varphi)$ be a representation of $\mathcal{G} = \{G^{[0]}, G^{[1]}\}$ such that the restrictions of φ to $G^{[0]}$ and $G^{[1]}$ afford $\chi^{[0]}$ and $\chi^{[1]}$, respectively. Then the equivalence classes of representations corresponding to the pair $(\chi^{[0]}, \chi^{[1]})$ are in one-to-one correspondence with the $(C^{[0]}, C^{[1]})$-double cosets in $C^{[01]}$, where*

$$C^{[x]} = C_{GL_m(\mathbf{C})}(\varphi(G^{[x]})) \quad \text{for } x = 0, 1, 01.$$

In particular there is precisely one equivalence class if and only if $C^{[01]} = C^{[0]}C^{[1]}$.

By Schur's lemma the isomorphism type of $C^{[x]}$ is determined by the multiplicities of the irreducible constituents of $\chi^{[x]}$. Suppose that $\chi^{[01]}$ is multiplicity-free. Then $C^{[01]}$ is the set of ordered sequences of elements from $\mathbf{C}^*$, whose components are indexed by the irreducible constituents of $\chi^{[01]}$ and the multiplication is componentwise. Then for $i = 0$ and 1 the subgroup $C^{[i]}$ consists of the sequences from $C^{[01]}$ subject to the condition that components, corresponding to $\chi^{[01]}$-constituents fused in $\chi^{[i]}$, are equal. In this case the number $k(\chi^{[0]}, \chi^{[1]})$ is uniquely determined by the characters and easy to calculate.

Next we are aiming to deduce an existence condition for representations in terms of the characters of $G^{[0]}$ only. If I is a conjugacy class of $G^{[0]}$ and χ is a character of $G^{[0]}$ then $\chi(I)$ is the value $\chi(i)$ for some (and hence for all) $i \in I$. Since $G^{[1]}$ contains $G^{[01]}$ as a normal subgroup (of index 2) the former acts on the conjugacy classes and on the irreducible characters of the latter. Let $t \in G^{[1]} \setminus G^{[01]}$, let I be a conjugacy class of $G^{[01]}$ and let λ be a character of $G^{[01]}$. Put

$$I^t = \{t^{-1}it \mid i \in I\}, \quad \lambda^t : g \mapsto \lambda(t^{-1}gt) \text{ for } g \in G^{[01]}.$$

Then $\{I, I^t\}$ and $\{\lambda, \lambda^t\}$ are $G^{[1]}$-orbits on classes and characters of $G^{[01]}$ of lengths 1 or 2. Clearly λ^t is irreducible if and only if λ is such. The following result is rather standard (cf. section 7.15 in Conway et al. (1985)).

Lemma 6.1.3 *Let μ be a character of $G^{[01]}$.*

(i) *If μ is the sum of two irreducibles transposed by $G^{[1]}$ then there exists a unique character ν of $G^{[1]}$ whose restriction to $G^{[01]}$ equals to μ, this character is defined by $\nu(a) = \mu(a)$ if $a \in G^{[01]}$ and $\nu(a) = 0$ if $a \in G^{[01]} \setminus G^{[1]}$.*

(ii) *If μ is irreducible stable under $G^{[1]}$, then there are exactly two characters ν of $G^{[1]}$ such that ν restricted to $G^{[01]}$ equals to μ; for these two characters the values on the elements from $G^{[1]} \setminus G^{[01]}$ are negatives of each other.*

Proof Let λ be an irreducible constituent of μ and θ be the character of $G^{[1]}$ induced from λ. By the Frobenius reciprocity ν is a constituent of θ. Evaluating the inner product of θ with itself, we deduce that in (i) θ is irreducible and hence equal to ν. In (ii) θ is the sum of two irreducibles, say θ_1 and θ_2 and the restriction of each to $G^{[01]}$ is μ. Since $\theta(a) = 0$ for every $a \in G^{[1]} \setminus G^{[01]}$, $\theta_1(a) = -\theta_2(a)$. ∎

Lemma 6.1.4 *Let $\chi^{[0]}$ be a character of $G^{[0]}$. Then for the existence of a representation $(GL_m(\mathbf{C}), \varphi)$, such that the restriction of φ to $G^{[0]}$ affords $\chi^{[0]}$ it is necessary and sufficient that for every conjugacy class I of $G^{[01]}$ the equality $\chi^{[0]}(K) = \chi^{[0]}(L)$ holds, where K and L are the conjugacy classes of $G^{[0]}$ containing I and I^t, respectively (where $t \in G^{[1]} \setminus G^{[01]}$).*

Proof Since the classes I and I^t are fused in $G^{[1]}$, the necessity is obvious. If $\chi^{[0]}$ satisfies the hypothesis, then the restriction of $\chi^{[0]}$ to $G^{[01]}$ is stable under the action of $G^{[1]}$. Therefore we can write this restriction as a sum of characters μ as in (6.1.3). By that lemma the restriction of $\chi^{[0]}$ to $G^{[01]}$ extends to a character of $G^{[1]}$ and the result follows. ∎

In practice the action of t on the classes of $G^{[01]}$ can be determined by calculating the automorphism group of the character table of $G^{[01]}$ and by identifying the image of t in this automorphism group.

6.2 Bounding the dimension

In this section we prove the following.

Proposition 6.2.1 *Let $m = m(\mathcal{G})$ be the smallest positive integer such that there exists a faithful completion $\varphi : \mathcal{G} \to A$ where $A = GL_m(\mathbf{C})$. Then*

$$m(\mathcal{G}) \geq 1333.$$

Further down in this chapter we will construct a completion of $\mathcal{G}$ in $GL_{1333}(\mathbf{C})$. This shows that in (6.2.1) the equality is attained. Originally this equality was established in Ivanov and Pasechnik (2004) by checking the necessary and sufficient conditions in (6.1.4) using explicit computer calculations with character tables.

Before proceeding to the proof of (6.2.1) we recall a few basic facts about representations and group characters. For a group F by $m(F)$ we denote the minimal degree of a faithful complex character of F (equivalently the smallest dimension of a faithful representation over the complex numbers).

Lemma 6.2.2 *Let Q be an elementary abelian 2-group and P be a subgroup of index 1 or 2 in Q. Let $\chi^{[P]}$ be the function on Q such that $\chi^{[P]}(q) = 1$ if $q \in P$ and $\chi^{[P]}(q) = -1$ otherwise. Then $\chi^{[P]}$ is an irreducible character (of degree one) of Q and all the irreducible characters of Q can be obtained in this way.*

The following result is a direct consequence of (6.2.2).

Lemma 6.2.3 *Let F be a group which contains an elementary abelian 2-group Q as a normal subgroup and l be the length of the shortest orbit on the set of hyperplanes (that are subgroups of index 2) in Q of F acting by conjugation. Then $l \leq m(F)$.*

Lemma 6.2.4 *Let $E \cong 2^{1+2m}_+$ be an extraspecial group of order 2^{2m+1} and of plus type. Let $Z = \langle z \rangle$ be the centre of E and $x \in E \setminus Z$ be an involution. Let χ be a faithful irreducible character of E. Then*

   (i) *the degree of χ is 2^m;*
   (ii) $\chi(z) = -2^m$;
   (iii) $\chi(x) = 0$;
   (iv) *χ is unique.*

Proof Let U be a maximal elementary abelian subgroup in E. By (1.6.7) $|U| = 2^{m+1}$ and U contains Z. Let ψ be an irreducible character of U of degree 1 whose kernel is disjoint from Z. Let χ be the character of E induced from ψ. Then it is well known and easy to check that χ is as stated. ∎

Proof of Proposition 6.2.1 Let $\varphi : \mathcal{G} \rightarrow A$ be a faithful completion of $\mathcal{G}$, where $A = GL_m(\mathbf{C})$ and let χ be the group character afforded by the restriction of φ to $Q^{[3]} \cong 2^{1+12}_+$. Then

$$\chi = \chi_0 + \chi_1 + \chi_2,$$

where χ_0 is the sum of the faithful constituents, χ_1 is the sum of the constituents with kernel $Z^{[3]}$ and χ_2 is the sum of the trivial constituents. Let z be the generator of $Z^{[3]}$ and $x \in Z^{[2]} \setminus Z^{[3]}$ (in particular x is an involution). Then $x \in Q^{[3]}$, z and x are conjugate in $G^{[2]}$ and hence $\chi(x) = \chi(z)$. By (6.2.4) the irreducible constituents in χ_0 are all equal to the unique irreducible faithful character of $Q^{[3]}$ of degree 64; if n is the multiplicity of this character, then

$$\chi_0(1) = 64n, \ \chi_0(z) = -64n, \ \chi_0(x) = 0.$$

Since χ_2 is a trivial character, $\chi_2(x) = \chi_2(z) = \chi(1)$. Thus for the equality $\chi(x) = \chi(z)$ to hold the constituent χ_1 must be non-zero. By (6.2.2) χ_1 is the sum of linear characters of $\bar{Q}^{[3]} = Q^{[3]}/Z^{[3]}$ (considered as characters of $Q^{[3]}$). By (5.2.4) the group $G^{[3]}$ induces on $\bar{Q}^{[3]}$ the group $3 \cdot \text{Aut}\,(M_{22})$, whose orbits on the non-zero vectors are described in (5.3.1). Notice that $\bar{Q}^{[3]}$ is self-dual and the orbit lengths on the hyperplanes are the same as on the set of non-zero vectors. Combining (6.2.3) and (the dual version of) (5.3.1) we observe that either $\chi_1(1) \geq 1386$ or $\chi_1(1) = 693$ and χ_1 is the sum of 693 linear characters corresponding to the orbit Ψ_1 in (5.3.1). Since $x \in Z^{[2]}$, we conclude from statement (5.3.1) and its proof that the image of x in $\bar{Q}^{[3]}$ belongs to Ψ_1. By (5.3.2) this gives

$$\chi_1(x) = 693 - 2 \cdot 320 = 53.$$

In this case, since $\chi_0(x) = 0$,

$$53 = \chi_0(z) + \chi_1(z) = -64n + 693,$$

which gives $n = 10$. Hence

$$\chi(1) \geq \chi_0(1) + \chi_1(1) \geq 64 \cdot 10 + 693 = 1333.$$

■

As an immediate consequence of the above proof we obtain the following.

Corollary 6.2.5 *Let $\varphi : \mathcal{G} \to A$ be a faithful completion, where $A = GL_{1333}(\mathbf{C})$ and χ is the corresponding character. Let z be a generator of $Z^{[3]}$. Then the following assertions hold:*

(i) *the restrictions of χ to $Q^{[3]}$ does not contain trivial constituents;*
(ii) *$\chi(z) = 53$.*

6.3   On irreducibility of induced modules

In this section we prove a standard result from the representation theory of finite groups.

Let F be a finite group, let Σ be an F-module over the complex numbers. Let Q be an elementary abelian normal 2-subgroup in F which acts fixed-point freely on Σ, that is, $C_\Sigma(Q) = 0$. Let $\mathcal{R}$ be the set of hyperplanes in Q having non-zero centralizers in Σ:

$$\mathcal{R} = \{R \mid R \leq Q, \ [Q : R] = 2, \ C_\Sigma(R) \neq 0\}.$$

For $R \in \mathcal{R}$ put $\Sigma_R = C_\Sigma(R)$.

Proposition 6.3.1 *In the situation described in the paragraph before the proposition suppose that F (acting by conjugation) transitively permutes the hyperplanes in $\mathcal{R}$. Then*

(i) *the module Σ is isomorphic to the one induced from the $N_F(R)$-module Σ_R;*
(ii) *Σ is irreducible provided that the action of $N_F(R)$ on Σ_R is irreducible;*
(iii) *suppose that Σ_R is the direct sum of k pairwise non-isomorphic constituents of dimensions $d_1, \ldots, d_k$, then Σ is a direct sum of k pairwise non-isomorphic irreducible constituents of dimensions $d_1 \cdot m, \ldots, d_k \cdot m$, respectively, where $m = |\mathcal{R}| = [F : N_F(R)]$.*

Proof By the definition Σ_R is a constituent of Σ restricted to $N_F(R)$. By the hypothesis the space Σ is spanned by the images of Σ_R under F. By the Frobenius reciprocity Σ is a quotient of the module I induced from the $N_F(R)$-module Σ_R. Since both I and Σ have dimension

$$\dim \Sigma_R \cdot [F : N_F(R)] = \dim \Sigma_R \cdot |\mathcal{R}|,$$

(i) follows.

Suppose that the action of $N_F(R)$ on Σ_R is irreducible. Let Θ be a non-zero F-submodule in Σ. By Maschke theorem Θ is a direct sum of irreducible Q-submodules. Since $C_\Sigma(Q) = 0$, by (6.2.2) each such submodule is non-trivial and its kernel is a hyperplane of Q. Therefore $\Theta_P : = C_\Theta(P)$ is non-trivial for some hyperplane P in Q. Since $C_\Theta(P) \le C_\Sigma(P)$, we have $P \in \mathcal{R}$ and without loss of generality we assume that $P = R$. Clearly $\Theta_R \le \Sigma_R$. Since Θ is an F-submodule in Σ, $N_F(R)$ stabilizes Θ_R. Since the action of $N_F(R)$ on Σ_R is irreducible, we conclude that $\Theta_R = \Sigma_R$ and (ii) follows.

To establish (iii) we consider an irreducible constituent of $N_F(R)$ in Σ_R, take its images under F and consider the subspace in Σ spanned by these images. By (i) and (ii) such a subspace is an irreducible F-submodule in Σ. Furthermore, arguing as in the above paragraph, it is easy to see that every constituent appears in this way. Finally the irreducible constituents of F in Σ are pairwise non-isomorphic since those of $N_F(R)$ in Σ_R are such. ∎

6.4 Representing $G^{[0]}$

In the next five sections (including this one) we construct a faithful completion $\varphi : \mathcal{G} \to GL_{1333}(\mathbf{C})$. Actually we will construct eight such completions and the result will be summarized in Section 6.9. Suppose first that such a completion exists. Let $\Pi^{[0]}$ be a 1333-dimensional vector space over the field $\mathbf{C}$ of complex numbers which supports the restriction of φ to $G^{[0]}$. By (6.2.5) the character χ of φ satisfies $\chi(z) = 53$, where z is the generator of $Z^{[3]}$. In terms of (3.6.2) the element z is in the orbit $\mathcal{Q}_1$ of length 155 of the action of $L^{[0]}$ on $Q^{[0]}$.

Consider the eigenspace decomposition of $\Pi^{[0]}$ with respect to $Q^{[0]}$. Let $\mathcal{P}$ be the set of hyperplanes (subgroups of index 2) in $Q^{[0]}$. Then by (3.6.1) under the natural action by conjugation $\mathcal{P}$ splits into two $G^{[0]}$-orbits $\mathcal{P}_1$ and $\mathcal{P}_2$ with lengths 155 and 868, respectively. For $P \in \mathcal{P}$ put

$$\Pi_P^{[0]} = C_{\Pi^{[0]}}(P).$$

Notice that $q \in Q^{[0]}$ acts on $\Pi_P^{[0]}$ as the identity operator if $q \in P$ and as the (-1)-scalar operator otherwise. Combining (3.6.3), (6.2.2), and (6.2.5) and applying direct calculations we obtain the following.

Lemma 6.4.1 *Let $\Pi^{[0]}$ be the $G^{[0]}$-module which supports the restriction to $G^{[0]}$ of a 1333-dimensional representation of $\mathcal{G}$. Then*

(i) $C_{\Pi^{[0]}}(Q^{[0]}) = 0$;

(ii) *if $P \in \mathcal{P}_1$ then* $\dim \Pi_P^{[0]} = 3$;

(iii) *if $P \in \mathcal{P}_2$ then* $\dim \Pi_P^{[0]} = 1$.

By (6.4.1) $\Pi^{[0]}$ is the direct sum of the $\Pi_P^{[0]}$'s taken for all $P \in \mathcal{P}$. Furthermore, if $\Pi_1^{[0]}$ is the sum of the subspaces $\Pi_P^{[0]}$ taken for all $P \in \mathcal{P}_1$ and $\Pi_2^{[0]}$ is the sum of such subspaces taken for all $P \in \mathcal{P}_2$ then both $\Pi_1^{[0]}$ and $\Pi_2^{[0]}$ are $G^{[0]}$-submodules in $\Pi^{[0]}$ and

$$\Pi^{[0]} = \Pi_1^{[0]} \oplus \Pi_2^{[0]}.$$

Finally, dim $\Pi_1^{[0]} = |\mathcal{P}_1| \cdot 3 = 465$ and dim $\Pi_2^{[0]} = |\mathcal{P}_2| = 868$ (notice that $465 + 868 = 1333$).

Let $P_1 \in \mathcal{P}_1$, $P_2 \in \mathcal{P}_2$. Let F_1 and F_2 be the stabilizers in $G^{[0]}$ of P_1 and P_2, respectively. Put $\Pi_1 = \Pi_{P_1}^{[0]}$ and $\Pi_2 = \Pi_{P_2}^{[0]}$. Then Π_1 is a 3-dimensional module for F_1, while Π_2 is a 1-dimensional module for F_2. By the (6.3.1 (i)) $\Pi_1^{[0]}$ is induced from Π_1 and $\Pi_2^{[0]}$ is induced from Π_2.

The above discussion suggests the following strategy to produce $\Pi^{[0]}$:

(a) calculate the stabilizers F_1 and F_2;

(b) describe a 3-dimensional **C**-module Π_1 of F_1 and a 1-dimensional module Π_2 of F_2;

(c) define $\Pi_1^{[0]}$ and $\Pi_2^{[0]}$ as the $G^{[0]}$-modules induced from Π_1 and Π_2, respectively;

(d) put $\Pi^{[0]} = \Pi_1^{[0]} \oplus \Pi_2^{[0]}$.

First we define Π_1 and Π_2 up to two possibilities each. The adjustment will be made when we restrict $\Pi^{[0]}$ to $G^{[01]}$ and require the restricted action to be extendable to an action of $G^{[1]}$.

In due course we will use the following direct consequence of (6.4.1) and (3.6.3).

Corollary 6.4.2 *Let y be an element from the orbit of length 868 of the action of $L^{[0]}$ on $Q^{[0]}$. Then*

$$\chi(y) = -11.$$

In terms introduced in the paragraph after (3.6.1), if $P_1 = P(U_3)$ then $F_1 = G^{[02]}$ and if $P_2 = P(U_1, f)$ then $F_2 = G^{[0]}(f)$ is the stabilizer in $G^{[0]}$ of a symplectic form f of rank 4 on U_5, whose radical is U_1. Equivalently, F_2 is the stabilizer in $G^{[04]}$ of the non-singular symplectic form $\bar{f}$ on U_5/U_1 induced by f. Therefore

$$[G^{[04]} : G^{[04]}(\bar{f})] = 28, \quad O_2(G^{[04]}) \le G^{[04]}(\bar{f}) \quad \text{and}$$

$$G^{[04]}(\bar{f})/O_2(G^{[04]}) \cong S_4(2) \cong \mathrm{Sym}_6.$$

Lemma 6.4.3 *The following assertions hold:*

(i) $G^{[02]}/O_2(G^{[02]}) \cong (L^{[0]} \cap G^{[2]})Q^{[2]}/Q^{[2]} \cong \mathrm{Sym}_3 \times L_3(2)$;

(ii) *if S_3 is a Sylow 3-subgroup of $O_{2,3}(G^{[02]})$, then $[O_2(G^{[02]}), S_3] = Q^{[2]} = O_2(N^{[2]})$;*

(iii) $Q^{[0]} \cap Q^{[2]} = P_1 = P(U_3)$.

Proof Statement (i) is immediate, since $G^{[02]}$ is the stabilizer of U_3 in $G^{[0]}$. In order to establish (ii), notice that

$$G^{[02]}/Q^{[2]} \cong (\mathrm{Sym}_3 \times 2) \times L_3(2),$$

and that $Z^{[2]}$ is the commutator subgroup of $Q^{[2]}$. Furthermore, $Q^{[2]}/Z^{[2]}$, as a module for

$$H^{[2]}/Q^{[2]} \cong O_4^+(2) \times L_3(2)$$

is isomorphic to the tensor product $W_4 \otimes U_3$, which shows that S_3 acts on $Q^{[2]}/Z^{[2]}$ is fixed-point freely. Without loss we assume that $S_3 \leq L^{[0]}$. Then elementary calculations in $Q^{[0]}$ (which is the exterior square of U_5) show that $C_{Q^{[0]}}(S_3)$ is 1-dimensional, which gives (iii). ∎

The following lemma was known since the time of A. Hurwitz and F. Klein.

Lemma 6.4.4 *The following assertions hold:*

(i) *the smallest faithful complex representation of $L_3(2)$ is 3-dimensional;*
(ii) *$L_3(2)$ possesses exactly two complex irreducible characters of degree 3, say λ_1 and λ_2, which are algebraically conjugate and transposed by the outer automorphism group of $L_3(2)$;*
(iii) *if $S \cong \mathrm{Sym}_4$ is a maximal parabolic subgroup in $L_3(2)$ then the restriction to S of a representation affording the character λ_1 or λ_2 is irreducible.*

For the remainder of the chapter we fix a complex character λ of $L^{[2]} \cong L_3(2)$ of degree 3. In terms of (6.4.4) λ is either λ_1 or λ_2.

Let us turn to the description of the modules Π_1 and Π_2 of $G^{[02]}$ and $G^{[04]}(f)$, respectively. First we consider the possibilities for the kernels K_1 and K_2 of these modules.

Lemma 6.4.5 *Let K_1 be a normal subgroup in $G^{[02]}$ such that*

$$G^{[02]}/K_1 = Q^{[0]}K_1/K_1 \times L^{[2]}K_1/K_1 \cong 2 \times L_3(2).$$

Then K_1 is one of two particular subgroups $K_1^{(1)}$ and $K_1^{(2)}$ distinguished by the condition that the equality $K_1^{(i)} = P_1(K_1^{(i)} \cap L^{[0]})$ holds for $i = 1$ but not for $i = 2$.

Proof Since K_1 is normal in $G^{[02]}$ and by considering the order of K_1 we conclude that K_1 contains a Sylow 3-subgroup S_3 of $O_{2,3}(G^{[02]})$. By (6.4.3 (iii)) $Q^{[2]} = [O_2(G^{[02]}), S_3]$. Therefore K_1 contains $Q^{[2]}$. We know that $G^{[02]}/Q^{[2]} \cong (\mathrm{Sym}_3 \times 2) \times L_3(2)$ and that $Q^{[0]}Q^{[2]}/Q^{[2]}$ is the centre of order 2 in $G^{[02]}/Q^{[2]}$. Therefore $K_1/Q^{[2]} \cong \mathrm{Sym}_3$, while by the above $G^{[02]}/Q^{[2]}$ contains exactly two normal subgroups isomorphic to Sym_3. On other hand, $(L^{[0]} \cap G^{[2]})Q^{[2]}/Q^{[2]} \cong \mathrm{Sym}_3 \times L_3(2)$ contains just one such normal subgroup. Hence the result. ∎

If $K_1^{(i)}$ is as in (6.4.5), then by (6.4.3 (iii))

$$K_1^{(i)} \cap Q^{[0]} = P_1 = P(U_3)$$

both for $i = 1$ and 2.

Let $\Pi_1 = \Pi_1^{(i)}$, where $i = 1$ or 2, be the 3-dimensional module for $G^{[02]}$ over the field of complex numbers whose kernel is $K_1^{(i)}$ and the action of $L^{[2]} \cong L_3(2)$ affords the character λ as defined after (6.4.4). Since $Q^{[0]} K_1^{(i)} / K_1^{(i)}$ is the centre of order 2 of $G^{[02]} / K_1^{(i)} \cong 2 \times L_3(2)$, every element from $Q^{[0]} \setminus K_1^{(i)} = Q^{[0]} \setminus P_1$ acts on $\Pi_1^{(i)}$ as the (-1)-scalar operator. Because of the factorization

$$G^{[02]} = K_1^{(i)} Q^{[0]} L^{[2]},$$

the above conditions define the module $\Pi_1^{(i)}$ up to isomorphism for $i = 1$ and 2.

Lemma 6.4.6 *The group* $G^{[04]}(\bar{f})$ *contains exactly two normal subgroups* $K_2^{(1)}$ *and* $K_2^{(2)}$ *of index* 2 *intersecting* $Q^{[0]}$ *in* P_2. *Furthermore the notation can chosen so that* $L^{[0]}(f) = S_f \cong 2^4 : S_4(2)$ *is contained in* $K_2^{(1)}$ *but not in* $K_2^{(2)}$.

Proof Since $G^{[04]}(\bar{f})/P_2 \cong 2 \times 2^4 : S_4(2)$, the quotient X of $G^{[04]}(\bar{f})$ over the commutator subgroup of $G^{[04]}(\bar{f})$ is elementary abelian of order 4. One of the order 2 subgroups in X is of course the image of $Q^{[0]}$, hence the result. ∎

Let $\Pi_2 = \Pi_2^{(i)}$ be a 1-dimensional module for $G^{[04]}(\bar{f})$ over the field of complex numbers, whose kernel is $K_2^{(i)}$ for $i = 1$ and 2.

There are just eight possibilities for the modules Π_1 and Π_2 (taking into account the two possibilities for the action of $L^{[2]}$ on Π_1). To specify these modules we have to choose K_1 between $K_1^{(1)}$ and $K_1^{(2)}$ and also to choose K_2 between $K_2^{(1)}$ and $K_2^{(2)}$. Inducing from $G^{[02]}$ and $G^{[04]}(\bar{f})$ to $G^{[0]}$ and taking the direct sum, we obtain a $G^{[0]}$-module $\Pi^{[0]}$, whose dimension is

$$1333 = 155 \times 3 + 868 = [G^{[0]} : G^{[02]}] \times \dim \Pi_1 + [G^{[0]} : G^{[04]}(\bar{f})] \times \dim \Pi_2.$$

6.5   Restricting to $G^{[01]}$

The $G^{[0]}$-module $\Pi^{[0]}$ from the previous section is 1333-dimensional over the field of complex numbers and it is the direct sum of two submodules $\Pi_1^{[0]}$ and $\Pi_2^{[0]}$ of dimension 465 and 868, respectively. The module $\Pi_1^{[0]}$ is induced from a 3-dimensional module $\Pi_1 = \Pi_1^{(i)}$ for $G^{[02]}$ and the module $\Pi_2^{[0]}$ is induced from a 1-dimensional module $\Pi_2 = \Pi_2^{(j)}$ of $G^{[04]}(f)$, where $i, j \in \{1, 2\}$. In this section we discuss the restriction of $\Pi^{[0]}$ from $G^{[0]}$ to $G^{[01]}$. Here the choice between possibilities for $\Pi^{[0]}$ is irrelevant.

If $P \in \mathcal{P}_1$ then $P = P(V_3)$ for a 3-dimensional $GF(2)$-subspace V_3 in U_5, and the module $\Pi_P^{[0]}$ is 3-dimensional (over the complex numbers) by (6.4.1 (ii)).

The stabilizer of V_3 in $G^{[0]}$ coincides with $N_{G^{[0]}}(\Pi_P^{[0]})$ and this stabilizer acts on $\Pi_P^{[0]}$ as $G^{[02]}$ does on Π_1. Namely, these two actions are conjugate by an element of $G^{[0]}$ which maps U_3 onto V_3. If $P \in \mathcal{P}_2$, then $P = P(V_1, f)$, $\Pi_P^{[0]}$ is 1-dimensional and the stabilizer of P in $G^{[0]}$ acts on $\Pi_P^{[0]}$ in the way $G^{[04]}(f)$ does on Π_2.

Lemma 6.5.1 *The $G^{[0]}$-modules $\Pi_1^{[0]}$ and $\Pi_2^{[0]}$ are irreducible.*

Proof The result is immediate by (6.3.1). ∎

Next we restrict $\Pi^{[0]}$ from $G^{[0]}$ to $G^{[01]}$. Recall that the latter is the stabilizer in $G^{[0]}$ of the hyperplane U_4 in U_5.

Lemma 6.5.2 *Put*

$$\mathcal{P}_{11}^{(1)} = \{P(V_3) \mid V_3 < U_4\}, \quad \mathcal{P}_{12}^{(1)} = \{P(V_3) \mid V_3 \not< U_4\},$$

$$\mathcal{P}_{21}^{(1)} = \{P(V_1, f) \mid V_1 < U_4\}, \quad \mathcal{P}_{22}^{(1)} = \{P(V_1, f) \mid V_1 \not< U_4\}.$$

Then

(i) *$\mathcal{P}_{11}^{(1)}$ and $\mathcal{P}_{12}^{(1)}$ are the $G^{[01]}$-orbits on $\mathcal{P}_1$;*

(ii) *$\mathcal{P}_{21}^{(1)}$ and $\mathcal{P}_{22}^{(1)}$ are the $G^{[01]}$-orbits on $\mathcal{P}_2$;*

(iii) *$|\mathcal{P}_{11}^{(1)}| = 15$, $|\mathcal{P}_{12}^{(1)}| = 140$, $|\mathcal{P}_{21}^{(1)}| = 420$, $|\mathcal{P}_{22}^{(1)}| = 448$.*

Proof Since $L_5(2)$ has two orbits on the set of pairs (V_3, V_4) where V_3 and V_4 are 3- and 4-dimensional subspaces in the natural module U_5, (i) follows. Let S_f be the stabilizer of f in $L^{[0]}$. Then by (1.5.1) $S_f \cong 2^4 : S_4(2)$. In order to establish (ii) we have to determine the orbits of S_f on the set of hyperplanes in U_5. By (1.5.3) the orbit of such a hyperplane V_4 is determined by whether or not V_4 contains V_1 and by the isomorphism type of $(V_4, f|_{V_4})$. Since V_1 is the radical of f, it is easy to see that the symplectic form $(V_4, f|_{V_4})$ has rank 2 if V_4 contains V_1 and it is non-singular otherwise. The assertion (iii) now follows from elementary geometric calculations. ∎

Lemma 6.5.3 *For a hyperplane $P \in \mathcal{P}$ of $Q^{[0]}$ the stabilizer $G^{[01]}(P)$ of P in $G^{[01]}$ acts non-trivially and irreducibly on $\Pi_P^{[0]}$.*

Proof An equivalent reformulation of the assertion is that for any pair (V_4, P) where V_4 is a hyperplane in U_5 and P is a hyperplane in $Q^{[0]}$, the stabilizer $G^{[0]}(V_4, P)$ of this pair in $G^{[0]}$ acts non-trivially and irreducibly on $\Pi_P^{[0]}$. By the proof of (6.5.2), $G^{[0]}$ has four orbits on the set of pairs (V_4, P) as above. We take P to be $P_1 = P(U_3)$ or $P_2 = P(U_1, f)$ and vary V_4. Thus we have to check that $G^{[0]}(V_4, P_i)$ acts irreducibly on Π_i. Since Π_2 is 1-dimensional and $Q^{[0]}$ acts on Π_2 non-trivially, and $G^{[0]}(V_4, P_i)$ contains $Q^{[0]}$, the result is obvious in the case $i = 2$. Therefore we consider the case $i = 1$. First suppose that

V_4 contains U_3. There are three hyperplanes in U_5 containing U_3. These hyperplanes are transitively permuted by a Sylow 3-subgroup S_3 of $O_{2,3}(G^{[02]})$. Hence $G^{[0]}(V_4, P_1)K_1 = G^{[02]}$, which implies that the actions induced on Π_1 by $G^{[02]}$ and $G^{[0]}(V_4, P_1)$ coincide, giving the assertion. If V_4 does not contain U_3, then as always we have $Q^{[0]} \leq G^{[0]}(V_4, P_1)$ and also without loss we may assume that $L^{[2]} \cap G^{[0]}(V_4, P_1)$ is the stabilizer of the 2-dimensional subspace $V_4 \cap U_3$ in the natural module U_3 of $L^{[2]}$. Therefore $G^{[0]}(V_4, P_1)$ induces $2 \times \mathrm{Sym}_4$ on Π_1 and the action is irreducible by (6.4.4 (iii)). ∎

Lemma 6.5.4 *For* $a, b \in \{1, 2\}$ *put*

$$\Pi_{ab}^{[01]} = \bigoplus_{P \in \mathcal{P}_{ab}^{(1)}} \Pi_P^{[0]}.$$

Then the $\Pi_{ab}^{[01]}$*'s are the irreducible* $G^{[01]}$*-submodules in* $\Pi^{[0]}$ *of dimensions 45, 420, 420, and 448 for* $ab = 11, 12, 21$*, and 22, respectively.*

Proof By (6.5.2) the $\Pi_{ab}^{[01]}$'s are $G^{[01]}$-submodules. Arguing as in the proof of (6.5.1) and making use of (6.5.3) it is easy to check that each of the four submodules is irreducible. ∎

6.6 Lifting $\Pi_{11}^{[01]}$

In terms of Section 3.7 we have $G^{[01]} = Q^{[1]}L^{[1]}$, where $Q^{[1]} = O_2(G^{[01]})$, $L^{[1]} \cong L_4(2)$, and U_4 is the natural module of $L^{[1]}$. Furthermore, $Q^{[1]} = Z^{[1]}A^{[1]}B^{[1]}$ and there are $L^{[1]}$-module isomorphisms $\alpha : U_4 \to A^{[1]}$, $\beta : U_4 \to B^{[1]}$ and $\zeta : \bigwedge^2 U_4 \to Z^{[1]}$, such that the key commutator relation reads as

$$[\alpha(w_1), \beta(w_2)] = \zeta(w_1 \wedge w_2),$$

where $w_1, w_2 \in U_4$. As above put $D^{[1]} = \langle \alpha(w)\beta(w) \mid w \in U_4 \rangle$ (recall that $D^{[1]}$ is elementary abelian of order 2^{10} and that $D^{[1]}$ contains $Z^{[1]}$).

The family $\mathcal{P}_{11}^{(1)}$ consists of 15 hyperplanes of $Q^{[0]}$ which are naturally indexed by the 3-dimensional subspaces in U_4. The group $G^{[01]}$ induces on this family the natural permutation action of $L_4(2) \cong L^{[1]}$ of degree 15 with kernel $Q^{[1]}$. The hyperplane P_1 corresponds to the subspace U_3. Let $L^{[1]}(U_3) \cong 2^3 : L_3(2)$ be the stabilizer of U_3 in $L^{[1]}$. Then

$$G^{[012]} = G^{[01]} \cap G^{[02]} = Q^{[1]}L^{[1]}(U_3)$$

is the stabilizer of U_3 in $G^{[01]}$.

Consider the action of $G^{[012]}$ on Π_{P_1} (the latter being canonically identified with Π_1). By (6.3.1 (i)) the action of $G^{[01]}$ on $\Pi_{11}^{[01]}$ is induced from that action. As above let K_1 denote the kernel of $G^{[02]}$ on Π_1 (so that K_1 is $K_1^{(1)}$ or $K_1^{(2)}$ in terms of (6.4.5)) and put $R_1 = K_1 \cap G^{[012]}$. By the proof of (6.5.3) we have $[K_1 : R_1] = 3$ and $K_1 = R_1 S_3$ for a Sylow 3-subgroup S_3 of $O_{2,3}(G^{[02]})$. On the other hand, R_1 contains $Q^{[2]}$ and by the structure of $G^{[01]}$, we have

$$Q^{[2]} = Z^{[1]}\alpha(U_3)\beta(U_3)O_2(L^{[1]}(U_3)).$$

Since $G^{[012]}/R_1 \cong G^{[02]}/K_1 \cong 2 \times L_3(2)$, while $G^{[012]}/Q^{[2]} \cong 2^2 \times L_3(2)$, we conclude that $R_1/Q^{[2]}$ is a normal subgroup of order 2 in $G^{[012]}/Q^{[2]}$. One the other hand, the normal subgroups of order 2 in $G^{[012]}/Q^{[2]}$ are $A^{[1]}Q^{[2]}/Q^{[2]}$, $B^{[1]}Q^{[2]}/Q^{[2]}$, and $D^{[1]}Q^{[2]}/Q^{[2]}$.

By the above $K_1 = R_1 S_3$ where S_3 is a Sylow 3-subgroup of $O_{2,3}(G^{[02]})$ and we will assume without loss of generality that $S_3 \leq L^{[0]}$. If $R_1 = A^{[1]}Q^{[2]}$ then R_1 (and hence K_1 as well) contains $Q^{[0]} = Z^{[1]}A^{[1]}$ which is not allowed. On the other hand $B^{[1]}S_3 \leq L^{[0]}$ and $B^{[1]}S_3 Q^{[2]}/Q^{[2]} \cong \mathrm{Sym}_3$. Therefore in terms of (6.4.5) we obtain:

$$R_1^{(1)} := K_1^{(1)} \cap G^{[012]} = Q^{[2]}B^{[1]},$$

$$R_1^{(2)} := K_1^{(2)} \cap G^{[012]} = Q^{[2]}D^{[1]}.$$

Suppose first that $R_1 = R_1^{(1)} = Q^{[2]}B^{[1]}$. Then $Z^{[1]}B^{[1]}$ is in K_1. Since $Z^{[1]}B^{[1]}$ is normal in $G^{[01]}$ and $G^{[01]}$ permutes transitively the subspaces $\Pi_P^{[0]}$ for $P \in \mathcal{P}_{11}^{(1)}$ the action of $Z^{[1]}B^{[1]}$ on the whole of $\Pi_{11}^{[01]}$ is trivial. Since the action of $Z^{[1]}A^{[1]}$ is non-trivial and $G^{[1]}$ contains an element t_1 which conjugates $Z^{[1]}A^{[1]}$ onto $Z^{[1]}B^{[1]}$, the action of $G^{[01]}$ on $\Pi_{11}^{[01]}$ cannot be lifted to that of $G^{[1]}$ in this case, and we have established the following result.

Lemma 6.6.1 *If the kernel of $G^{[02]}$ on Π_1 is $K_1^{(1)}$, then the action of $G^{[01]}$ on $\Pi_{11}^{[01]}$ cannot be lifted to an action of $G^{[1]}$.*

Now suppose that $R_1 = R_1^{(2)} = Q^{[2]}D^{[1]}$. Then $D^{[1]} \leq K_1$. Since $D^{[1]}$ is normal in $G^{[01]}$, we conclude that $D^{[1]}$ acts trivially on the whole of $\Pi_{11}^{[01]}$. Therefore the action of $G^{[01]}$ on $\Pi_{11}^{[01]}$ is isomorphic to

$$G^{[01]}/D^{[1]} \cong 2^4 : L_4(2).$$

On the other hand, $D^{[1]}$ stays normal in $G^{[1]}$ and the element t_1 (defined in Section 3.7 which extends $G^{[01]}$ to $G^{[1]}$) centralizes $G^{[01]}$ modulo $D^{[1]}$. Therefore

$$G^{[1]}/D^{[1]} \cong 2^4 : L_4(2) \times 2$$

and the action of $G^{[01]}$ on $\Pi_{11}^{[01]}$ can be lifted to that of $G^{[1]}$ simply by declaring that t_1 acts as a (± 1)-scalar operator.

Proposition 6.6.2 *If the kernel of $G^{[02]}$ on Π_1 is $K_2^{(2)}$ then there are exactly two ways to lift the action of $G^{[01]}$ on $\Pi_{11}^{[01]}$ to an action of $G^{[1]}$. There is an element $t_1 \in G^{[1]} \setminus G^{[01]}$ commuting with $D^{[1]}$, which acts as the identity operator in one of the liftings and as the (-1)-scalar operator in the other one.*

6.7 Lifting $\Pi_{22}^{[01]}$

Recall that for $P \in \mathcal{P}_{22}^{(1)}$ the stabilizer of P in $G^{[01]}$ coincides with the stabilizer of a pair (V_1, f), where V_1 is a 1-dimensional subspace in U_5, disjoint from U_4 and f is a non-singular symplectic form on U_4 (notice that in the considered case U_4 is canonically isomorphic to U_5/V_1). In terms of the development of Sections 3.4 and 3.7 we assume that $V_1 = U_5 \cap W_2$ and $V_1 = \langle u \rangle$, so that the stabilizer of V_1 in $G^{[01]}$ is $Z^{[1]}A^{[1]}L^{[1]}$ and

$$G^{[01]}(P) = G^{[01]}(V_1, f) = Z^{[1]}A^{[1]}S$$

where $S \cong S_4(2) \cong \mathrm{Sym}_6$ is the stabilizer of f in $L^{[1]}$. As an $L^{[1]}$-module the subgroup $Z^{[1]}$ is isomorphic to the exterior square of U_4 and also to the exterior square of U_4^*. By (1.4.3) the elements of $Z^{[1]}$ are the symplectic forms on U_4. Therefore f is an element of $Z^{[1]}$ and its perp (with respect to the non-singular symplectic form on $Z^{[1]}$ preserved by $L^{[1]} \cong \Omega_6^+(2)$) is an S-invariant hyperplane in $Z^{[1]}$. We denote this hyperplane by $Z_f^{[1]}$. Then $P = Z_f^{[1]}A^{[1]}$, and if C is the commutator subgroup of $G^{[01]}(V_1, f)$, then

$$C = Z_f^{[1]}A^{[1]}T,$$

where $T \cong \mathrm{Alt}_6$ is the commutator subgroup of $S \cong \mathrm{Sym}_6$. Let R_2 be the kernel of the action of $G^{[01]}(V_1, f)$ on the subspace $\Pi_P^{[0]}$. Then R_2 contains C with index 2 and R_2 does not contain $Z^{[1]}$. This leaves us with two possibilities for R_2 (notice that the choice between these two possibilities specifies the choice between $K_2^{(1)}$ and $K_2^{(2)}$ in (6.4.6)).

In any event we have $R_2 \cap Z^{[1]} = Z_f^{[1]}$. Let

$$\Pi^{(f)} = C_{\Pi_{22}^{[01]}}(Z_f^{[1]}).$$

By the above, $\Pi^{(f)}$ is the sum of the subspaces $\Pi_Q^{[0]}$ taken for all hyperplanes $Q = P(V_1, f)$, where V_1 is a 1-dimensional subspace in U_5 disjoint from U_4 and f is the above form on $U_5/V_1 \cong U_4$.

Lemma 6.7.1 $\dim \Pi^{(f)} = 16$ *and the action of $G^{[01]}$ on $\Pi_{22}^{[01]}$ is induced from the action of*

$$C_{G^{[01]}}(f) = Z^{[1]}A^{[1]}B^{[1]}S$$

on $\Pi^{(f)}$.

Proof Since there are exactly 16 1-dimensional subspaces in U_5 disjoint from U_4, we get the dimension of $\Pi^{(f)}$. Since

$$448 = \dim \Pi_{22}^{[01]} = \dim \Pi^{(f)} \times [G^{[01]} : C_{G^{[01]}}(f)] = 16 \times 28,$$

the second assertion is by (6.3.1 (i)).   ∎

Put $\bar{F}^{[01]} = C_{G^{[01]}}(f)/Z_f^{[1]}$ and adopt the bar notation for the images in $\bar{F}^{[01]}$ of elements and subgroups of $C_{G^{[01]}}(f)$, and also for the homomorphisms induced by α, β and ζ.

Lemma 6.7.2 *The following assertions hold:*

 (i) $\bar{Z}^{[1]}\bar{A}^{[1]}\bar{B}^{[1]} = O_2(\bar{F}^{[01]})$ *is extraspecial of type* 2_+^{1+8} *with centre* $\bar{Z}^{[1]}$;
 (ii) $\bar{S} \cong S_4(2) \cong \mathrm{Sym}_6$ *is a complement to* $O_2(\bar{F}^{[01]})$ *in* $\bar{F}^{[01]}$;
 (iii) $\bar{Z}^{[1]}\bar{A}^{[1]}$, $\bar{Z}^{[1]}\bar{B}^{[1]}$ *and* $\bar{D}^{[1]}$ *are the elementary abelian normal subgroups of order* 2^5 *in* $\bar{F}^{[01]}$; *the former two are semisimple while the latter is indecomposable as* $\bar{S}$-*modules.*

Proof The key commutation of $G^{[01]}$ in $\bar{F}^{[01]}$ becomes

$$[\bar{\alpha}(w_1), \bar{\beta}(w_2)] = \bar{z}^{f(w_1, w_2)},$$

where $\bar{z}$ is the generator of $\bar{Z}^{[1]}$, which gives (i). The remaining assertions are obvious and/or follow from the above commutator relation. ∎

Consider the action of $O_2(\bar{F}^{[01]})$ on $\Pi^{(f)}$. This action is induced from the linear representation of the subgroup $\bar{Z}^{[1]}\bar{A}^{[1]}$ (of index 16 in $O_2(\bar{F}^{[01]})$) on Π_P with kernel $\bar{A}^{[1]}$. Therefore the action is faithful and since $16 = \dim \Pi^{(f)}$ is the dimension of the unique faithful irreducible representation of $O_2(\bar{F}^{[01]}) \cong 2_+^{1+8}$, we have the following lemma.

Lemma 6.7.3 *As a module for* $O_2(\bar{F}^{[01]}) \cong 2_+^{1+8}$, *the space* $\Pi^{(f)}$ *is isomorphic to the unique irreducible faithful module of that group.*

Let $\bar{F}^{[1]} = C_{G^{[1]}}(f)/Z_f^{[1]}$. Then $\bar{F}^{[1]}$ is a semidirect product of $\bar{F}^{[01]}$ and a group of order 2 generated by the image $\bar{t}_1$ of t_1 in $\bar{F}^{[1]}$, where t_1 is the involution which extends $G^{[01]}$ to $G^{[1]}$ as in Section 3.7. We require a detailed information about the action of $\bar{t}_1$ on $\bar{F}^{[01]}$.

The action of $\bar{t}_1$ on $O_2(\bar{F}^{[01]})$ is clear: it transposes $\bar{\alpha}(w)$ onto $\bar{\beta}(w)$ for every $w \in U_4^{\#}$.

Lemma 6.7.4 *The element* t_1 *can be chosen so that* $\bar{t}_1$ *centralizes* $\bar{S}$.

Proof Consider $L^{[1]} \cong \mathrm{Alt}_8$ as the alternating group of an 8-element set P_8. Then $Z^{[1]}$ is the heart $\mathcal{P}_8^e/\mathcal{P}_8^c$ of the $GF(2)$-permutation module of $L^{[1]}$ on P_8 (compare Section 3.3). Since

$$\langle Z^{[1]}L^{[1]}, t_1 \rangle \cong (\mathcal{P}_8/\mathcal{P}_8^c) : L^{[1]},$$

there are 64 candidates for t_1 and these candidates are indexed by the partitions of P_8 into two odd subsets. In these terms the subgroup $S \cong \mathrm{Sym}_6$ is the stabilizer in $L^{[1]}$ of a 2-element subset $\{a, b\} \subset P_8$. The partition $\rho = (\{a, b\}, P_8 \backslash \{a, b\})$ (considered as an element of $Z^{[1]}$) is the only non-identity element in $Z^{[1]}$ stabilized by S and therefore ρ corresponds to the symplectic form f on U_4, stabilized by S; in particular $\rho \in Z_f^{[1]}$. Suppose that t_1 corresponds to the partition $(\{a\}, P_8 \setminus \{a\})$. Then by (3.3.3)

$$C_{L^{[1]}}(t_1) = L^{[1]}(a) \cong \mathrm{Alt}_7$$

and $[\bar{t}_1, \bar{T}] = 1$, where $\bar{T} \cong \mathrm{Alt}_6$ is the commutator subgroup of $\bar{S}$. Let π be the transposition (a, b). Then $S = \langle T, \pi \rangle$ and $[t_1, \pi] = \rho$. Since $\rho \in Z_f^{[1]}$; the result follows.   ∎

At this point we feel it is appropriate to explain our further strategy. By (6.7.3) the action of $O_2(\bar{F}^{[01]})$ on $\Pi^{(f)}$ is determined uniquely (up to isomorphism). Our goal is to lift this action to that of the whole of $\bar{F}^{[1]}$, subject to the condition that the restriction to $\bar{F}^{[01]}$ is known (up to the choices). Since $\bar{D}^{[1]}$ is normal in $\bar{F}^{[1]}$ it is natural to study the eigenspace decomposition of $\Pi^{(f)}$ with respect to $\bar{D}^{[1]}$. Since the action of $O_2(\bar{F}^{[01]})$ on $\Pi^{(f)}$ is faithful irreducible and $\bar{z}$ is an involution in the centre of $O_2(\bar{F}^{[01]})$, we observe (by Schur's lemma or otherwise) that $\bar{z}$ acts on $\Pi^{(f)}$ as the (-1)-scalar operator. Since $\bar{D}^{[1]}$ contains $\bar{z}$, this implies that $\bar{D}^{[1]}$ acts fixed-point freely on $\Pi^{(f)}$. Hence the kernel of every eigenspace is a hyperplane $\bar{C}$ in $\bar{D}^{[1]}$ which does not contain $\bar{z}$. There are 16 such hyperplanes and it is easy to deduce from (1.6.7 (iii)) that $O_2(\bar{F}^{[01]})$ acting on $\bar{D}^{[1]}$ by conjugation permutes these hyperplanes transitively. Since 16 is also the dimension of $\Pi^{(f)}$, every hyperplane not containing $\bar{z}$ is the kernel of an action on an eigenspace and every eigenspace is 1-dimensional.

Let $\bar{C}$ be a hyperplane in $\bar{D}^{[1]}$ which does not contain $\bar{z}$. Then by the above the action of $O_2(\bar{F}^{[01]})$ on $\Pi^{(f)}$ is induced from the linear action of $\bar{D}^{[1]}$ with kernel $\bar{C}$. Put

$$\bar{I} = N_{\bar{F}^{[1]}}(\bar{C}).$$

Then by (6.3.1 (i)) the action of $\bar{F}^{[1]}$ on $\Pi^{(f)}$ is induced from a linear representation of $\bar{I}$. Therefore our immediate goal is to calculate $\bar{I}$. By (1.6.7 (ii)), $\bar{I} \cap O_2(\bar{F}^{[01]}) = \bar{D}^{[1]}$ and since $O_2(\bar{F}^{[01]})$ acts transitively on the set of hyperplanes in $\bar{D}^{[1]}$ not containing $\bar{z}$, $\bar{I}O_2(\bar{F}^{[01]}) = \bar{F}^{[01]}$ and so by (6.7.4)

$$\bar{I}/\bar{D}^{[1]} \cong S_4(2) \times 2.$$

In order to accomplish our goal we study the classes of Alt_6-subgroups in $\bar{F}^{[01]}$ and the action of $\bar{t}_1$ on these classes.

For $X = A$, B, or D, the group $Y = \bar{Z}^{[1]}\bar{X}^{[1]}\bar{T}$ is the semidirect product of $\bar{T} \cong \mathrm{Alt}_6$ and an extension of the trivial 1-dimensional module $\bar{Z}^{[1]}$ by the natural symplectic module $\bar{Z}^{[1]}\bar{X}^{[1]}/\bar{Z}^{[1]}$ of $\bar{S}$ restricted to $\bar{T}$ (compare (6.7.2)). By (3.2.6 (iv)) Y contains two classes of Alt_6-subgroups. Let $\bar{T}^X$ denote such a subgroup not conjugate to $\bar{T}$.

Lemma 6.7.5 *The following assertions hold:*

(i) $\bar{F}^{[01]}$ *contains exactly four classes of* Alt_6*-subgroups with representatives* $\bar{T}$, $\bar{T}^A$, $\bar{T}^B$, *and* $\bar{T}^D$;

(ii) $\bar{D}^{[1]}$ *is semisimple as a module for* $\bar{T}^A$ *or* $\bar{T}^B$ *and indecomposable as a module for* $\bar{T}$ *or* $\bar{T}^D$;

(iii) *the representatives* $\bar{T}^A$ *and* $\bar{T}^B$ *can be chosen in such a way that*

$$\bar{D}^{[1]}\bar{T}^A = \bar{D}^{[1]}\bar{T}^B;$$

(iv) *the element* $\bar{t}_1$ *stabilizes the classes containing* $\bar{T}$ *and* $\bar{T}^D$ *and transposes the classes of* $\bar{T}^A$ *and* $\bar{T}^B$.

Proof By (6.7.2) $O_2(\bar{F}^{[01]}/\bar{Z}^{[1]})$ is the direct sum of two copies of the natural symplectic module of $\bar{S}$ and by (3.2.6 (iv)) $H^1(\bar{T}, O_2(\bar{F}^{[01]}/\bar{Z}^{[1]})) \cong 2^2$ which gives (i). Now (ii) is by (3.2.7 (iii)) while (iii) follows from the definition. Finally (iv) is by (6.7.4) and the paragraph before that lemma.   ∎

By (6.7.5 (ii)), $\bar{T}^A$ stabilizes a direct sum decomposition

$$\bar{D}^{[1]} = \bar{Z}^{[1]} \oplus \bar{C},$$

where $\bar{C}$ is a hyperplane in $\bar{D}^{[1]}$ disjoint from $\bar{Z}^{[1]}$. Clearly this decomposition is also stabilized by $N_{\bar{F}^{[01]}}(\bar{T}^A)$. By (3.2.6) we have $H^1(\bar{S}, \bar{A}^{[1]}) \cong H^1(\bar{T}, \bar{A}^{[1]}) \cong 2$, and therefore

$$N_{\bar{F}^{[01]}}(\bar{T}^A) \cong \mathrm{Sym}_6 \times 2.$$

And now by the order reason the following hold.

Lemma 6.7.6 $\bar{I} = \bar{D}^{[1]}N_{\bar{F}^{[01]}}(\bar{T}^A)\langle\bar{t}_1\rangle.$

Thus the quotient of $\bar{I}$ over the commutator subgroup of $\bar{I}$ is elementary abelian of order 2^3 and hence there are exactly four linear representations of $\bar{I}$ whose kernels do not contain $\bar{Z}^{[1]}$. Therefore we can construct four required representations of $\bar{F}^{[1]}$ on $\Pi^{(f)}$. For each such representation we consider the eigenspace decomposition with respect to $\bar{Z}^{[1]}\bar{A}^{[1]}$ and observe that the restriction to $\bar{F}^{[01]}$ is induced from a linear representation of

$$\bar{Z}^{[1]}\bar{A}^{[1]}N_{\bar{F}^{[01]}}(\bar{T}^A) = \bar{Z}^{[1]}\bar{A}^{[1]}N_{\bar{F}^{[01]}}(\bar{T}) \cong (2 \times 2^4) : \mathrm{Sym}_6$$

as required. Thus we obtain the final result of the section.

Proposition 6.7.7 *For each of the two choices of the kernel* K_2 *of* $G^{[04]}(f)$ *on* Π_2, *there are two ways to lift the action of* $G^{[01]}$ *on* $\Pi_{22}^{[01]}$ *to an action of* $G^{[1]}$.

6.8   Lifting $\Pi_{12}^{[01]} \oplus \Pi_{21}^{[01]}$

At this stage it is appropriate to calculate the eigenspace decomposition of Π with respect to $Z^{[1]}$.

Lemma 6.8.1 *Let* f *and* h *be elements in* $Z^{[1]}$ *contained, respectively in the orbits of length 28 and 35 of the action of* $L^{[1]}$ *on* $Z^{[1]}$. *Let* $Z_f^{[1]}$ *and* $Z_h^{[1]}$ *be the hyperplanes in* $Z^{[1]}$ *formed by the vectors perpendicular to* f *and* h, *respectively,* with respect to the unique non-zero symplectic form on $Z^{[1]}$ preserved by $L^{[1]}$. Then the following assertions hold:

(i) $C_{\Pi}(Z^{[1]}) = \Pi_{11}^{[01]}$ and $[\Pi, Z^{[1]}] = \Pi_{12}^{[01]} \oplus \Pi_{21}^{[01]} \oplus \Pi_{22}^{[01]}$;

(ii) $C_{[\Pi, Z^{[1]}]}(Z_f^{[1]}) \leq \Pi_{22}^{[01]}$ and $\dim C_{[\Pi, Z^{[1]}]}(Z_f^{[1]}) = 16$;

(iii) $C_{[\Pi, Z^{[1]}]}(Z_h^{[1]}) \leq \Pi_{12}^{[01]} \oplus \Pi_{21}^{[01]}$;

(iv) $\dim C_{\Pi_{12}^{[01]}}(Z_h^{[1]}) = \dim C_{\Pi_{21}^{[01]}}(Z_h^{[1]}) = 12$.

Proof The orbit of length 35 of $L^{[1]}$ on $Z^{[1]}$ is the intersection $\mathcal{Q}_1 \cap Z^{[1]}$ and the elements in this orbit are indexed by the 2-dimensional subspaces contained in U_4, while the orbit of length 28 is the intersection $\mathcal{Q}_2 \cap Z^{[1]}$. Keeping in mind this observation it is easy to deduce the assertions from (3.6.3) in view of (6.5.2) and (6.5.4). Alternatively one can make use of the equalities $\chi(h) = 53$ and $\chi(f) = -11$ implied by (6.2.5) and (6.4.2). ∎

Let h be as in (6.8.1) so that h is a symplectic form of rank 2 on U_4, whose radical will be denoted by W. Notice that h is uniquely determined by W and that

$$C_{G^{[01]}}(h) = Z^{[1]} A^{[1]} B^{[1]} M,$$

where $M := L^{[1]}(W) \cong 2^4 : (\mathrm{Sym}_3 \times \mathrm{Sym}_3)$ is the stabilizer of W in $L^{[1]}$. With $Z_h^{[1]}$ as in (6.8.1) put

$$\Pi_{12}^{(h)} = C_{\Pi_{12}^{[01]}}(Z_h^{[1]}), \quad \Pi_{21}^{(h)} = C_{\Pi_{21}^{[01]}}(Z_h^{[1]}),$$

and

$$\Pi^{(h)} = \Pi_{12}^{(h)} \oplus \Pi_{21}^{(h)}.$$

Since

$$420 = \dim \Pi_{\alpha}^{[01]} = \dim \Pi_{\alpha}^{(h)} \times [G^{[01]} : C_{G^{[01]}}(h)] = 12 \times 35,$$

for $\alpha = 12$ and 21, we have the following result analogous to (6.7.1).

Lemma 6.8.2 *For $\alpha = 12$ and 21 the action of $G^{[01]}$ on $\Pi_{\alpha}^{[01]}$ is induced from the action of $C_{G^{[01]}}(h)$ on $\Pi_{\alpha}^{(h)}$.*

6.8.1 24-dim representations of $C_{G^{[1]}}(h)$

In this subsection we put

$$\bar{E}^{[1]} = C_{G^{[1]}}(h)/Z_h^{[1]},$$

and describe a class of 24-dimensional representation of $\bar{E}^{[1]}$ over the field of complex numbers. These representations can be considered as that of $C_{G^{[1]}}(h)$ with kernel $Z_h^{[1]}$. After that we restrict a representation from this class to $C_{G^{[01]}}(h)$ and compare the restriction with the action of $C_{G^{[01]}}(h)$ on $\Pi_{12}^{(h)} \oplus \Pi_{21}^{(h)}$.

First of all we investigate the structure of $\bar{E}^{[1]}$. Similarly to the previous section we adopt the bar convention for the images in $\bar{E}^{[1]}$ of elements and subgroups of $C_{G^{[01]}}(h)$, and also for the homomorphisms induced by α, β, and ζ. Then in $\bar{E}^{[1]}$ the key commutator relation of $G^{[01]}$ becomes

$$[\bar{\alpha}(w_1), \bar{\beta}(w_2)] = \bar{z}^{h(w_1, w_2)},$$

where $w_1, w_2 \in U_4$, $\bar{z}$ is the generator of $\bar{Z}^{[1]} = Z^{[1]}/Z_h^{[1]}$ and h is treated as a symplectic form on U_4.

We have seen in Section 3.7 that $G^{[1]}$ is a semidirect product of $Q^{[m]} \cong 2^{11}$ and $A^{[1]}L^{[1]} \cong 2^4 : L_4(2)$. The relevant action is described in (3.8.1) in terms of the Mathieu group M_{24} and its irreducible Todd module $\bar{C}_{11}$. Since $Z_h^{[1]} \leq Q^{[m]}$, $\bar{E}^{[1]}$ is a semidirect product of $\bar{Q}^{[m]} = Q^{[m]}/Z_h^{[1]} \cong 2^6$ and $\bar{A}^{[1]}\bar{M} \cong A^{[1]}M$, where

$$M = L^{[1]}(W) = N_{L^{[1]}}(Z_h^{[1]}) \cong 2^4 : (\mathrm{Sym}_3 \times \mathrm{Sym}_3).$$

Thus the structure of $\bar{E}^{[1]}$ is completely determined by the action of $\bar{A}^{[1]}\bar{M}$ on $\bar{Q}^{[m]}$. One could determine the action implementing (3.8.1) and conducting relevant calculations in the Todd module, although here we follow a more elementary approach. First we describe the action of $\bar{A}^{[1]}$ on $\bar{Q}^{[m]}$.

Lemma 6.8.3 *The following assertions hold where $\bar{z}$ denotes the generator of $\bar{Z}^{[1]}$:*

(i) $\bar{Q}^{[m]} = \langle \bar{z}, \bar{\alpha}(w)\bar{\beta}(w), \bar{t}_1 \mid w \in U_4 \rangle \cong 2^6$ *and* $[\bar{Z}^{[1]}, \bar{A}^{[1]}] = 1$;
(ii) *if* $v, w \in U_4$ *then* $[\bar{\alpha}(v), \bar{\alpha}(w)\bar{\beta}(w)]$ *is trivial unless* $\langle v, w \rangle$ *is a 2-dimensional subspace in U_4 disjoint from W, in which case the commutator is $\bar{z}$;*
(iii) $[\bar{\alpha}(v), \bar{t}_1] = \bar{\alpha}(v)\bar{\beta}(v)$ *for every* $v \in U_4^{\#}$.

Proof (i) is immediate from the definition of $\bar{Q}^{[m]}$ before (3.7.1), while (ii) follows from the basic commutator relation. Since t_1 (acting by conjugation) transposes $\alpha(v)$ and $\beta(v)$ for every $v \in U_4^{\#}$, (iii) follows. ∎

We follow the notation for subgroups and elements of $M = L^{[1]}(W)$ introduced between the proofs of (3.3.5) and (3.3.6) (we still follow the bar convention). In particular $T \cong T^{(1)} \times T^{(2)}$ is a Sylow 3-subgroup of M, which stabilizes a direct sum decomposition $U_4 = W^{(1)} \oplus W^{(2)}$, such that $W^{(1)} = W$ and $W^{(i)} = C_{U_4}(T^{(3-i)})$; R is a Sylow 2-subgroup of $N_M(T)$, $R^{\#} = \{r^{(1)}, r^{(2)}, r^{(12)}\}$ and $\langle r^{(i)}, T^{(3-i)} \rangle$ is abelian of order 6 for $i = 1$ and 2. The following lemma describes the action of $\bar{M}$ on $\bar{Q}^{[m]}$.

Lemma 6.8.4 *The following assertions hold:*

(i) $[\bar{Z}^{[1]}, \bar{M}] = 1$;
(ii) *if* $v \in U_4$, $m \in M$, *and* $m(v)$ *is the image of v under m, then $\bar{m}$ conjugates $\bar{\alpha}(v)\bar{\beta}(v)$ onto $\bar{\alpha}(m(v))\bar{\beta}(m(v))$;*

(iii) $C_{\bar{M}}(\bar{t}_1) = O_2(\bar{M})\bar{T}\langle \bar{r}^{(12)}\rangle \cong 2^4 : 3^2 : 2$, *in particular* $[\bar{M} : C_{\bar{M}}(\bar{t}_1)] = 2$;
(iv) $[\bar{r}^{(1)}, \bar{t}_1] = [\bar{r}^{(2)}, \bar{t}_1] = \bar{z}$.

Proof Statement (i) is quite clear since $Z^{[1]}$ is normal in $G^{[1]}$, while (ii) is by the definition of $\bar{Q}^{[m]}$. Finally (iii) and (iv) follow from (3.3.6). Notice that t_1 commutes with $O_2(M)$ modulo $Z_h^{[1]}$. ∎

The following corollary is a direct consequence of (6.8.3) and (6.8.4).

Corollary 6.8.5 *The group* $\bar{A}^{[1]}\bar{M}$ *acts on the non-identity elements of* $\bar{Q}^{[m]}$ *with five orbits of lengths 1, 3, 3, 24, and 32. The orbits are the following:*

$$\Upsilon_1 = \{\bar{z}\}, \quad \Upsilon_2 = \{\bar{\alpha}(w)\bar{\beta}(w) \mid w \in W^{\#}\},$$

$$\Upsilon_3 = \{\bar{\alpha}(w)\bar{\beta}(w)\bar{z} \mid w \in W^{\#}\},$$

$$\Upsilon_4 = \{\bar{\alpha}(v)\bar{\beta}(v), \bar{\alpha}(v)\bar{\beta}(v)\bar{z} \mid v \in U_4 \setminus W\} \text{ and}$$

$$\Upsilon_5 = \{\bar{\alpha}(v)\bar{\beta}(v)\bar{t}_1, \bar{\alpha}(v)\bar{\beta}(v)\bar{t}_1\bar{z} \mid v \in U_4\}.$$

If $V = C_{\bar{Q}^{[m]}}(O_2(\bar{A}^{[1]}\bar{M}))$, *then* $V^{\#} = \Upsilon_1 \cup \Upsilon_2 \cup \Upsilon_3$; $\bar{D}^{[1]}$ *is the only index 2 subgroup in* $\bar{Q}^{[m]}$ *normalized by* $\bar{A}^{[1]}\bar{M}$ *and* $\bar{D}^{[1]} = \bar{Q}^{[m]} \setminus \Upsilon_5$.

Next we would like to enumerate the hyperplanes in $\bar{Q}^{[m]}$ (treating $\bar{Q}^{[m]}$ as a 6-dimensional $GF(2)$-vector space) disjoint from $\bar{Z}^{[1]}$ and calculate their stabilizers.

Lemma 6.8.6 *Let* q *be an orthogonal form on* U_4 *associated with* h *in the sense that*

$$h(w, v) = q(w) + q(v) + q(w + v)$$

for all $w, v \in U_4$ *and let* $p \in \{0, 1\}$. *Then every hyperplane in* $\bar{Q}^{[m]}$ *disjoint from* $\bar{Z}^{[1]}$ *is of the form*

$$\{0\} \cup \{\bar{\alpha}(v)\bar{\beta}(v)\bar{z}^{q(v)}, \bar{\alpha}(v)\bar{\beta}(v)\bar{t}_1\bar{z}^{q(v)+p} \mid v \in U_4\}.$$

Proof Let $\bar{P}$ be a hyperplane in $\bar{Q}^{[m]}$ disjoint from $\bar{Z}^{[1]}$. Since $\bar{Q}^{[m]} = \bar{P}\bar{Z}^{[1]}$, for every $v \in U_4$ exactly one of $\bar{\alpha}(v)\bar{\beta}(v)$ and $\bar{\alpha}(v)\bar{\beta}(v)\bar{z}$ is in $\bar{P}$. Since

$$\bar{\alpha}(v)\bar{\beta}(v)\bar{\alpha}(w)\bar{\beta}(w) = \bar{\alpha}(v+w)\bar{\beta}(v+w)\bar{z}^{h(v,w)}$$

the mapping $q : U_4 \to GF(2)$ defined by

$$q(v) = \begin{cases} 0, & \text{if } \bar{\alpha}(v)\bar{\beta}(v) \in \bar{P}; \\ 1, & \text{if } \bar{\alpha}(v)\bar{\beta}(v)\bar{z} \in \bar{P}. \end{cases}$$

is an orthogonal form on U_4 associated with h. This orthogonal form uniquely determines the intersection $\bar{P} \cap \bar{D}^{[1]}$. Since $\{\bar{z}, \bar{t}_1, \bar{t}_1\bar{z}\}$ is the set of non-zero vectors of a 2-dimensional subspace in $\bar{Q}^{[m]}$, $\bar{P}$ contains either $\bar{t}_1$ or $\bar{t}_1\bar{z}$ but not both. Since there are $16 = 2^{\dim U_4}$ choices for q (compare (1.1.1)) and two choices for p, the above construction gives the totality of 32 hyperplanes in $\bar{Q}^{[m]}$ disjoint from $\bar{Z}^{[1]}$. ∎

Lemma 6.8.7 *The group $\bar{A}^{[1]}\bar{M}$ has exactly two orbits $\bar{\mathcal{P}}^{(0)}$ and $\bar{\mathcal{P}}^{(1)}$ on the set of hyperplanes in $\bar{Q}^{[m]}$ disjoint from $\bar{Z}^{[1]}$, described as follows:*

(i) $|\bar{\mathcal{P}}^{(0)}| = 8$, *if $q^{(0)}$ is the form on U_4, such that $q^{(0)}(v) = 0$ if and only if $v \in W$, then the two hyperplanes corresponding to $q^{(0)}$ are in $\bar{\mathcal{P}}^{(0)}$ and the stabilizer of every one of them is*

$$\bar{S}^{(0)} = \bar{\alpha}(W)O_2(\bar{M})\bar{T}\langle \bar{r}^{(12)} \rangle;$$

(ii) $|\bar{\mathcal{P}}^{(1)}| = 24$, *if $q^{(1)}$ is the form on U_4, such that there is exactly one $w_1^{(1)} \in W$ such that $q^{(1)}(w_1^{(1)}) = 0$ and $q^{(1)}(v) = 1$ for every $v \in W^{(2)}$, then the two hyperplanes corresponding to $q^{(1)}$ are in $\bar{\mathcal{P}}^{(1)}$ and (assuming that $r^{(12)}$ stabilizes $w_1^{(1)}$) the stabilizer of each of the two hyperplanes is*

$$\bar{S}^{(1)} = \bar{\alpha}(W)O_2(\bar{M})\bar{T}_2\langle \bar{r}^{(12)} \rangle;$$

(iii) *the stabilizer of the pair of hyperplanes in (ii) as a whole is*

$$\bar{S}^{(2)} = \bar{\alpha}(W)O_2(\bar{M})\bar{T}_2\bar{R}.$$

Proof It is straightforward to check that both $q^{(0)}$ and $q^{(1)}$ are orthogonal forms on U_4 associated with h. Therefore by (6.8.6) in order to specify the corresponding hyperplane we only have to decide whether $\bar{t}_1$ or $\bar{z}\bar{t}_1$ is in the hyperplane. The stabilizers are easy to calculate using (6.8.3) and (6.8.4). Since then $|\bar{\mathcal{P}}^{(i)}| = [\bar{A}^{[1]}\bar{M} : \bar{S}^{(i)}]$ for $i = 0$ and 1 and $|\bar{\mathcal{P}}^{(0)}| + |\bar{\mathcal{P}}^{(1)}| = 8 + 24 = 32$ is the total number of hyperplanes in question, (i) and (ii) follow. The statement (iii) is also quite clear. ∎

Let Σ be a 24-dimensional **C**-module for $\bar{E}^{[1]} = C_{G^{[1]}}(h)/Z_h^{[1]}$ such that $C_\Sigma(\bar{Q}^{[m]}) = 0$ and $C_\Sigma(\bar{P})$ is 1-dimensional for every $\bar{P} \in \bar{\mathcal{P}}^{(1)}$. Since

$$24 = \dim \Sigma = |\bar{\mathcal{P}}^{(1)}| \times \dim C_\Sigma(\bar{P}),$$

by (6.3.1 (i)) Σ is induced from a 1-dimensional linear representation of the stabilizer of $\bar{P}$ in $\bar{E}^{[1]}$. By (6.8.7) the stabilizer is the semidirect product of $\bar{Q}^{[m]}$ and $\bar{S}^{(1)}$. Therefore the structure of Σ is uniquely determined by the kernel $\bar{K}_{\bar{P}}$ of the action of $\bar{Q}^{[m]}\bar{S}^{(1)}$ on $C_\Sigma(\bar{P})$. A subgroup $\bar{K}_{\bar{P}}$ in $\bar{Q}^{[m]}\bar{S}^{(1)}$ is suitable for being the kernel if and only if

$$[\bar{Q}^{[m]}\bar{S}^{(1)} : \bar{K}_{\bar{P}}] = 2,$$

$$\bar{K}_{\bar{P}} \cap \bar{Q}^{[m]} = \bar{P}.$$

By the former of the two conditions $\bar{K}_{\bar{P}}$ contains the commutator subgroup of $\bar{Q}^{[m]}\bar{S}^{(1)}$. The next lemma, which is a direct consequence of (6.8.7), describes this commutator subgroup.

Lemma 6.8.8 *Suppose that $\bar{P} \in \bar{\mathcal{P}}^{(1)}$ is determined by the orthogonal form $q^{(1)}$ as in (6.8.7 (ii)). Then the elements $\bar{z}$, $\bar{t}_1$, $\bar{\alpha}(w_1^{(1)})$, $\bar{r}^{(12)}$ generate in $\bar{Q}^{[m]}\bar{S}^{(1)}$ an elementary abelian subgroup of order 16 which complements the commutator subgroup of $\bar{Q}^{[m]}\bar{S}^{(1)}$.*

By the following lemma there are four possibilities for the kernel $\bar{K}_{\bar{P}}$.

Lemma 6.8.9 *Suppose that $\bar{P}$ is as in (6.8.7 (ii)) and that $\bar{P}$ contains $\bar{t}_1$. Then in order to generate $\bar{K}_{\bar{P}}$ one should take*

(a) *the commutator subgroup of $\bar{Q}^{[m]}\bar{S}^{(1)}$;*
(b) *the element $\bar{t}_1$; and*
(c) *one of the four subgroups of order 2^2 in $\langle \bar{z}, \bar{\alpha}(w_1^{(1)}), \bar{r}^{(12)} \rangle$ which are disjoint from $\bar{z}$.*

6.8.2 Structure of $\Pi_{12}^{(h)}$

In this section we study the action of $C_{G^{[01]}}(h)$ on $\Pi_{12}^{(h)}$ and relate it to the action of $C_{G^{[1]}}(h)$ on Σ described in Section 6.8.1.

We follow notation introduced above in this section. To refresh the memory we suggest the reader to take a note of the paragraph before (6.8.4). In addition recall that d denotes the $L^{[0]}$-invariant injection of the set of 2-dimensional subspaces of U_5 into $Q^{[0]} \cong \bigwedge^2 U_5$ and that u is the unique vector from $U_5 \setminus U_4$ which is stabilized $L^{[1]} \cong L_4(2)$.

Let $V_3 = \langle W, u \rangle$. Then V_3 is 3-dimensional and $V_3 \cap U_4 = W$. Let $P(V_3)$ be the hyperplane in $Q^{[0]}$ which corresponds to V_3 and let $z = d(W^{(2)})$. Since $V_3 \cap W^{(2)} = 0$, (3.6.3 (iii)) implies that $z \notin P(V_3)$. According to the bar convention adopted since the beginning of Section 6.8.1 $\bar{z}$ is the generator of $\bar{Z}^{[1]}$.

Notice that $\langle T^{(2)}, r^{(2)} \rangle \cong L_2(2) \cong \mathrm{Sym}_3$ is the largest subgroup in $L^{[0]}$ which stabilizes the subspace V_3 vectorwise and the subspace $W^{(2)}$ as a whole. Let $M^{(3)} \cong L_3(2)$ be the subgroup of $L^{[0]}(V_3)$ which stabilizes $W^{(2)}$ vectorwise.

Let $K_1(V_3)$ be the kernel of the action of $G^{[0]}(V_3)$ on $\Pi_{P(V_3)}^{[0]}$. Then by (6.4.5) and (6.6.2) we have the following:

$$K_1(V_3) = P(V_3)O_2(L^{[0]}(V_3))T^{(2)}\langle zr^{(2)} \rangle.$$

Furthermore both z and $r^{(2)}$ act on $\Pi_{P(V_3)}^{[0]}$ as the (-1)-scalar operator, while $M^{(3)}$ acts irreducibly. This implies the following.

Lemma 6.8.10 *Let W be a 2-dimensional subspace in U_4 and let h be the symplectic form on U_4 whose radical is W. Let X_3 be a 3-dimensional subspace in U_5 such that $X_3 \cap U_4 = W$. Then*

$$K_1(X_3) \cap Z^{[1]} = P(X_3) \cap Z^{[1]} = Z_h^{[1]},$$

in particular $\Pi_{P(X_3)}^{[0]} < \Pi_{12}^{(h)}$.

Elementary calculations in the projective geometry of U_5 show that for a fixed 2-dimensional subspace W in U_4 there are exactly four 3-dimensional subspaces X_3 which satisfy the hypothesis of (6.8.10). Since $\Pi_{P(X_3)}^{[0]}$ is 3-dimensional over **C** and since by (6.8.1 (iv)) $\Pi_{12}^{(h)}$ is 12-dimensional, we obtain the following.

Lemma 6.8.11 *The space* $\Pi_{12}^{(h)}$ *is the direct sum of four 3-dimensional* **C**-*subspaces* $\Pi_{P(X_3)}^{[0]}$ *taken for all 3-dimensional* $GF(2)$-*subspaces* X_3 *in* U_5 *such that* $X_3 \cap U_4 = W$.

So far we were able to get away without giving explicit names to all the elements of $W = W^{(1)}$ and $W^{(2)}$. It is getting to be more and more difficult, so we are giving up. For $i = 1$ and 2 let

$$\{w_j^{(i)} \mid 1 \le j \le 3\}$$

be the set of non-zero vectors of the 2-dimensional subspace $W^{(i)}$.

Let $V_3^{(0)} = V_3$ and $V_3^{(j)} = \langle W, u + w_j^{(2)} \rangle$ for $j = 1, 2, 3$. Then it is easy to see that

$$\mathcal{X} = \{V_3^{(j)} \mid 0 \le j \le 3\}$$

is the set of 3-dimensional subspaces in U_5 intersecting U_4 in W. Notice that every vector from $U_5 \setminus U_4$ is contained in exactly one of the $V_3^{(j)}$'s. For $0 \le k \le 3$ put $\Phi^{(k)} = \Pi_{P(V_3^{(k)})}^{[0]}$. Then by (6.8.11) we have

$$\Pi_{12}^{(h)} = \Phi^{(0)} \oplus \Phi^{(1)} \oplus \Phi^{(2)} \oplus \Phi^{(3)}.$$

Our next objective is to describe the action of $\bar{D}^{[1]}$ on $\Pi_{12}^{(h)}$ and to calculate the eigenspaces of $\bar{D}^{[1]}$ in $\Pi_{12}^{(h)}$. We know that $\bar{z}$ acts as the (-1)-scalar operator. Next lemma describes the action of $\bar{A}^{[1]}$ on $\Pi_{12}^{(h)}$.

Lemma 6.8.12 *The following assertions hold:*

(i) *for* $1 \le j \le 3$ *the element* $\bar{\alpha}(w_j^{(1)})$ *acts trivially on* $\Pi_{12}^{(h)}$;
(ii) *for* $1 \le j \le 3$ *the element* $\bar{\alpha}(w_j^{(2)})$ *acts trivially on* $\Phi^{(0)} \oplus \Phi^{(j)}$ *and as the* (-1)-*scalar operator on* $\Phi^{(k)} \oplus \Phi^{(l)}$, *where* $\{j, k, l\} = \{1, 2, 3\}$.

Proof By the definition, for $w \in U_4^{\#}$ we have

$$\alpha(w) = d(\langle u, w \rangle).$$

On the other hand, if $X_3 \in \mathcal{X}$ and $\langle u, w \rangle \cap X_3$ is non-zero then $\alpha(w) \in P(X_3)$ and $\alpha(w)$ acts trivially on $\Pi_{P(X_3)}^{[0]}$; if $\langle u, w \rangle \cap X_3 = 0$ then $\alpha(w)$ acts on $\Pi_{P(X_3)}^{[0]}$ as the (-1)-scalar operator. Now the result follows from the definition of the $w_j^{(l)}$'s. $\blacksquare$

Next we turn to the action of $\bar{B}^{[1]}$ on $\Pi_{12}^{(h)}$.

Lemma 6.8.13 *Let $1 \leq j \leq 3$. Then*

(i) $\bar{\beta}(w_j^{(1)})$ *stabilizes* $\Phi^{(k)}$ *as a whole and acts on it with one eigenvalue 1 and two eigenvalues* -1 *for every* $0 \leq k \leq 3$;

(ii) $\bar{\beta}(w_j^{(2)})$ *transposes* $\Phi^{(0)}$ *and* $\Phi^{(j)}$, *and also* $\Phi^{(k)}$ *and* $\Phi^{(l)}$, *where* $\{j, k, l\} = \{1, 2, 3\}$.

Proof If $w \in U_4^{\#}$ then $\beta(w)$ is the transvection from $L^{[0]}$ whose centre is w and whose axis U_4. Put

$$Y = \langle T^{(1)}, \{\beta(w_j^{(1)}) \mid 1 \leq j \leq 3\}, r^{(1)} \rangle.$$

Then $Y \cong \mathrm{Sym}_4$ and for every $X_3 \in \mathcal{X}$ the subgroup Y stabilizes X_3 as a whole and acts faithfully on the vector-set of X_3. This means that Y acts faithfully on $\Phi^{(k)}$ for every $0 \leq k \leq 3$. In addition, Y is a maximal parabolic subgroup in $M^{(3)}$. Now (i) follows from elementary properties of the 3-dimensional **C**-module $\Phi^{(k)}$ of $M^{(3)}$ restricted to Y. The element $\beta(w_j^{(2)})$ is the transvection of U_5 with centre $w_j^{(2)}$ and axis U_4. Therefore it transposes u and $u + w_j^{(2)}$, and also $u + w_k^{(2)}$ and $u + w_l^{(2)}$, which gives (ii). ■

Since $r^{(1)}$ and $\beta(w_1^{(1)})$ are conjugate in $M^{(3)}$ we observe that $r^{(1)}$ acts on $\Phi^{(k)}$ with one eigenvalue 1 and two eigenvalues -1 for every $0 \leq k \leq 3$.

For $1 \leq j \leq 3$ and $0 \leq k \leq 3$ put $\Phi_j^{(k)} = C_{\Phi^{(k)}}(\beta(w_j^{(1)}))$. Then by the proof of (6.8.13 (i)) for $0 \leq k \leq 3$

$$\Phi^{(k)} = \Phi_1^{(k)} \oplus \Phi_2^{(k)} \oplus \Phi_3^{(k)}$$

is the eigenspace decomposition of $\Phi^{(k)}$ with respect to $\langle \beta(w_j^{(1)}) \mid 1 \leq j \leq 3 \rangle$.

For $1 \leq j \leq 3$ put

$$\Phi_j = \Phi_j^{(0)} \oplus \Phi_j^{(1)} \oplus \Phi_j^{(2)} \oplus \Phi_j^{(3)}.$$

Since $\bar{B}^{[1]}$ is abelian, Φ_j is a $\bar{B}^{[1]}$-submodule in $\Pi_{12}^{(h)}$, and the dimension of Φ_j is 4 (because of (6.8.11 (ii), (iii))). This gives the following.

Lemma 6.8.14 *As a module for* $\bar{B}^{[1]}$, *the subspace* $\Pi_{12}^{(h)}$ *possesses the direct sum decomposition*

$$\Pi_{12}^{(h)} = \Phi_1 \oplus \Phi_2 \oplus \Phi_3,$$

whose summands are 4-dimensional.

We choose a basis in $\Pi_{12}^{(h)}$ with respect to which the elements of $\bar{D}^{[1]}$ are presented by monomial matrices. We achieve this by choosing consistently a non-zero vector $e_j^{(k)}$ from each of the $\Phi_j^{(k)}$'s. We choose the $e_j^{(0)}$ arbitrary and take $e_j^{(k)}$ to be the image of $e_j^{(0)}$ under $\beta(w_k^{(2)})$.

The next lemma describes the action of $\bar{D}^{[1]}$ on Φ_1.

Lemma 6.8.15 *The following assertions hold:*

(i) $\bar{\alpha}(w_1^{(1)})\bar{\beta}(w_1^{(1)})$ *acts on* Φ_1 *trivially;*

(ii) $\bar{\alpha}(w_j^{(1)})\bar{\beta}(w_j^{(1)})$ *acts on* Φ_1 *as the* (-1)*-scalar operator for* $j = 2, 3$;

(iii) *the actions of the* $\bar{\alpha}(w_j^{(2)})\bar{\beta}(w_j^{(2)})$*'s for* $j = 1, 2$, *and* 3 *in the basis* $\{e_1^{(k)} \mid 0 \leq k \leq 3\}$ *are described, respectively, by the following matrices:*

$$\begin{pmatrix} 0 & 1 & 0 & 0 \\ 1 & 0 & 0 & 0 \\ 0 & 0 & 0 & -1 \\ 0 & 0 & -1 & 0 \end{pmatrix}, \quad \begin{pmatrix} 0 & 0 & 1 & 0 \\ 0 & 0 & 0 & -1 \\ 1 & 0 & 0 & 0 \\ 0 & -1 & 0 & 0 \end{pmatrix} \text{ and } \begin{pmatrix} 0 & 0 & 0 & 1 \\ 0 & 0 & -1 & 0 \\ 0 & -1 & 0 & 0 \\ 1 & 0 & 0 & 0 \end{pmatrix}.$$

Proof Statements (i) and (ii) follow from (6.8.12 (i)) and (6.8.13 (i)), while (iii) follows from (6.8.12 (ii)), (6.8.13 (ii)) and the definition of the $e_j^{(k)}$'s. ∎

Calculating with the matrices in (6.8.15 (iii)) one checks that

$$f_1^{(0)} = -e_1^{(0)} + e_1^{(1)} + e_1^{(2)} + e_1^{(3)}$$

is an eigenvector of the action of $\bar{D}^{[1]}$ on $\Pi_{12}^{(h)}$. Generalising this observation we obtain the following.

Lemma 6.8.16 *For* $1 \leq j \leq 3$ *and* $0 \leq k \leq 3$ *put* $f_j^{(k)} = \sum_{l=0}^{3}(-1)^{\delta(l,k)}e_j^{(k)}$, *where* δ *is the Kronecker delta function. Then* $\Pi_{12}^{(h)} = \bigoplus_{j=1}^{3}\bigoplus_{k=0}^{3}\langle f_j^{(k)}\rangle$ *is the eigenspace decomposition of* $\Pi_{12}^{(h)}$ *with respect to the action of* $\bar{D}^{[1]}$.

The next and final lemma in this subsection relates the action of $C_{G^{[01]}}(h)$ on $\Pi_{12}^{(h)}$ with that of $C_{G^{[1]}}(h)$ on Σ. Notice that $\bar{A}^{[1]}\bar{M}$ is both a complement to $\bar{D}^{[1]}$ in $C_{G^{[01]}}(h)$ and a complement to $\bar{Q}^{[m]}$ in $C_{G^{[1]}}(h)$ (compare (6.8.7), (6.8.8), and (6.8.9)). In the next lemma we follow notations used in (6.8.7)

Lemma 6.8.17 *Let* $\bar{P}^{(1)}$ *and* $\bar{P}^{(2)}$ *be the hyperplanes of* $\bar{Q}^{[m]}$ *as in* (6.8.7 (ii)) *and let* $\bar{C}$ *be the commutator subgroup of* $\bar{D}^{[1]}\bar{S}^{(1)}$. *Then*

(i) $C_{\bar{D}^{[1]}}(f_1^{(0)}) = \bar{P}^{(1)} \cap \bar{D}^{[1]} = \bar{P}^{(2)} \cap \bar{D}^{[1]}$;

(ii) *the normalizer of* $C_{\bar{D}^{[1]}}(f_1^{(0)})$ *in* $\bar{A}^{[1]}\bar{M}$ *is* $\bar{S}^{(2)}$;

(iii) $C_{\bar{A}^{[1]}\bar{M}}(f_1^{(0)}) = \langle \bar{C}, \bar{\alpha}(w_1^{(1)}), \bar{r}^{(12)}, \bar{z}\bar{r}^{(1)}\rangle$.

Proof By (6.8.15) if $1 \leq l \leq 2$ and $1 \leq j \leq 3$ then $\bar{\alpha}(w_j^{(l)})\bar{\beta}(w_j^{(l)})$ negates $f_1^{(0)}$ unless $l = j = 1$ which gives (i). Clearly $\bar{C}$ centralizes $f_1^{(0)}$. In addition, $\bar{\alpha}(w_1^{(1)})$ centralizes the whole of $\Pi_{12}^{(h)}$, in particular it centralizes $f_1^{(0)}$. It was mentioned at the beginning of this subsection that $r^{(2)}$ acts on $\Pi_{12}^{(h)}$ as the (-1)-scalar operator, in particular $r^{(2)}$ negates $f_1^{(0)}$. By the remark after the proof of (6.8.13), it follows that for $0 \leq k \leq 3$ the element $r^{(1)}$ acts on $\Phi^{(k)}$ with one eigenvalue 1 and two eigenvalues -1. Since $r^{(1)}$ transposes $\langle f_2^{(k)}\rangle$ and $\langle f_3^{(k)}\rangle$, $r^{(1)}$ must negate $f_1^{(1)}$.

Hence $r^{(12)} = r^{(1)}r^{(2)}$ centralizes $f_1^{(0)}$. Keeping in mind that $\bar{z}$ negates every vector in $\Pi_{12}^{(h)}$ we observe that the proof is complete. ∎

6.8.3 *Structure of* $\Pi_{21}^{(h)}$

In this subsection we turn our attention to the action of $C_{G^{[01]}}(h)$ on $\Pi_{21}^{(h)}$. This action will also be related to that of $C_{G^{[1]}}(h)$ on Σ as in Section 6.8.1.

In the above terms let V_1 be a 1-dimensional subspace in U_4 and let g be a non-singular symplectic form on U_5/V_1, which we consider as a symplectic form on U_5 with radical V_1 (compare Section 1.4). Then the restriction $g|_{U_4}$ of g to U_4 is a non-zero symplectic from, whose radical is also non-zero (clearly this radical contains V_1). By (1.1.5 (i)) the radical of $g|_{U_4}$ is 2-dimensional. In this subsection we consider the situation when this 2-dimensional subspace is W or, equivalently, that $g|_{U_4} = h$.

Lemma 6.8.18 *Let V_1 be a 1-dimensional subspace in W and g be a symplectic form on U_5 such that V_1 is the radical of g and W is the radical of the restriction of g to U_4. Then*

$$P(V_1, g) \cap Z^{[1]} = Z_h^{[1]},$$

in particular $\Pi_{P(V_1,g)}^{[0]} < \Pi_{21}^{(h)}$.

Proof The result follows from (3.6.3 (iv)) along with elementary geometrical considerations. ∎

Let us describe the symplectic forms g on U_5 such that the restriction of g to U_4 coincides with h. As above let u denote the vector in $U_5 \setminus U_4$ stabilized by $L^{[1]}$. Then g is uniquely determined by the values

$$g(u, w_j^{(l)}) \text{ for } 1 \le l \le 2,\ 1 \le j \le 3.$$

Lemma 6.8.19 *Let g be a symplectic form on U_5 whose radical is a 1-dimensional subspace contained in W and whose restriction to U_4 coincides with h. Then*

(i) $g(u, w_j^{(1)}) = 0$ *for exactly one* $j \in \{1, 2, 3\}$;

(ii) $g(u, w_j^{(2)}) = 0$ *either for all or for exactly one* $j \in \{1, 2, 3\}$.

Proof Since $\{w_j^{(l)} \mid 1 \le j \le 3\}$ is the set of non-zero vectors in a 2-dimensional subspace, either all or exactly one of the three values $g(u, w_j^{(l)})$ for $1 \le j \le 3$ is zero (here $l \in \{1, 2\}$). Only if $g(u, w_j^{(1)})$ would be zero for all $j \in \{1, 2, 3\}$, the whole of $W = W^{(1)}$ would be in the radical of g, which is not allowed. ∎

Thus there are 12 forms g satisfying the hypothesis of (6.8.19) (and 12 is also the dimension of $\Pi_{21}^{(h)}$). For $0 \le k \le 3$ and $1 \le j \le 3$ let $g_j^{(k)}$ denote a form satisfying the hypothesis of (6.8.19) and subject to the following:

$$g_j^{(k)}(u, w_i^{(1)}) = 0 \quad \text{if and only if } i = j;$$

$$g_j^{(0)}(u, w_i^{(2)}) = 0 \quad \text{for all } i \in \{1, 2, 3\};$$

$$\text{if } k \neq 0 \text{ then } g_j^{(k)}(u, w_i^{(2)}) = 0 \quad \text{if and only if } i = k.$$

For k and j from the above ranges put

$$\Psi_j^{(k)} = \Pi_{P(V_1, g_j^{(k)})}^{[0]},$$

where $V_1 = \langle w_j^{(1)} \rangle$ is the radical of $g_j^{(k)}$. Then by (6.8.18) $\Pi_{21}^{(h)}$ is the direct sum of the twelve $\Psi_j^{(k)}$'s. We are going to use this decomposition to describe the action of $\bar{D}^{[1]}$ on $\Pi_{21}^{(h)}$. We proceed in the usual manner starting with the remark that $\bar{z}$ negates every vector in $\Pi_{21}^{(h)}$.

Lemma 6.8.20 *If $v \in U_4^\#$ then the element $\bar{\alpha}(v)$ centralizes $\Psi_j^{(k)}$ if $g_j^{(k)}(u, v) = 0$ and negates the vectors in $\Psi_j^{(k)}$ otherwise.*

Proof Recall that $\alpha(v)$ is the image in $Q^{[0]}$ of the 2-dimensional subspace $\langle u, v \rangle$ under the $L^{[0]}$-invariant injection d. By (3.6.3 (iv)) $\alpha(v) \in P(V_1, g)$ if and only if $\langle u, v \rangle$ is totally singular with respect to g. Hence the result. ∎

Lemma 6.8.21 *The following assertions hold:*

(i) *if $1 \leq j \leq 3$ then $\bar{\beta}(w_j^{(1)})$ stabilizes $\Psi_i^{(k)}$ as a whole for every $0 \leq k \leq 3$ and $1 \leq i \leq 3$;*

(ii) *if $j \in \{1, 2, 3\}$ and $i \in \{1, 2, 3\}$ then $\bar{\beta}(w_j^{(2)})$ transposes $\Psi_i^{(0)}$ and $\Psi_i^{(j)}$, and also $\Psi_i^{(l)}$ and $\Psi_i^{(k)}$, where $\{j, l, k\} = \{1, 2, 3\}$.*

Proof Recall that for $v \in U_4^\#$ the element $\beta(v)$ is the transvection from $L^{[0]}$ whose centre is v and whose axis is U_4. Then the action of $\beta(v)$ on the set of symplectic forms (satisfying the hypothesis of (6.8.19)) is quite clear. In fact, if g is such a form and $g^{(v)}$ is the image of g under $\beta(v)$, then

$$g^{(v)}(u, w) = g(u + v, w) = g(u, w) + g(v, w) = g(u, w) + h(v, w)$$

for every $w \in U_4$. Since W is the radical of h we immediately get (i). Now (ii) follows easily from the definition of the $g_j^{(k)}$'s. ∎

To spare a letter let $g_j^{(k)}$ denote also a non-zero vector in $\Psi_j^{(k)}$. We assume that for every $1 \leq j \leq 3$ the vector $g_j^{(k)}$ is the image under $\beta(w_k^{(2)})$ of the vector $g_j^{(0)}$ (compare the paragraph after (6.8.14)).

Recall that there are actually two possibilities for the module $\Pi_2^{[0]}$ corresponding to the two possible kernels $K_2^{(1)}$ and $K_2^{(2)}$ of $F_1 = G^{[04]}(f)$ on Π_2. Furthermore in (6.4.6) we have chosen our notation so that

$$K_2^{(1)} = P(V_1, f)L^{[0]}(V_1, f).$$

For $1 \leq i \leq 3$ let $\Psi_i = \Psi_i^{(0)} \oplus \Psi_i^{(1)} \oplus \Psi_i^{(2)} \oplus \Psi_i^{(3)}$. Then the Ψ_i's are $\bar{D}^{[1]}$-modules and the next lemma (which is analogous to (6.8.15)) describes the action of $\bar{D}^{[1]}$ on Ψ_1.

Lemma 6.8.22 *Suppose that the kernel of $G^{[04]}(f)$ on Π_2 is $K_2^{(\varepsilon+1)}$ where $\varepsilon = 0$ or 1. Then the following assertions hold:*

(i) *$\bar{\alpha}(w_1^{(1)})\bar{\beta}(w_1^{(1)})$ acts on Ψ_1 as $(-1)^\varepsilon$-scalar operator;*

(ii) *$\bar{\alpha}(w_j^{(1)})\bar{\beta}(w_j^{(1)})$ acts on Ψ_1 as the $(-1)^{\varepsilon+1}$-scalar operator for $j = 2, 3$;*

(iii) *the actions of the $\bar{\alpha}(w_j^{(2)})\bar{\beta}(w_j^{(2)})$'s for $j = 1, 2$, and 3 in the basis $\{g_1^{(k)} \mid 0 \leq k \leq 3\}$ are described by the matrices as in (6.8.15 (iii)).*

Proof The assertions follow from (6.8.20) and (6.8.21) in view of the observation that the elements $\beta(v)$ for $v \in U_4^\#$ are transvections from $L^{[0]}$ and therefore they are contained in a $G^{[0]}$-conjugate of $K_2^{(1)}$, but not of $K_2^{(2)}$. ∎

Comparing (6.8.15) and (6.8.22) we notice that if $\varepsilon = 1$, then the element $\bar{\alpha}(w_1^{(1)})\bar{\beta}(w_1^{(1)})$ has different characters on $\Pi_{12}^{(h)}$ and $\Pi_{21}^{(h)}$. Since this elements is centralized by t_1 and t_1 must transpose $\Pi_{12}^{(h)}$ and $\Pi_{21}^{(h)}$, we arrive to the following conclusion.

Proposition 6.8.23 *In the considered situation, unless the kernel of F_2 on Π_2 is $P(V_1, f)L^{[0]}(V_1, f)$, the action of $G^{[01]}$ on $\Pi_{12}^{[01]} \oplus \Pi_{21}^{[01]}$ cannot be extended to an action of $G^{[1]}$.*

From now on we assume that

$$K_2 = P(V_1, g)L^{[0]}(V_1, f).$$

Then in (6.8.22) we have $\varepsilon = 0$, which makes the analogy between (6.8.15) and (6.8.22) is even more striking. The vector

$$h_1^{(0)} = -g_1^{(0)} + g_1^{(1)} + g_1^{(2)} + g_1^{(3)}$$

is an eigenvector of the action of $\bar{D}^{[1]}$ on $\Pi_{21}^{(h)}$ and we are ready to prove the following result analogous to (6.8.17).

Lemma 6.8.24 *Let $\bar{P}^{(1)}$ and $\bar{P}^{(2)}$ be the hyperplanes of $\bar{Q}^{[m]}$ as in (6.8.7 (ii)) and let $\bar{C}$ be the commutator subgroup of $\bar{D}^{[1]}\bar{S}^{(1)}$. Then*

(i) *$C_{\bar{D}^{[1]}}(h_1^{(0)}) = \bar{P}^{(1)} \cap \bar{D}^{[1]} = \bar{P}^{(2)} \cap \bar{D}^{[1]}$;*

(ii) *the normalizer of $C_{\bar{D}^{[1]}}(h_1^{(0)})$ in $\bar{A}^{[1]}\bar{M}$ is $\bar{S}^{(2)}$;*

(iii) $C_{\bar{A}^{[1]}\bar{M}}(h_1^{(0)}) = \langle \bar{C}, \bar{\alpha}(w_1^{(1)}), \bar{r}^{(12)}, \bar{r}^{(1)} \rangle.$

Proof Statements (i) and (ii) are immediate from (6.8.22). The element $r^{(1)}$ is a transvection from $L^{[0]}(g_j^{(k)})$ for $0 \le k \le 3$, therefore it centralizes $h_1^{(0)}$, because of (6.8.23). We assume that $r^{(2)}$ stabilizes $w_1^{(2)}$ and transposes $w_2^{(2)}$ and $w_3^{(2)}$. Then $r^{(2)}$ is a transvection from $L^{[0]}$ which stabilises $g_1^{(0)}$ and $g_1^{(1)}$. Since $r^{(2)}$, acting by conjugation, transposes $\beta(w_2^{(2)})$ and $\beta(w_3^{(2)})$, we conclude the $r^{(2)}$ transposes $g_1^{(2)}$ and $g_1^{(3)}$. Overall this shows that $r^{(2)}$ centralizes $h_1^{(0)}$ and the result follows. $\blacksquare$

6.8.4   *Gluing*

In this section we show that the action of $C_{G^{[01]}}(h)$ on $\Pi_{12}^{(h)} \oplus \Pi_{21}^{(h)}$ extends uniquely to an action of $C_{G^{[1]}}(h)$. With (6.8.17) and (6.8.24) in hands it is rather easy to achieve.

Lemma 6.8.25 *As above let $t_1 \in Q^{[m]} \setminus D^{[1]}$. Then the following assertions hold:*

   (i) *the subgroup $\bar{Q}^{[m]}\bar{S}^{(2)}$ is normalized by $\bar{t}_1$;*

   (ii) *acting by conjugation, $\bar{t}_1$ transposes the centralizers in $C_{G^{[01]}}(h)$ of $f_1^{(0)}$ and $h_1^{(0)}$;*

   (iii) *$\bar{t}_1$ transposes (the characters of) the actions of $C_{G^{[01]}}(h)$ on $\Pi_{12}^{(h)}$ and $\Pi_{21}^{(h)}$.*

Proof The action of $\bar{t}_1$ on $C_{G^{[01]}}(h)$, described in (6.8.3) and (6.8.4), gives (i). Compare (6.8.17 (iii)) and (6.8.24 (iii)). By (6.8.4 (iv)) $\bar{t}_1$ conjugates $\bar{r}^{(1)}$ onto $\bar{r}^{(1)}\bar{z}$; by (6.8.4 (iii)) $\bar{t}_1$ centralizes $\bar{r}^{(12)}$ and finally, by (6.8.3 (iii)) $\bar{t}_1$ conjugates $\bar{\alpha}(w_1^{(1)})$ onto $\bar{\beta}(w_1^{(1)})$. Since $\bar{\alpha}(w_1^{(1)})\bar{\beta}(w_1^{(1)})$ centralizes both $f_1^{(0)}$ and $h_1^{(0)}$ (compare (6.8.7), (6.8.17 (i)), and (6.8.24 (i))), we obtain (ii). Now the actions of $C_{G^{[01]}}(h)$ on $\Pi_{12}^{(h)}$ and $\Pi_{21}^{(h)}$ are induced from 1-dimensional representations of $\bar{D}^{[1]}\bar{S}^{(2)}$ and by (ii) the element $\bar{t}_1$ acting by conjugation transposes the kernels of these 1-dimensional representations. $\blacksquare$

Alternatively we can arrive to the same conclusion from the other side.

Proposition 6.8.26 *Let Σ be a 24-dimensional module for $C_{G^{[1]}}(h)$ induced from a 1-dimensional module of $\bar{Q}^{[m]}\bar{S}^{(1)}$, whose kernel contains $\bar{t}_1$, $\bar{\alpha}(w_1^{(1)})$ and $\bar{r}^{(12)}$. Then the restriction of this module to $C_{G^{[01]}}(h)$ is isomorphic to the direct sum $\Pi_{12}^{(h)} \oplus \Pi_{21}^{(h)}$.*

Proof Just compare (6.8.9), (6.8.17 (iii)), and (6.8.24 (iii)). $\blacksquare$

Corollary 6.8.27 *There is a unique $G^{[1]}$-module $\Pi_2^{[1]}$, whose restriction to $G^{[01]}$ is isomorphic to $\Pi_{12}^{[01]} \oplus \Pi_{21}^{[01]}$. Furthermore the module $\Pi_2^{[1]}$ is induced from the 24-dimensional module Σ of $C_{G^{[1]}}(h)$ as in (6.8.26).*

Proof The result is immediate from (6.8.2), (6.8.25), and (6.8.26). ∎

6.9   The minimal representations of $\mathcal{G}$

Above in this section we have classified the representation of the amalgam $\mathcal{G} = \{G^{[0]}, G^{[1]}\}$ of minimal dimension 1333. Before stating the result of the classification we recall a few definitions.

Let U_3 be the 3-dimensional subspace of U_5 stabilized by $G^{[02]}$; let f be a symplectic form on U_5 whose radical is the 1-dimensional subspace U_1 stabilized by $G^{[04]}$. Let $P_1 = P(U_3)$ and $P_2 = P(U_1, f)$ be the corresponding hyperplanes in $Q^{[0]}$. Let $L^{[2]} \cong L_3(2)$ be a complement to $Q^{[2]}$ in $N^{[2]}$, let $z \in Q^{[0]} \setminus P(U_3)$ and let r be an involution from $L^{[0]}(U_3) \cong 2^6 : (L_3(2) \times \mathrm{Sym}_3)$ which commutes with $L^{[2]}$.

Let $\Pi_1 = \Pi_1(\lambda)$ be a 3-dimensional module for $G^{[02]}$ whose kernel is

$$K_1 = P(U_3)O_{2,3}(L^{[0]}(U_3))\langle zr \rangle,$$

and in which the action of $L^{[2]}$ affords the character λ. Let $\Pi_1^{[0]} = \Pi_1^{[0]}(\lambda)$ be the $G^{[0]}$-module induced from the $G^{[02]}$-module Π_1. Let Π_2 be the 1-dimensional module for $G^{[0]}(f) = G^{[04]}(f)$ whose kernel is

$$K_2 = P(U_1, f)L^{[0]}(f).$$

Let $\Pi_2^{[0]}$ be the $G^{[0]}$-module induced from the $G^{[0]}(f)$-module Π_2.

Next we are going to describe a few $G^{[1]}$-modules. We refer to Sections 3.7 and 4.1 for definition and basic properties of some important subgroups in $G^{[1]}$. Notice that there is an action of $G^{[1]}$ on U_4, where

$$O_2(G^{[1]}) = Q^{[1]}\langle t_1 \rangle = Q^{[m]}A^{[1]} = Q^{[m]}B^{[1]}$$

is the kernel and $L^{[1]} \cong L_4(2)$ acts naturally.

Recall that U_3 is the 3-dimensional subspace in U_4, stabilized by $G^{[12]}$ and $v \in U_4 \setminus U_3$. The commutator subgroup of $G^{[12]}$ is $N^{[2]} = Q^{[2]}L^{[2]}$, where

$$Q^{[2]} = Z^{[1]}\alpha(U_3)\beta(U_3)O_2(L^{[1]}),$$

and the elements $\alpha(v)$, $\beta(v)$, and t_1 generate a D_8-subgroup which complements $N^{[2]}$ in $G^{[12]}$.

For $\varepsilon_1 \in \{0, 1\}$ let $\Sigma_1 = \Sigma_1(\mu, \varepsilon_1)$ be a 3-dimensional module for $G^{[02]}$ whose kernel is

$$\langle Q^{[2]}, \alpha(v)\beta(v), t_1^{1-\varepsilon_1}, (\beta(v)t_1)^{\varepsilon_1} \rangle$$

and the action of $L^{[2]}$ affords the character μ (the latter must be an irreducible character of degree 3). Let $\Pi_1^{[1]} = \Pi_1^{[1]}(\mu, \varepsilon_1)$ be the $G^{[1]}$-module induced from the $G^{[12]}$-module Σ_1.

It is worth mentioning that t_1 acts on $\Pi_1^{[1]}(\mu, \varepsilon_1)$ as the $(-1)^{\varepsilon_1}$-scalar operator.

Let h be a symplectic form on U_4 whose radical is 2-dimensional. We consider h as an element of $Z^{[1]} \cong \bigwedge^2 U_4$. Let Σ be a 24-dimensional module for $G^{[1]}(h) = C_{G^{[1]}}(h)$ described in Section 6.8.1, such that in terms of (6.8.9) the kernel $\bar{K}_{\bar{P}}$ contains $\bar{a}(w_1^{(1)})$ and $\bar{r}^{(12)}$. Notice that the latter condition means that $\bar{K}_{\bar{P}}$ is the semidirect product of $\bar{P}$ and the stabilizer of $\bar{P}$ in $\bar{A}^{[1]}\bar{L}^{[1]}$. Let $\Pi_2^{[1]}$ be the $G^{[1]}$-module induced from the $G^{[1]}(h)$-module Σ. Since Σ is induced from a 1-dimensional module for the stabilizer in $Q^{[m]}$ of a hyperplane, the whole module $\Pi_2^{[1]}$ is also induced from this representation.

Let f be a non-singular symplectic form on U_4. Then f also can be considered as an element of $Z^{[1]} \cong \bigwedge^2 U_4$. The subgroup $G^{[1]}(f)$ stabilizes in $Q^{[m]}$ exactly two hyperplanes $Q_f^{(0)}$ and $Q_f^{(1)}$ disjoint from f. For $\varepsilon_2 \in \{0,1\}$ let $\Pi_3^{[1]} = \Pi_3^{[1]}(\varepsilon_2)$ be the $G^{[1]}$-module induced from the 1-dimensional $G^{[1]}(f)$-module, whose kernel is

$$Q_f^{(\varepsilon_2)} A^{[1]} L^{[1]}(f).$$

Proposition 6.9.1 *If $\lambda = \mu$ then the $G^{[0]}$-module*

$$\Pi_1^{[0]}(\lambda) \oplus \Pi_2^{[0]}$$

and the $G^{[1]}$-module

$$\Pi_1^{[1]}(\mu, \varepsilon_1) \oplus \Pi_2^{[1]} \oplus \Pi_3^{[1]}(\varepsilon_2)$$

have isomorphic restrictions to $G^{[01]}$.

Proof The result is by (6.6.2), (6.7.7), and (6.8.26). ∎

Proposition 6.9.2 *Let $m = m(\mathcal{G})$ be the smallest positive integer such that there exists a faithful completion $\varphi : \mathcal{G} \to A$, where $A = GL_m(\mathbb{C})$. Then*

(i) *$m = 1333$;*
(ii) *for every completion φ as above the restrictions of φ to $G^{[0]}$ and $G^{[1]}$ turn the underlying vector space into the modules*

$$\Pi_1^{[0]}(\lambda) \oplus \Pi_2^{[0]} \quad and \quad \Pi_1^{[1]}(\lambda, \varepsilon_1) \oplus \Pi_2^{[1]} \oplus \Pi_3^{[1]}(\varepsilon_2),$$

respectively, where λ is one of the two 3-dimensional irreducible characters of $L^{[2]} \cong L_3(2)$ and $\varepsilon_1, \varepsilon_2 \in \{0,1\}$;
(iii) *up to conjugacy there are exactly eight representations of $\mathcal{G}$ of the minimal dimension 1333.*

Proof Combining (6.2.1) and (6.9.1) we obtain (i). The statement (ii) follow from (6.4.1), (6.6.1), and (6.8.23). Finally (iii) follows from the above in view of (6.4.2). In fact, in the considered situation

(a) $\chi^{[0]}$ is a sum of two irreducible characters with degrees 465 and 868;
(b) $\chi^{[1]}$ is a sum of three irreducible characters with degrees 45, 840, and 448;
(c) $\chi^{[01]}$ is a sum of four distinct irreducible characters with degrees 45, 420, 420, and 448.

The obvious fusion pattern immediately shows that $C^{[01]}$ is factorized by $C^{[0]}$ and $C^{[1]}$. ∎

6.10　The action of $G^{[2]}$

Let $\varphi : \mathcal{G} \to GL_{1333}(\mathbf{C})$ be a representation of $\mathcal{G}$ of the minimal degree 1333. By (6.9.1) and (6.9.2) we assume that $\varphi = \varphi(\lambda, \varepsilon_1, \varepsilon_2)$ in the sense that the restrictions of φ to $G^{[0]}$ and $G^{[1]}$ turn the underlying 1333-dimensional vector space Π (over the complex numbers) into the modules $\Pi^{[0]} = \Pi_1^{[0]}(\lambda) \oplus \Pi_2^{[0]}$ and $\Pi^{[1]} = \Pi_1^{[1]}(\lambda, \varepsilon_1) \oplus \Pi_2^{[1]} \oplus \Pi_3^{[1]}(\varepsilon_2)$, respectively. As usual put

$$G^{[2]} = \langle \varphi(G^{[02]}), \varphi(G^{[12]}) \rangle.$$

In the next section we show that under a suitable choice of ε_1 and ε_2 the completion $\varphi = \varphi(\lambda, \varepsilon_1, \varepsilon_2)$ is constrained at level 2. Recall that by the definition the latter means that

$$C_{G^{[2]}}(N^{[2]}) \leq N^{[2]}.$$

Since the centre of $N^{[2]}$ is trivial the completion is constrained at level 2 if and only if

$$C^{[2]} := C_{G^{[2]}}(N^{[2]})$$

is the identity subgroup.

As in (4.2.2) put

$$\widehat{G^{[2]}} = G^{[2]}/(N^{[2]}C_{G^{[2]}}(N^{[2]})).$$

Since φ is a completion of $\mathcal{G}$, by (4.8.5 (iii)), independently of the particular choice of λ, ε_1 and ε_2, we have $\widehat{G^{[2]}} \cong \mathrm{Sym}_5$. Let us adopt the hat convention for the images in $\widehat{G^{[2]}}$ of elements and subgroups of $G^{[2]}$.

Lemma 6.10.1 *The following assertions hold:*

(i) $\widehat{G^{[02]}} \cong G^{[02]}/N^{[2]} \cong \mathrm{Sym}_3 \times 2 \cong D_{12}$;

(ii) $\widehat{G^{[12]}} \cong G^{[12]}/N^{[2]} \cong D_8$;

(iii) $\widehat{G^{[012]}} \cong G^{[012]}/N^{[2]} \cong 2^2$;

(iv) *if $t_1 \in Q^{[m]} \setminus D^{[1]}$ then $\hat{t}_1$ belongs to the commutator subgroup of $\widehat{G^{[2]}}$ isomorphic to Alt_5.*

Proof Statements (i) to (iii) are implicit in the first paragraph of Section 4.7. Statement (iv) is an immediate consequence of the following elementary observation: every involution which stabilizes as a whole but not vertexwisely an edge of the Petersen graph is an even permutation (of the underlying 5-element set). ∎

We are going to study the restriction to $G^{[02]}$ of the module $\Pi^{[0]} = \Pi^{[0]}(\lambda, \varepsilon_1)$. First we consider the action of $G^{[02]}$ on the set of hyperplanes in $Q^{[0]}$ and prove the following result, which is analogous to (6.5.2).

Lemma 6.10.2 *Put*

$$\mathcal{P}_{11}^{(2)} = \{P(U_3)\}, \quad \mathcal{P}_{12}^{(2)} = \{P(V_3) \mid \dim (U_3 \cap V_3) = 2\},$$

$$\mathcal{P}_{13}^{(2)} = \{P(V_3) \mid \dim (U_3 \cap V_3) = 1\};$$

$$\mathcal{P}_{21}^{(2)} = \{P(V_1, f) \mid f|_{U_3} = 0\}, \quad \mathcal{P}_{22}^{(2)} = \{P(V_1, f) \mid V_1 < U_3 \text{ and } f|_{U_3} \neq 0\},$$

$$\mathcal{P}_{23}^{(2)} = \{P(V_1, f) \mid V_1 \not< U_3\}.$$

Then

(i) $\mathcal{P}_{11}^{(2)}$, $\mathcal{P}_{12}^{(2)}$ *and* $\mathcal{P}_{13}^{(2)}$ *are the* $G^{[02]}$*-orbits on* $\mathcal{P}_1$;

(ii) $|\mathcal{P}_{11}^{(2)}| = 1$, $|\mathcal{P}_{12}^{(2)}| = 42$, $|\mathcal{P}_{13}^{(1)}| = 112$;

(iii) *the subgroup* $N^{[2]}$

    (a) *acts transitively on* $\mathcal{P}_{11}^{(2)}$;

    (b) *has three orbits on* $\mathcal{P}_{12}^{(2)}$ *of length 14 each;*

    (c) *acts transitively on* $\mathcal{P}_{13}^{(2)}$;

(iv) $\mathcal{P}_{21}^{(2)}$, $\mathcal{P}_{22}^{(2)}$, *and* $\mathcal{P}_{23}^{(2)}$ *are the* $G^{[02]}$*-orbits on* $\mathcal{P}_2$;

(v) $|\mathcal{P}_{21}^{(2)}| = 84$, $|\mathcal{P}_{22}^{(2)}| = 112$, $|\mathcal{P}_{23}^{(2)}| = 672$.

Proof It is a direct consequence of the main theorem of projective geometry that the orbit under a general linear group L of a pair of subspaces from the natural module of L is uniquely determined by the dimensions of the subspaces together with the dimension of their intersection. This gives (i) and reduces (ii) to direct calculations.

Let us turn to (iii). Statement (a) is obvious. Let M be the kernel of the action of $L^{[0]}(U_3)$ on the quotient U_5/U_3. Then $N^{[2]}Q^{[0]} = MQ^{[0]}$ and hence in order to establish (b) and (c) it is sufficient to calculate the orbits of M on the set of 3-dimensional subspaces V_3 in U_5. First suppose that $U_3 \cap V_3$ is 2-dimensional. Then $\langle U_3, V_3\rangle$ is a hyperplane in U_5 containing U_3. On the other hand, it is easy to see that M stabilizes each of the three hyperplanes in U_5 containing U_3. Therefore $N^{[2]}$ has at least three orbits on $\mathcal{P}_{12}^{(2)}$. Suppose that $\langle U_3, V_3\rangle = U_4$. The subgroup M induces on U_4 the maximal parabolic subgroup isomorphic to $2^3 : L_3(2)$. Clearly this subgroup acts transitively on $U_4 \setminus U_3$. Since V_3 is generated by $U_3 \cap V_3$ and a vector from $V_3 \setminus (V_3 \cap U_3)$, we conclude that there are exactly three orbits of $N^{[2]}$ on $\mathcal{P}_{12}^{(2)}$ and (b) follows. Finally, suppose that $U_3 \cap V_3$ is 1-dimensional. Then $V_3 = \langle v, V_2\rangle$, where v is the non-zero vector in $U_3 \cap V_3$ and V_2 is a complement to $\langle v \rangle$ in V_3 which is also a complement to U_3 in U_5. It is easy to check that $O_2(M) \cong 2^6$ fixes all the vectors in U_3 and acts regularly on the set of 2-dimensional subspaces in U_5 disjoint from U_3. This gives (c).

By the version (1.5.3) of Witt's theorem the orbit of a subspace W under the stabilizer of a symplectic form f is determined by the intersection of W with the radical of f together with the restriction $f|_W$ of f to W. This gives (iv) and reduces (v) to some straightforward calculations (one might find it even easier to compute the orbits of $G^{[0]}(f)$ on the set of 3-dimensional subspaces in U_5). ∎

For $a \in \{1, 2\}$ and $b \in \{1, 2, 3\}$ put

$$\Pi_{ab}^{[02]} = \bigoplus_{P \in \mathcal{P}_{ab}^{(2)}} \Pi_P^{[0]}.$$

It is clear that $\Pi_{ab}^{[02]}$ is a $G^{[02]}$-module whose dimension is $3^{\delta(1,a)} \cdot |\mathcal{P}_{ab}^{(2)}|$.

Lemma 6.10.3 *The following assertions holds:*

(i) $\Pi_{11}^{[02]}$ *is stable under* $G^{[2]}$;
(ii) *both* $G^{[02]}$ *and* $G^{[2]}$ *induce on* $\Pi_{11}^{[02]}$ *the group* $L_3(2) \times 2$;
(iii) *if* $\varepsilon_1 = 1$ *then the completion* φ *is not constrained at level* 2.

Proof By the paragraph before the lemma the subspace $\Pi_{11}^{[02]}$ is stable under $G^{[02]}$. On the other hand, if t_1 is as in (6.10.1 (iv)) then $G^{[2]} = \langle G^{[02]}, t_1 \rangle$. By (6.6.2) t_1 acts on $\Pi_{11}^{[01]}$ as a (± 1)-scalar operator. Since $\Pi_{11}^{[02]} \leq \Pi_{11}^{[01]}$, (i) follows. Since $\Pi_{11}^{[02]}$ is canonically isomorphic to the $G^{[02]}$-module Π_1 (from which the module $\Pi_1^{[0]}$ was induced), we have (ii). Suppose that the completion is constrained at level 2. Then

$$G^{[2]} \cong 2^{3+12} \cdot (L_3(2) \times \mathrm{Sym}_5)$$

(the pentad group). Hence in this case $G^{[2]}$ possesses a unique homomorphism onto a group of order 2 and by (6.10.1 (iv)) the element t_1 is in the kernel of this homomorphism. On the other hand, if $\varepsilon_1 = 1$ then t_1 acts on $\Pi_{11}^{[02]}$ as the (-1)-scalar operator. Hence (iii) follows. ∎

Lemma 6.10.4 *Let* $b = 2$ *or* 3. *Let* $V_3^{(b)} \in \mathcal{P}_{1b}^{(2)}$, *so that*

$$\dim (U_3 \cap V_3^{(b)}) = 4 - b,$$

and let $P^{(b)} = P(V_3^{(b)})$. *Then*

(i) *both* $G^{[02]}(V_3^{(2)})$ *and* $N^{[2]}(V_3^{(2)})$ *induce on* $\Pi_{P^{(2)}}$ *an irreducible action of* $\mathrm{Sym}_4 \times 2$;
(ii) $G^{[02]}(V_3^{(3)})$ *induces on* $\Pi_{P^{(3)}}$ *an irreducible action of* $\mathrm{Sym}_4 \times 2$, *while* $N^{[2]}(V_3^{(3)})$ *induces on* $\Pi_{P^{(3)}}$ *an elementary abelian group of order* 2^3;
(iii) *under the action of* $G^{[02]}$ *the module* $\Pi_{1b}^{[02]}$ *is irreducible of dimension* $3 \cdot |\mathcal{P}_{1b}^{(2)}|$;

(iv) *under the action of $N^{[2]}$ the module $\Pi_{1b}^{[02]}$ is the direct sum of three irreducible pairwise non-isomorphic irreducible modules of dimension $|\mathcal{P}_{1b}^{(2)}|$ each.*

Proof Let $G^{(b)} = G^{[0]}(V^{(b)})$, let $N^{(b)}$ be the largest normal subgroup in $G^{(b)}$ such that $G^{(b)}/N^{(b)}$ is solvable, and let $Q^{(b)} = O_2(N^{(b)})$. Then an element of $G^{[0]}$ which maps U_3 onto $V_3^{(b)}$ conjugates $N^{[2]}$ onto $N^{(b)}$ and $Q^{[2]}$ onto $Q^{(b)}$.

Let $K_d^{(b)}$, $K_u^{(b)}$, and $K_l^{(b)}$ be the kernels of the actions of $G^{(b)}$ on $V_3^{(b)}$, $U_5/V_3^{(b)}$, and $\Pi_{P^{(b)}}$, respectively. By the structure of the parabolic subgroups in $L^{[0]} \cong L_5(2)$ we have

$$G^{(b)}/K_d^{(b)} \cong L_3(2), \quad G^{(b)}/K_u^{(b)} \cong L_2(2) \cong \mathrm{Sym}_3,$$

while

$$G^{(b)}/K_l^{(b)} \cong L_3(2) \times 2, \quad G^{(b)}/N^{(b)} \cong \mathrm{Sym}_3 \times 2$$

(compare (6.4.5) and the first paragraph of Section 4.7). Since

$$G^{(b)}/Q^{(b)} \cong \mathrm{Sym}_3 \times 2 \times L_3(2)$$

and $Q^{(b)}$ is contained in each of $K_d^{(b)}$, $K_u^{(b)}$, $K_l^{(b)}$, and $N^{(b)}$, we conclude that $K_d^{(b)}$ contains $K_l^{(b)}$ with index 2, and $K_u^{(b)}$ contains $N^{(b)}$ with index 2. In addition, since

$$N^{(b)} \cap Q^{[0]} = K_l^{(b)} \cap Q^{[0]} = P^{(b)},$$

we have

$$K_d^{(b)} = Q^{[0]}K_l^{(b)} \quad \text{and} \quad K_u^{(b)} = Q^{[0]}N^{(b)}.$$

In the above terms in order to establish (i) and (ii) we have to calculate

$$(G^{[02]} \cap G^{(b)})/(G^{[02]} \cap K_l^{(b)}) \quad \text{and} \quad (N^{[2]} \cap G^{(b)})/(N^{[2]} \cap K_l^{(b)})$$

for $b = 2$ and 3. Since $P(U_3)$ and $P^{(b)}$ are different hyperplanes in $Q^{[0]}$ and $P(U_3) < N^{[2]} < G^{[2]}$, by the above for $X = G^{[02]}$ or $N^{[2]}$

$$(K_d^{(b)} \cap X)/(K_l^{(b)} \cap X) \cong (K_u^{(b)} \cap X)/(N^{(b)} \cap X) \cong 2.$$

In view of the obvious symmetry between $V_3^{(b)}$ and U_3 we observe that $N^{[2]}$ is a subgroup of index 2 in the kernel of the action of $G^{[02]}$ on U_5/U_3. Now, easy geometrical considerations show that for $b = 2$ and 3 the subgroup $G^{[02]}$ induces Sym_4 on the vector-set of $V_3^{(b)}$, while $N^{[2]}$ induces on this set Sym_4 if $b = 2$ and the elementary abelian group of order 2^2 if $b = 3$. This gives the assertions (i) and (ii).

Now (iii) and (iv) follow from (i), (ii), (6.10.2 (i), (ii), (iii)), and (6.3.1 (ii), (iii)). Notice that the modules in (iv) have pairwise different kernels, therefore they are clearly pairwise non-isomorphic. ∎

Lemma 6.10.5 *For every $b \in \{1, 2, 3\}$ the action of $G^{[02]}$ on $\Pi_{2b}^{[02]}$ is irreducible.*

Proof The result immediately follows from (6.10.2 (iv), (v)) and (6.3.1 (ii)). ∎

Let us take a closer look at Π, considering it as a 1333-dimensional module for $N^{[2]}$. Let us discuss the irreducible constituents of $N^{[2]}$ involved in Π and their characters. By (6.10.3) $\Pi_{11}^{[02]}$ is such a constituent of dimension 3. By (6.10.4 (iv)) each of $\Pi_{12}^{[02]}$ and $\Pi_{13}^{[02]}$ is the sum of three pairwise non-isomorphic irreducible constituents of dimension 42 and 112, respectively. The group $G^{[2]}$ acts on $N^{[2]}$ by automorphisms. This action induces an action on the set of conjugacy classes and hence also on the set of irreducible characters of $N^{[2]}$ (clearly $N^{[2]}$ is the kernel of the latter action).

For $b = 2$ and 3 let $\chi^{(b)}$ be the character of an irreducible constituent of $N^{[2]}$ in $\Pi_{1b}^{[02]}$. Then by (6.10.4 (iv))

$$\chi^{(2)}(1) = 42 \text{ and } \chi^{(3)}(1) = 112.$$

Let $X^{(b)}$ be the set of images of $\chi^{(b)}$ under the action of $G^{[2]}$. Since the action of $N^{[2]}$ on Π is the restriction of an action of $G^{[2]}$ we conclude that every character from $X^{(2)} \cup X^{(3)}$ is afforded by an irreducible constituent of $N^{[2]}$ in Π. Furthermore the irreducible constituents in $\Pi_{1b}^{[02]}$ correspond to three different characters in $X^{(b)}$. Comparing (6.10.4 (iii)) and (6.10.4 (iv)), we conclude that these three characters in $X^{(b)}$ are transitively permuted by $G^{[02]}$. Thus $\widehat{G^{[2]}} \cong \text{Sym}_5$ acts faithfully on $X^{(b)}$ for $b = 2$ and 3. In particular each of $X^{(2)}$ and $X^{(3)}$ contains at least 5 elements. By (6.10.5) for $c \in \{1, 2, 3\}$ and $b \in \{2, 3\}$ either all or none of the characters of irreducible constituent of $N^{[2]}$ in $\Pi_{2c}^{[02]}$ are contained in $X^{(b)}$. Since $|X^{(3)}| \geq 5$ and the characters in $X^{(3)}$ have degree 112, we conclude that the restriction of $\Pi_{23}^{[02]}$ to $N^{[2]}$ is the sum of six irreducible constituents of degree 112, whose characters are contained in $X^{(3)}$. By a divisibility condition and since $|X^{(2)}| \geq 5$ we conclude that $\Pi_{22}^{[02]}$ is the sum of two irreducible constituents, whose characters are contained in $X^{(2)}$. Finally it is easy to conclude that the character of $\Pi_{22}^{[02]}$ must also be in $X^{(3)}$. In fact, otherwise it would be impossible to define an action of $\widehat{G^{[2]}} \cong \text{Sym}_5$ on $X^{(3)}$, whose restriction to $\widetilde{G^{[02]}} \cong \text{Sym}_3 \times 2$ has an orbit of length 3 and an orbit of length 6. Thus we have established the following main result of the section.

Proposition 6.10.6 *Define the following $G^{[02]}$-submodules in Π:*

$$\Pi_1^{[2]} = \Pi_{11}^{[02]}, \quad \Pi_2^{[2]} = \Pi_{12}^{[02]} \oplus \Pi_{21}^{[02]}, \text{ and } \Pi_3^{[2]} = \Pi_{13}^{[02]} \oplus \Pi_{22}^{[02]} \oplus \Pi_{23}^{[02]}.$$

Let Ω be a 5-element set on which $\widehat{G^{[2]}} \cong \text{Sym}_5$ acts naturally as the symmetric group. Then

$$\Pi = \Pi_1^{[2]} \oplus \Pi_2^{[2]} \oplus \Pi_3^{[2]}$$

is a decomposition of Π into irreducible $G^{[2]}$-submodules. Furthermore,

(i) $\Pi_1^{[2]}$ *restricted to $N^{[2]}$ stays irreducible;*

(ii) $\Pi_2^{[2]}$ *restricted to $N^{[2]}$ is the direct sum of five pairwise non-isomorphic irreducible constituents of dimension 42 each;*

(iii) $\Pi_3^{[2]}$ *restricted to $N^{[2]}$ is the direct sum of ten pairwise non-isomorphic irreducible constituents of dimension 112 each;*

(iv) $G^{[2]}$ *permutes the irreducible constituents in $\Pi_2^{[2]}$ as it permutes the elements of Ω;*

(v) $G^{[2]}$ *permutes the irreducible constituents in $\Pi_3^{[3]}$ as it permutes the 2-element subsets of Ω.*

One can conclude from the above proposition that $N^{[2]}$

(a) has two irreducible constituents in $\Pi_{21}^{[02]}$ of dimension 42 each;

(b) acts irreducibly on $\Pi_{22}^{[02]}$;

(c) has six irreducible constituents in $\Pi_{23}^{[02]}$ of dimension 112 each.

We could have deduced this conclusion earlier by computing the orbits of $N^{[2]}$ on $\mathcal{P}_2$ (compare Exercise 3 at the end of the chapter).

6.11   The centralizer of $N^{[2]}$

The completion $\varphi(\lambda, \varepsilon_1, \varepsilon_2)$ is constrained at level 2 if and only if the centralizer $C^{[2]}$ of $N^{[2]}$ in $G^{[2]}$ is trivial. In order to get hold of $C^{[2]}$ it is useful to calculate the centralizer of $N^{[2]}$ in the general linear group of Π or even in the ring of all (possibly non-invertible) linear transformations of Π. The structure of these centralizers is immediate from (6.10.6) in view of Schur's lemma. Before stating the result we slightly improve our notation.

As in (6.10.6) let Ω be a set of 5-elements on which $\widehat{G^{[2]}} \cong \mathrm{Sym}_5$ acts as the symmetric group. Let

$$\Pi_2^{[2]} = \bigoplus_{\omega \in \Omega} \Upsilon_\omega^{(2)} \text{ and } \Pi_3^{[2]} = \bigoplus_{\pi \in \binom{\Omega}{2}} \Upsilon_\pi^{(3)}$$

be the direct sum decompositions into (pairwise non-isomorphic) irreducible constituents of $N^{[2]}$ as in (6.10.6 (iv)) and (6.10.6 (v)), respectively. Also put $\Pi_1^{[2]} = \Upsilon_0^{(1)}$. Then

$$\mathcal{I} = \{0\} \cup \Omega \cup \binom{\Omega}{2}$$

is the index-set of the irreducible constituents of $N^{[2]}$ in Π and every such component is $\Upsilon_\iota^{(a)}$ for some $a \in \{1, 2, 3\}$ and some $\iota \in \mathcal{I}$ (notice that the possible values of ι depend on a). There is a natural action of $G^{[2]}$ on $\mathcal{I}$, where $N^{[2]}C^{[2]}$ is the kernel and $\widehat{G^{[2]}} = G^{[2]}/N^{[2]}C^{[2]} \cong \mathrm{Sym}_5$ acts naturally.

Lemma 6.11.1 *Let $\mathcal{C}^L$ be the centralizer of $N^{[2]}$ in the general linear group of Π and $\mathcal{C}^M$ be the centralizer of $N^{[2]}$ in the ring of all (possibly non-invertible) linear transformations of Π. Then*

(i) *as a module for $\widetilde{G^{[2]}}$ the centralizer $\mathcal{C}^M$ is isomorphic to the space of complex-valued functions on $\mathcal{I}$;*

(ii) *$\mathcal{C}^L$ consists of the invertible elements in $\mathcal{C}^M$;*

(iii) *$C^{[2]}$ is a subgroup of $\mathcal{C}^L$.*

Proof As an $N^{[2]}$-module, Π is the sum of $16 = 1 + 5 + 10$ pairwise non-isomorphic irreducible modules indexed by the set $\mathcal{I}$. Hence the result is immediate from Schur's lemma. ∎

In what follows an element f of $\mathcal{C}^M$ which acts on $\Upsilon_\iota^{(a)}$ as the $f(\iota)$-scalar operator will be identified with $\mathbf{C}$-valued function on $\mathcal{I}$ which sends ι onto $f(\iota)$.

Below we indicate the explicit fusion of the irreducible constituents of $N^{[2]}$ into those of $G^{[01]}$. For this purpose put $\Omega = \{1, 2, 3, 4, 5\}$ and suppose that $G^{[02]}$ is the stabilizer in $G^{[2]}$ of $\{1, 2\}$ while $G^{[12]}$ is the stabilizer of $\{\{1, 2\}, \{3, 4\}\}$. Then by (6.10.4), (6.10.5), (6.10.6), and the remark after (6.10.6) we have the following:

$$\Pi_{11}^{[01]} = \Upsilon_0^{(1)} \oplus \Upsilon_5^{(2)};$$

$$\Pi_{12}^{[01]} = \Upsilon_3^{(2)} \oplus \Upsilon_4^{(2)} \oplus \Upsilon_{\{34\}}^{(3)} \oplus \Upsilon_{\{35\}}^{(3)} \oplus \Upsilon_{\{45\}}^{(3)};$$

$$\Pi_{21}^{[01]} = \Upsilon_1^{(2)} \oplus \Upsilon_2^{(2)} \oplus \Upsilon_{\{12\}}^{(3)} \oplus \Upsilon_{\{15\}}^{(3)} \oplus \Upsilon_{\{25\}}^{(3)};$$

$$\Pi_{22}^{[01]} = \Upsilon_{\{13\}}^{(3)} \oplus \Upsilon_{\{14\}}^{(3)} \oplus \Upsilon_{\{23\}}^{(3)} \oplus \Upsilon_{\{24\}}^{(3)}.$$

The elements t_1 as in (6.10.1 (iv)) can be taken to act on Ω as the permutation $(1, 3)(2, 4)$.

6.12 The fundamental group of the Petersen graph

As above let $\varphi = \varphi(\lambda, \varepsilon_1, \varepsilon_2)$ be the completion map of $\mathcal{G}$ into $A = GL_{1333}(\mathbf{C})$. Let G be the subgroup of A generated by the image of φ, and let

$$\Gamma = \Lambda(\mathcal{G}, \varphi, G)$$

the corresponding coset graph. Then Γ is connected of valency 31 and the natural action of G on Γ is locally projective of type $(5, 2)$. Let $\Gamma^{[2]}$ be the geometric subgraph in Γ of valency 3, stabilized by $G^{[2]}$ (this subgraph is connected by the definition).

The subgroup $N^{[2]}$ is the kernel of the action of $G^{[2]}$ on $\Gamma^{[2]}$. Put $\widetilde{G^{[2]}} = G^{[2]}/N^{[2]}$ and adopt the tilde convention for the images in $\widetilde{G^{[2]}}$ of elements and subgroups of $G^{[2]}$. Let $\widehat{\Gamma^{[2]}}$ be the graph on the set of orbits of $\widetilde{C^{[2]}}$ on $\Gamma^{[2]}$ in which two orbits are adjacent if there is an edge of $\Gamma^{[2]}$ which joins them. Let

$$\psi : \Gamma^{[2]} \to \widehat{\Gamma^{[2]}}$$

be the mapping which assigns to a vertex of $\Gamma^{[2]}$ its orbit under $\widetilde{C^{[2]}}$.

Lemma 6.12.1 *Let* $\widetilde{\mathcal{E}} = \{\widetilde{G^{[02]}}, \widetilde{G^{[12]}}\}$ *and* $\widehat{\mathcal{E}} = \{\widehat{G^{[02]}}, \widehat{G^{[12]}}\}$ *be subamalgams in* $\widetilde{G^{[2]}}$ *and* $\widehat{G^{[2]}}$, *respectively. Then*

(i) $\widetilde{\mathcal{E}} \cong \widehat{\mathcal{E}}$;

(ii) $\widetilde{\Gamma^{[2]}} \cong \Lambda(\widetilde{\mathcal{E}}, \mathrm{id}, \widetilde{G^{[2]}})$ *and* $\widehat{\Gamma^{[2]}} \cong \Lambda(\widehat{\mathcal{E}}, \mathrm{id}, \widehat{G^{[2]}})$, *where* id *denotes the identity map;*

(iii) $\widehat{\Gamma^{[2]}}$ *is isomorphic to the Petersen graph;*

(iv) ψ *is a covering of graphs which commutes with the action of* $G^{[2]}$.

Proof Statement (i) follows from (6.10.1), while (ii) follows directly from the definition. Since $\widehat{G^{[2]}} \cong \mathrm{Sym}_5$, (iii) follows from (ii) and the definition of the Petersen graph. Finally, (iv) is by (i) and (ii). ∎

The subgroup $\widetilde{C^{[2]}} \cong C^{[2]}$ is the kernel of the natural homomorphism of $\widetilde{G^{[2]}}$ onto $\widehat{G^{[2]}}$ associated with the covering ψ. By (6.11.1 (iii)) $\widetilde{C^{[2]}}$ is abelian. On the other hand by (6.12.1) $\widetilde{C^{[2]}}$ is a factor group of the fundamental group of the Petersen graph. Thus we have the following.

Lemma 6.12.2 *The subgroup* $\widetilde{C^{[2]}}$ *is an abelian factor group of the fundamental group of the Petersen graph.*

Let us recall some basic facts about fundamental groups of graphs and their quotients over commutator subgroups. These standard results can be found in any textbook on algebraic topology, for instance in (Vick 1994), although I prefer the brief summary in (Venkatesh 1998).

Let Ξ be an undirected connected graph which has $n = |V(\Xi)|$ vertices and $m = |E(\Xi)|$ edges. Consider every edge $\{x, y\}$ as a pair of directed arcs (x, y) and (y, x). Let Θ be a *spanning tree* of Ξ. By the definition Θ is a connected subgraph in Ξ with no cycles which contains all the vertices. Then Θ contains exactly $n - 1$ edges. Let x be a vertex of Ξ called a *base vertex*. With every edge $\{y, z\}$ of Ξ which is not in Θ we associate a cycle T by the following rule:

(a) choose an arc corresponding to $\{y, z\}$, say (y, z);
(b) take the only path $(x = x_0, x_1, \ldots, x_k = y)$ which joins x with y in Θ;
(c) take the only path $(x_{k+1} = z, x_{k+2}, \ldots, x_l = x)$ which joins z with x;
(d) the resulting cycle is $T = (x_0, x_1, \ldots, x_k, x_{k+1}, \ldots, x_l = x_0)$ (T contains the arcs (x_i, x_{i+1}) for $0 \leq i \leq l - 1$).

The cycles obtained by the above procedure are called *fundamental cycles*. Thus there are

$$|E(\Xi)| - |V(\Xi)| + 1 = m - n + 1$$

fundamental cycles.

The so-called Hurewicz homomorphism (cf. proposition 4.21 in Vick (1994)) establishes an isomorphism between the quotient of the fundamental group of Ξ

over its commutator subgroup and the first homology group $H_1(\Xi)$ of Ξ. This isomorphism commutes with the natural action of the automorphism group of Ξ. In turn the homology group $H_1(\Xi)$ can be defined in the following way. Let $E_{\mathbf{Z}}$ be the free abelian group on the set of arcs of Ξ modulo the identification $(x, y) = -(y, x)$. Let $V_{\mathbf{Z}}$ be the free abelian group on the set of vertices of Ξ. Let

$$\partial : E_{\mathbf{Z}} \to V_{\mathbf{Z}}$$

be the homomorphism defined by $\partial : (x, y) \mapsto y - x$. The following result is standard.

Proposition 6.12.3 *The following assertions hold:*

   (i) *the kernel of ∂ is (isomorphic to) the homology group $H_1(\Xi)$ of Ξ;*
  (ii) *the sum h_T of arcs over a cycle T in Ξ is an element of $\ker \partial$;*
 (iii) *the elements h_T taken for all fundamental cycles T (with respect to a spanning tree and a base vertex) freely generate $H_1(\Xi) \cong \ker \partial$.*

From now on (till the end of the section) we assume that Ξ is the Petersen graph and that $X = \mathrm{Aut}\ (\Xi) \cong \mathrm{Sym}_5$ is the automorphism group of Ξ (thus Ξ and X are aliases of $\widehat{\Gamma^{[2]}}$ and $\widehat{G^{[2]}}$, respectively).

If x is a (base) vertex of Ξ then a spanning tree Θ in Ξ can be obtained by removing the six edges contained in $\Xi_2(x)$. This way we obtain six fundamental cycles of length 5. By (6.12.3) this means that $H_1(\Xi)$ is a free abelian group of rank 6 with free generators indexed by the cycles of length 5 in Ξ passing through x. In order to recover the X-module structure of $H_1(\Xi)$ we take all the twelve cycles of length 5 in Ξ as generators (these generators are no longer free, of course). For each such 5-cycle T we fix one of the two possible orientations of T and obtain an element h_T of $H_1(\Xi) \cong \ker \partial$ as in (6.12.3 (ii)). The subgroup H_T in $H_1(\Xi)$ generated by h_T is the infinite cyclic group (isomorphic to the additive group $\mathbf{Z}$ of integers). The group X acts on $H_1(\Xi)$ permuting the subgroups H_T in the way it permutes the corresponding cycles of length 5 in Ξ. The stabilizer of H_T with respect to this action coincides with $X(T) \cong D_{10}$. The subgroup $O^2(X(T))$ has order 5 and it centralizes H_T; an element from $X(T) \setminus O^2(X(T))$ negates $H_T \cong \mathbf{Z}$. This gives the following result.

Lemma 6.12.4 *Let T be a 5-cycle in the Petersen graph and $X(T) \cong D_{10}$ be the stabilizer of T in*

$$X = \mathrm{Aut}\ (\Xi) \cong \mathrm{Sym}_5.$$

Let H_T be the 1-dimensional $\mathbf{Z}$-module for $X(T)$ such that $O^2(X(T))$ acts trivially, while every element from $X(T) \setminus O^2(X(T))$ negates H_T. Let $\mathcal{T}$ be the X-module induced from the $X(T)$-module H_T. Then the following assertions hold *(where H_T is canonically identified with a $X(T)$-submodule in $\mathcal{T}$):*

(i) *there is an X-homomorphism*

$$\lambda : \mathcal{T} \to H_1(\Xi);$$

(ii) *the restriction of λ to H_T is an isomorphism.*

By (6.12.2) there is an X-homomorphism

$$\mu : H_1(\Xi) \to \mathcal{C}^L,$$

whose image is $C^{[2]} \cong \widetilde{C^{[2]}}$. Let ν be the composition of λ and μ.

Lemma 6.12.5 *If T is a 5-cycle in Ξ and $f = \nu(h_T)$ then $f(\iota) = \pm 1$ for every $\iota \in \mathcal{I}$, in particular the order of $\nu(H_T)$ is at most 2.*

Proof By (6.11.1) $\mathcal{C}^M$ is the **C**-permutation module of $X = \widehat{G^{[2]}} \cong \mathrm{Sym}_5$ on $\mathcal{I}$. It is elementary to check that both $X(T)$ and $O^2(X(T))$ act on $\mathcal{I}$ with three orbits of length 5 and one fixed element (the latter element is of course 0). Therefore we have the equality

$$C_{\mathcal{C}^M}(X(T)) = C_{\mathcal{C}^M}(O^2(X(T))).$$

In plain words the equality means that whenever an element from $\mathcal{C}^M$ is centralized by $O^2(X(T))$, it is centralized by the whole of $X(T)$. On the other hand, every element $g \in X \setminus O^2(X)$ negates h_T and therefore it sends f onto its inverse in $\mathcal{C}^L$. By the above g also centralizes f. Hence

$$f(\iota) = f(\iota)^{-1}$$

for every $\iota \in \mathcal{I}$ and the result follows. ∎

Since $\mathcal{T}$ is generated by the images of H_T under X, as an immediate consequence of (6.12.5) we obtain the following crucial result.

Corollary 6.12.6 *Let $\mathcal{P}$ be the subgroup in $\mathcal{C}^L$ generated by the (± 1)-valued functions. Then*

(i) *$\mathcal{P}$ is an elementary abelian 2-group;*

(ii) *as a $GF(2)$-module for $\widehat{G^{[2]}}$, $\mathcal{P}$ is isomorphic to*

$$\mathcal{P}_1 \oplus \mathcal{P}_5 \oplus \mathcal{P}_{10},$$

where $\mathcal{P}_1$, $\mathcal{P}_5$ and $\mathcal{P}_{10}$ are the $GF(2)$-permutation modules of $\widehat{G^{[2]}}$ on $\{0\}$, Ω and $\binom{\Omega}{2}$, respectively;

(iii) *$C^{[2]}$ is isomorphic to a $\widehat{G^{[2]}}$-submodule in $\mathcal{P}$.*

In order to restrict $C^{[2]}$ even further we once again make use of the fact that it is the image of $H_1(\Xi)$ under an X-homomorphism μ. By (6.12.5) and (6.12.6) $\mu(h^2)$ is the identity element for every $h \in H_1(\Xi)$. Put

$$H_1^{[2]}(\Xi) = H_1(\Xi)/\langle h^2 \mid h \in H_1(\Xi)\rangle.$$

Then $H_1^{[2]}(\Xi)$ is the largest elementary abelian factor-group of $H_1(\Xi)$ and in view of (6.12.6) μ induces an X-homomorphism

$$\mu^{[2]} : H_1^{[2]}(\Xi) \to \mathcal{P},$$

where $\mathcal{P} \cong \mathcal{P}_1 \oplus \mathcal{P}_5 \oplus \mathcal{P}_{10}$.

The above definition of $H_1(\Xi)$ in terms of the map ∂ can be easily transfered into the following definition of $H_1^{[2]}(\Xi)$. Let $2^{E(\Xi)}$ be the power-space of $E(\Xi)$. This means that the elements of $2^{E(\Xi)}$ are the subsets of $E(\Xi)$ and the addition is performed by the symmetric difference operator. Let $2^{V(\Xi)}$ be the power-space of $V(\Xi)$ and $\partial^{[2]}$ be the map defined by

$$\partial^{[2]} : \{x, y\} \mapsto \{x\} \cup \{y\}.$$

Then $H_1^{[2]}(\Xi) = \ker \partial^{[2]}$.

Lemma 6.12.7 *The following assertions hold:*

(i) $H_1^{[2]}(\Xi)$ *is elementary abelian of order* 2^6;

(ii) *considering* $H_1^{[2]}(\Xi)$ *as an* X-*module,*

    (a) *there are five orbits on the non-zero elements with lengths 6, 10, 12, 15 and 20;*

    (b) *there is a unique composition series*

$$0 < M^{(1)} < M^{(2)} < H_1^2(\Xi),$$

    where $M^{(1)}$ is the natural module of $X \cong P\Sigma L_2(4)$ and the remaining two composition factors are trivial 1-dimensional modules;

    (c) *the orbit of length 15 is in $M^{(1)}$ while the orbits of length 6 and 10 constitute $M^{(2)} \setminus M^{(1)}$.*

Proof From the definition of $H_1^{[2]}(\Xi)$ it is easy to deduce that a collection F of edges of Ξ is contained in $\ker \partial^{[2]} = H_1^{[2]}(\Xi)$ if and only if every vertex $x \in V(\Xi)$ is incident to an even number $n_F(x)$ of edges from F. Since Ξ is cubic, $n_F(x) \in \{0, 2\}$ and therefore F is either empty or the edge-set of a union of disjoint cycles. Elementary calculations in the Petersen graph Ξ show that F (if non-empty) is one of the following:

(1) a 5-cycle (there are 12 of them);

(2) a union of two disjoint 5-cycles (there are 6 of them);

(3) a 6-cycle (there are 10 of them);

(4) an 8-cycle (there are 15 of them);

(5) a 9-cycle (there are 20 of them).

Furthermore, each of (1) to (5) is an X-orbit, which gives (ii) (a). With the explicit form of $H_1^{[2]}(\Xi)$ in hand the remaining statements are easy to check. ∎

It is worth mentioning that $H_1^{[2]}(\Xi)$ is dual to the universal representation module of the geometry of edges and vertices of the Petersen graph (compare section 3.9 in Ivanov and Shpectorov (2002)).

Lemma 6.12.8 *Let* $\mu^{[2]} : H_1^{[2]}(\Xi) \to \mathcal{P}$ *be an X-homomorphism. Then* $\mathrm{Im}\ \mu^{[2]}$ *is either zero or a 1-dimensional submodule in*

$$\mathcal{R} := \mathcal{P}_1^{(c)} \oplus \mathcal{P}_5^{(c)} \oplus \mathcal{P}_{10}^{(c)}$$

(where $\mathcal{P}_m^{(c)}$ is the 1-dimensional submodule in $\mathcal{P}_m$ consisting of the constant-valued functions).

Proof Let T_1 and T_2 be disjoint 5-cycles in Ξ, $T_0 = T_1 \cup T_2$. Let F_i be the edge-set of T_i for $i = 0, 1, 2$. Then the F_i's are elements of $H_1^{[2]}(\Xi)$. Let X_i be the setwise stabilizer of F_i in $X \cong \mathrm{Sym}_5$ for $i = 0, 1, 2$. Then

$$X_1 = X_2 \cong D_{10}, \ X_0 \cong F_5^4 \ \text{and} \ X_1, X_2 \le X_0.$$

Therefore F_1 and F_2 are in the orbit of length 12 of X on $H_1^{[2]}(\Xi)$, while F_0 is in the 6-orbit. Since X_0 acts transitively both on Ω and on $\binom{\Omega}{2}$, we conclude that

$$\mu^{[2]}(F_0) \le \mathcal{R}.$$

By (6.12.7 (ii)) this implies that $\mu^{[2]}(M^{(2)}) \le \mathcal{R}$ and either $\mathrm{Im}\ \mu^{[2]} \le \mathcal{R}$ or $(\mathrm{Im}\ \mu^{[2]})\mathcal{R}/\mathcal{R}$ is a 1-dimensional X-submodule in $\mathcal{P}/\mathcal{R}$. It is easy to check that there are no 1-dimensional submodules in $\mathcal{P}/\mathcal{R}$. Hence $\mu^{[2]}(F_1) \le \mathcal{R}$. Since $H_1^{[2]}(\Xi)$ is generated by the orbit of F_1 under X and X centralizes $\mathcal{R}$ we conclude that $\mathrm{Im}\ \mu^{[2]}$ is generated by $\mu^{[2]}(F_1)$. ∎

The final result of the section is the following proposition.

Proposition 6.12.9 *Suppose that $G^{[2]}$ is defined with respect to $\varphi = \varphi(\lambda, \varepsilon_1, \varepsilon_2)$. Then (depending on the particular choice of φ) one of the following holds:*

(i) $\widetilde{G^{[2]}} \cong \mathrm{Sym}_5$ *and φ is constrained at level 2;*

(ii) $\widetilde{G^{[2]}} \cong \mathrm{Sym}_5 \times 2$ *and $C^{[2]}$ is a 1-dimensional submodule in $\mathcal{R}$.*

Proof By the definition $\widetilde{G^{[2]}}$ is an extension of $\widetilde{C^{[2]}}$ by $\widetilde{G^{[2]}} \cong \mathrm{Sym}_5$. By (6.12.8) $\widetilde{C^{[2]}}$ is of order 1 or 2. Since $\widetilde{G^{[12]}} \cong \widetilde{G^{[2]}} \cong D_8$ is a Sylow 2-subgroup of $\widetilde{G^{[2]}}$ the extension splits by Gaschütz's theorem. ∎

6.13 Completion constrained at level 2

In this section we complete the analysis of the subgroup $G^{[2]}$ corresponding to a completion $\varphi = \varphi(\lambda, \varepsilon_1, \varepsilon_2)$ of $\mathcal{G}$. We are interested in completions which are constrained at level 2. Therefore in view of (6.10.3 (iii)) we assume that $\varepsilon_1 = 0$. Let us make explicit a remark made in Section 6.9 after defining $\Pi_1^{[1]}$.

Lemma 6.13.1 *Let $t_1 \in Q^{[m]} \setminus D^{[1]}$ and suppose that $\varepsilon_1 = 0$. Then t_1 centralizes $\Pi_1^{[1]}$.*

Since $N^{[2]} \leq G^{[1]}$, $\Pi_1^{[1]}$ is an $N^{[2]}$-module. Since $\dim \Pi_1^{[1]} = 45$, we deduce from (6.10.6) that

$$\Pi_1^{[1]} = \Upsilon_0^{(1)} \oplus \Upsilon_\omega^{(2)}$$

for some $\omega \in \Omega$.

Lemma 6.13.2 *There is a unique element $\omega \in \Omega$ such that $\Upsilon_\omega^{(2)}$ is normalized by t_1. Furthermore, $\Upsilon_\omega^{(2)}$ is centralized by t_1 if $\varepsilon_1 = 0$ and it is negated by t_1 otherwise.*

Proof By (6.10.1 (iv)) t_1 acts on Ω as an even involution, therefore it has a unique fixed element. The second assertion is by (6.13.1). ∎

At this stage it is convenient to make use of the fact that $G^{[2]}$ splits over $N^{[2]}$. Let

$$N_7 \cong F_7^3$$

be the normalizer of a Sylow 7-subgroups in $N^{[2]}$. For a subgroup Y of $G^{[2]}$ containing $N^{[2]}$ let Y_7 denote a complement to N_7 in $N_Y(N_7)Y_7$.

By the proof of (4.9.1) N_7 is self-normalized in $N^{[2]}$. By a Frattini argument this implies that $Y \cong Y_7$. By (6.12.9) this gives the following result.

Lemma 6.13.3 *If φ is constrained at level 2 then $G_7^{[2]} \cong \mathrm{Sym}_5$, otherwise $G_7^{[2]} \cong \mathrm{Sym}_5 \times 2$.*

Thus $C_7^{[2]}$ is of order 1 or 2. Our next goal is to show that provided $\varepsilon_1 = 0$ the action of $C_7^{[2]}$ on

$$\Upsilon_0^{(1)} \bigoplus_{\omega \in \Omega} \Upsilon_\omega^{(2)}$$

is trivial. Without loss of generality we assume that t_1 commutes with N_7. We need the following preliminary result.

Lemma 6.13.4 *Suppose that $G_7^{[2]} \cong \mathrm{Sym}_5 \times 2$. Then*

(i) $G_7^{[02]} \cong \mathrm{Sym}_3 \times 2$ *is contained in a* Sym_5*-complement to* $C_7^{[2]}$ *in* $G_7^{[2]}$;

(ii) t_1 *is not contained in a* Sym_5*-complement to* $C_7^{[2]}$ *in* $G_7^{[2]}$.

Proof First notice that $\mathrm{Sym}_5 \times 2$ contains exactly two subgroups isomorphic to Sym_5. Let $\mathcal{E}_7 = \{G_7^{[02]}, G_7^{[12]}\}$ be the subamalgam in $G_7^{[2]}$. Then $\Gamma^{[2]}$ is canonically isomorphic to $\Lambda(\mathcal{E}_7, \mathrm{id}, G_7^{[2]})$. Since $G_7^{[2]} \cong \mathrm{Sym}_5 \times 2$, $\Gamma^{[2]}$ is the standard bipartite doubling of Ξ. This means that $V(\Gamma^{[2]}) = V(\Xi) \times \{0, 1\}$ and $\{(x, \alpha), (y, \beta)\} \in E(\Gamma^{[2]})$ if and only if $\{x, y\} \in E(\Xi)$ and $\alpha \neq \beta$. This graph is bipartite and the stabilizer of a part is a Sym_5-complement, hence (i) follows. By (6.10.1 (iv)) t_1 is contained in $O^2(G_7^{[2]}) \cong \mathrm{Alt}_5$, therefore it is contained in either both or none of the Sym_5-complements. Since $G_7^{[2]} = \langle G_7^{[02]}, t_1 \rangle$, (ii) follows. ∎

Lemma 6.13.5 *If* $\varepsilon_1 = 0$ *then the subgroup* $C_7^{[2]}$ *acts trivially on* $\Upsilon_0^{(1)}$.

Proof By (6.10.3 (ii)) we know that $G^{[2]}$ induces on $\Upsilon_0^{(1)}$ the group $L_3(2) \times 2$. Clearly $N^{[2]}$ induces $L_3(2)$, therefore $G_7^{[2]}$ induces a group of order 2. Suppose that $G_7^{[2]} \cong \mathrm{Sym}_5 \times 2$ (otherwise the claim is obvious). Then the kernel of the action of $G_7^{[2]}$ on $\Upsilon_0^{(1)}$ is isomorphic either to Sym_5 or to $\mathrm{Alt}_5 \times 2$. Since t_1 is contained in the kernel by (6.13.1) and by (6.13.4 (ii)) it is not contained in a Sym_5-subgroup, the kernel is $\mathrm{Alt}_5 \times 2$ and in particular it contains $C_7^{[2]}$. ∎

Lemma 6.13.6 *If* $\varepsilon_1 = 0$ *then the subgroup* $C_7^{[2]}$ *acts trivially on*

$$\Upsilon^{(2)} := \bigoplus_{\omega \in \Omega} \Upsilon_\omega^{(2)}.$$

Proof Suppose the contrary. Then in view of (6.13.3) $G_7^{[2]} \cong \mathrm{Sym}_5 \times 2$ and the generator σ of $C_7^{[2]}$ acts on $\Upsilon^{(2)}$ as the (-1)-scalar operator.

We follow the explicit form of Ω and of the actions on Ω of $G^{[02]}$, $G^{[12]}$ and t_1 as at the end of Section 6.12. In particular

$$\widehat{t_1} = (1, 3)(2, 4)$$

(we follow the hat convention for the images in $\widehat{G^{[2]}}$). Let s_1 be the unique element from $G_7^{[02]}$ such that

$$\widehat{s_1} = (1, 2)(3, 5).$$

Let us identify t_1 and s_1 with their images under φ and put $r = s_1 t_1$. Then $r \in G^{[2]}$,

$$\widehat{r} = (1, 5, 3, 2, 4)$$

and exactly one of the following holds:

(a) r^5 centralizes $\Upsilon^{(1)}$, r^3 conjugates the action of t_1 on $\Upsilon^{(2)}$ onto that of s_1, $C^{[2]}$ acts trivially on $\Upsilon^{(2)}$;

(b) $r^5 = \sigma$ acts on $\Upsilon^{(2)}$ as the (-1)-scalar operator, r^3 conjugates the action of t_1 on $\Upsilon^{(2)}$ onto that of $s_1\sigma$, $C^{[2]} \cong 2$ and $C^{[2]}$ acts faithfully on $\Upsilon^{(2)}$.

By (6.13.2) $\Upsilon_5^{(2)}$ is the only irreducible constituent of $N^{[2]}$ in $\Upsilon^{(2)}$ which is stabilized by t_1 and it is centralized by t_1 (since ε_1 is assumed to be 0). Therefore a conjugate of t_1 in $G^{[2]}$ centralizes the only irreducible constituent of $N^{[2]}$ which it stabilizes, while a conjugate of $t_1\sigma$ negates every vector in the constituent it stabilizes.

On the other hand, $\Upsilon_4^{(2)}$ is the only irreducible constituent in $\Upsilon^{(2)}$ stabilized by s_1. Comparing (a) and (b) above, we conclude that s_1 either centralizes or negates every vector in $\Upsilon_4^{(1)}$ in the respective cases (a) and (b). Thus in order to establish the claim it is sufficient to show that s_1 centralizes at least one vector from $\Upsilon_4^{(1)}$.

Since $\widehat{t_1}$ and $\widehat{s_1}$ are conjugate in $\widehat{G^{[2]}}$, comparing the paragraph before (6.6.2) we conclude that s_1 is in the kernel K_1 of $G^{[02]}$ on Π_1. Let V_2 be a 2-dimensional subspace which complements U_3 in U_5 and let v be a non-zero vector from V_2. Suppose that N_7 stabilizes the direct sum decomposition

$$U_5 = U_3 \oplus V_2.$$

Then V_2 is the only 2-dimensional subspace in U_5 which is stabilized by N_7. The element s_1 is the product $\tau\delta$, where τ is the transvection from $L^{[0]}$ with centre v and axis $\langle v, U_3 \rangle$ while δ is the image of V_2 under the $L^{[0]}$-invariant mapping d of the set of 2-dimensional subspaces of U_5 into $Q^{[0]}$ (compare the beginning of Section 6.9).

Now we need to find a 3-dimensional subspace V_3 in U_5 such that

(1) $\Pi_{P(V_3)}^{[0]}$ is contained in $\Upsilon_4^{(2)}$;

(2) s_1 centralizes $\Pi_{P(V_3)}^{[0]}$.

By (6.10.2) in order to satisfy (1) we must have $\dim (U_3 \cap V_3) = 2$ and $V_4 := \langle U_3, V_3 \rangle$ must be an s_1-invariant hyperplane in U_5. Put $V_4 = \langle U_3, v \rangle$ and take V_3 to be a 3-dimensional subspace in V_4 which does not contain v and which intersects U_3 in a 2-dimensional subspace. An easy counting shows that such a 3-dimensional subspace V_3 exists. Under our choice V_2 is disjoint from U_3. By (3.6.3 (iii)) this means that $\delta = d(V_2)$ is in $Q^{[0]} \setminus P(V_3)$. Since $V_3 \le V_4$ and V_4 is the centre of the transvection τ, we have $s_1 = \tau\delta \in G^{[0]}(V_3)$. Furthermore, s_1 is in the kernel of the action of $G^{[0]}(V_3)$ on $\Pi_{P(V_3)}^{[0]}$ (compare the beginning of Section 6.9). Hence the condition (2) is also satisfied and the result follows. ∎

Now it only remains to show that for at least one of the completions $\varphi(\lambda, 0, \varepsilon_2)$ the centre $C^{[2]}$ of $G^{[2]}$ does not project onto

$$\Upsilon^{(3)} := \bigoplus_{\pi \in \binom{\Omega}{2}} \Upsilon_\pi^{(3)}.$$

For accomplishing this we still have one life line left in the sense that we can choose ε_2 between 0 and 1.

As a direct consequence of (6.4.3 (ii)) and (6.7.7) we have the following useful result.

Lemma 6.13.7 *Let ρ denote the element which centralizes*

$$\Pi_{11}^{[01]} \oplus \Pi_{12}^{[01]} \oplus \Pi_{21}^{[01]}$$

and negates every vector in $\Pi_{22}^{[01]}$. Let $t_1^{(\varepsilon_2)}$ denote the image of t_1 under $\varphi(\lambda, 0, \varepsilon_2)$. Then

$$t_1^{(1-\varepsilon_2)} = t_1^{(\varepsilon_2)}\rho.$$

Lemma 6.13.8 *There is exactly one $\varepsilon_2^{(0)} \in \{0,1\}$ such that the completion*

$$\varphi(\lambda, 0, \varepsilon_2^{(0)})$$

is constrained at level 2.

Proof Let $s_1 \in G^{[0]}$ be as in the proof of (6.13.6). As above we identify s_1 with its image under the completion $\varphi(\lambda, 0, \varepsilon_2)$. Let γ be the element from $\mathcal{P}$ which negates every vector from $\Upsilon^{(3)}$ and centralizes $\Upsilon_0^{(1)} \oplus \Upsilon^{(2)}$. Then, arguing as in the proof of (6.13.6), we observe that

$$(s_1 t_1^{(\varepsilon_2)})^5 = 1$$

if the completion $\varphi(\lambda, 0, \varepsilon_2)$ is constrained at level 2 and

$$(s_1 t_1^{(\varepsilon_2)})^5 = \gamma$$

otherwise. Applying (6.13.7) we check that

$$(s_1 t_1^{(1-\varepsilon_2)})^5 = (s_1 t_1^{\varepsilon_2})^5 \cdot \prod_{\widehat{p} \in \langle \widehat{r} \rangle} \rho^{\widehat{p}} = (s_1 t_1^{(\varepsilon_2)})^5 \gamma.$$

This immediately gives the result. ∎

Thus we have a pair of completions $\varphi(\lambda, 0, \varepsilon^{(0)})$ which are constrained at level 2. These completion differ by the choice of the irreducible character λ of $L^{[2]} \cong L_3(2)$ of degree 3 among two algebraically conjugate such characters. Thus we have arrived to the following main result of the chapter.

Proposition 6.13.9 *The amalgam $\mathcal{G} = \{G^{[0]}, G^{[1]}\}$ possesses a faithful generating completion which is constrained at level 2, that is a completion*

$$\varphi : \mathcal{G} \to G,$$

such that

$$G^{[2]} := \langle \varphi(G^{[02]}), \varphi(G^{[12]}) \rangle \cong 2^{3+12} \cdot (L_3(2) \times \mathrm{Sym}_5).$$

The completion in (6.13.9) is $\varphi(\lambda, 0, \varepsilon^{(0)})$ for λ being either λ_1 or λ_2. By the Main Theorem the completion group of a generating completion of $\mathcal{G}$ which is constrained at level 2 is isomorphic to the fourth Janko group J_4. Thus $\varphi(\lambda_1, 0, \varphi^{(0)})$ and $\varphi(\lambda_2, 0, \varphi^{(0)})$ induce the algebraically conjugate pair of faithful complex representations of J_4 of the minimal degree 1333. This implies that the image of $\varphi(\lambda, 1, 1 - \varepsilon_2^{(0)})$ is $J_4 \times 2$ (compare Exercise 4 below).

Exercises

1. Prove (6.1.2), (6.4.1) and (6.4.2).
2. Complete the proof of (6.2.4).
3. Extend argument in the proof of (6.10.2) to calculate the $N^{[2]}$-orbits on $\mathcal{P}_2$.
4. With $\varepsilon_2^{(0)}$ as in (6.13.8) decide what is the group generated by the image of $\varphi(\lambda, 0, 1 - \varepsilon_2^{(0)})$.
5. Determine the dimension of the minimal representation of the amalgam $\mathcal{H} = \{H^{[0]}, H^{[1]}\}$ from $O_{10}^+(2)$. What are the groups generated by the corresponding images?

7

GETTING THE PARABOLICS TOGETHER

From now on we assume that G is the completion group of $\mathcal{G}$ which is constrained at level 2. We identify the third geometric subgroup with the famous involution centralizer

$$2^{1+12}_{+} \cdot 3 \cdot \text{Aut } (M_{22})$$

in J_4. We also recover another important subgroup in G which is

$$2^{11} : M_{24}.$$

This enables us to associate with G a coset geometry $\mathcal{D}(G)$ which eventually will be identified with the Ronan–Smith geometry for J_4.

7.1   Encircling $2^{1+12}_{+} \cdot 3 \cdot \text{Aut } (M_{22})$

Let $\varphi : \mathcal{G} \to G$ be a faithful generating completion of the amalgam $\mathcal{G}$ which is constrained at level 2. The existence of such a completion is guaranteed by (6.13.9). First we assume that $\varphi : \mathcal{G} \to G$ is universal among the completions which are constrained at level 2. Since the centre of $N^{[2]} = K^{[2]}$ is trivial we can define such a completion in the following way (compare Section 2.5).

Let $\widetilde{\varphi} : \mathcal{G} \to \widetilde{G}$ be the universal completion of $\mathcal{G}$, $\varphi : \mathcal{G} \to G$ be an arbitrary completion which is contrained at level 2, $\psi : \widetilde{G} \to G$ be the corresponding homomorphism of completions and Y be the kernel of ψ. Then the restriction of ψ to

$$\widetilde{G}^{[2]} = \langle \widetilde{\varphi}(G^{[02]}), \widetilde{\varphi}(G^{[12]}) \rangle$$

is a homomorphism onto

$$G^{[2]} = \langle \varphi(G^{[02]}), \varphi(G^{[12]}) \rangle$$

with kernel $Y^{[2]} = Y \cap \widetilde{G}^{[2]}$. Since $\varphi : \mathcal{G} \to G$ is constrained at level 2, we have

$$C_{G^{[2]}}(\varphi(N^{[2]})) \leq Z(\varphi(N^{[2]})) = 1.$$

On the other hand, the restriction of ψ to $\widetilde{\varphi}(N^{[2]})$ is an isomorphism onto $\varphi(N^{[2]})$ and therefore,

$$Y^{[2]} = C_{\widetilde{G}^{[2]}}(\widetilde{\varphi}(N^{[2]})).$$

If we want G to be the 'largest' completion group subject to the property that it is constrained at level 2 we must take Y to be the smallest normal subgroup in $\widetilde{G}$ which intersects $\widetilde{G}^{[2]}$ in $Y^{[2]}$. This means that Y should be taken to be the normal closure in $\widetilde{G}$ of $C_{\widetilde{G}^{[2]}}(\widetilde{\varphi}(N^{[2]}))$.

Alternatively we can define G to be the universal completion of the rank 3 amalgam

$$\mathcal{J} = \{G^{[0]}, G^{[1]}, G^{[2]}\}.$$

It is worth mentioning that the existence of the amalgam $\mathcal{J}$ is independent of the existence of completions of $\mathcal{G}$ which are constrained at level 2. In fact, $\mathcal{J}$ is the amalgam

$$\{\widetilde{\varphi}(G^{[0]}), \widetilde{\varphi}(G^{[1]}), \widetilde{G}^{[2]}\}$$

factorised over $C_{\widetilde{G}^{[2]}}(\widetilde{\varphi}(N^{[2]}))$. On the hand, $\mathcal{J}$ possesses a faithful completion if and only if $\mathcal{G}$ possesses a completion constrained at level 2.

From now on (unless explicitly stated otherwise)

$$\varphi : \mathcal{G} \to G$$

is assumed to be an *arbitrary* faithful completion of $\mathcal{G}$ which is constrained at level 2. The amalgam $\mathcal{G}$ will be identified with its image in G under φ, so that we can plainly write

$$G^{[i]} = \langle G^{[0i]}, G^{[1i]} \rangle.$$

By (4.2.6) and (4.2.8) $N^{[2]}$ and $N^{[3]}$ are non-trivial, so by (4.2.1 (iv)) $G^{[2]}$ and $G^{[3]}$ are proper subgroups in G. On the other hand by (5.4.1) $N^{[4]} = 1$ and in Section 7.4 we will show that $G^{[4]}$ is in fact the whole of G.

Let $\Gamma = \Lambda(\mathcal{G}, \varphi, G)$ be the coset graph corresponding to the completion $\varphi : \mathcal{G} \to G$. Let x and $\{x, y\}$ be defined as in the paragraph before (4.1.1) so that

$$\mathcal{G} = \{G(x), G\{x, y\}\}.$$

Let $\Gamma^{[2]}$ and $\Gamma^{[3]}$ be the geometric subgraphs in Γ induced by the images of x under $G^{[2]}$ and $G^{[3]}$, respectively (compare (4.2.1 (iii))). Since $\Gamma^{[2]}$ is of valency 3 and $G^{[2]}$ induces on the vertex set of $\Gamma^{[2]}$ an action of $G^{[2]}/N^{[2]} \cong \mathrm{Sym}_5$ on the cosets of $G^{[02]}/N^{[2]} \cong \mathrm{Sym}_3 \times \mathrm{Sym}_2$ the following statement is an immediate consequence of the definition of the Petersen graph.

Lemma 7.1.1 $\Gamma^{[2]}$ *is isomorphic to the Petersen graph.*

Since the action of $G^{[3]}$ on $\Gamma^{[3]}$ is locally projective of type $(3, 2)$ and $\Gamma^{[2]}$ is a geometric cubic subgraph in $\Gamma^{[3]}$, (7.1.1), (5.2.3), and (11.4.3) imply the following.

Lemma 7.1.2 *One of the following two possibilities takes place:*

(i) $\Gamma^{[3]}$ *is the octet graph* $\Gamma(M_{22})$, $C_{G^{[3]}}(N^{[3]}) = Z(N^{[3]}) = Z^{[3]} \cong 2$, $G^{[3]}/N^{[3]} \cong \mathrm{Aut}\,(M_{22})$ *and*

$$G^{[3]} \cong 2_+^{1+12} \cdot 3 \cdot \mathrm{Aut}\,(M_{22});$$

(ii) $\Gamma^{[3]}$ *is the Ivanov–Ivanov–Faradjev graph* $\Gamma(3 \cdot M_{22})$, $C_{G^{[3]}}(N^{[3]}) \cong 2 \times 3$, $G^{[3]}/N^{[3]} \cong 3 \cdot \operatorname{Aut}(M_{22})$ *and*

$$G^{[3]} \cong (2_+^{1+12} \times 3) \cdot 3 \cdot \operatorname{Aut}(M_{22}).$$

It will be proved in Section 7.4 that the possibility (7.1.2 (i)) takes place. Clearly $G^{[3]}$ is a completion of the amalgam

$$\mathcal{G}^{[3]} = \{G^{[03]}, G^{[13]}, G^{[23]}\}.$$

Since the completion $\varphi : \mathcal{G} \to G$ is constrained at level 2, it is rather straightforward to check that

$$C_{G^{[2]}}(N^{[3]}) \leq Z(N^{[3]}) = Z^{[3]}$$

and therefore the amalgam $\bar{\mathcal{G}}^{[3]} = \{G^{[03]}/N^{[3]}, G^{[13]}/N^{[3]}, G^{[23]}/N^{[3]}\}$ is isomorphic to the amalgam $\widehat{\mathcal{A}^{[3]}}$ defined before (5.2.1). Hence by (5.2.1) $\bar{\mathcal{G}}^{[3]}$ is isomorphic to the amalgam $\mathcal{Z}$ as in (11.4.1).

Lemma 7.1.3 *Let* $C^{[3]}$ *be the universal completion of* $\mathcal{G}^{[3]}$. *Then* $C^{[3]}/N^{[3]}$ *is the universal completion of* $\bar{\mathcal{G}}^{[3]} \cong \mathcal{Z}$, *therefore* $C^{[3]}/N^{[3]} \cong 3 \cdot \operatorname{Aut}(M_{22})$.

Proof Let $\widetilde{K} \cong 3 \cdot \operatorname{Aut}(M_{22})$ be the universal completion of $\mathcal{Z}$ and let us identify $\mathcal{Z}$ with its image in $\widetilde{K}$. Let α be a homomorphism of $\mathcal{G}^{[3]}$ onto $\mathcal{Z}$ which is the composition of the canonical homomorphism $g \mapsto gN^{[3]}$ of $\mathcal{G}^{[3]}$ onto $\bar{\mathcal{G}}^{[3]}$ and an isomorphism of $\bar{\mathcal{G}}^{[3]}$ onto $\mathcal{Z}$. Let

$$C^{[3]} \times \widetilde{K} = \{(c, k) \mid c \in C^{[3]}, k \in \widetilde{K}\}$$

be the direct product of $C^{[3]}$ and $\widetilde{K}$ and let $\mathcal{X}$ be the subset of $C^{[3]} \times \widetilde{K}$ consisting of the pairs (c, k), such that $\alpha(c) = k$. Then $\mathcal{X}$ is isomorphic to $\mathcal{G}^{[3]}$. Furthermore if X is the subgroup in $C^{[3]} \times \widetilde{K}$ generated by $\mathcal{X}$ then the restriction to X of the canonical homomorphism of $C^{[3]} \times \widetilde{K}$ onto $\widetilde{K}$ is surjective and the claim follows. ∎

By (7.1.3) if $G^{[3]}$ is the universal completion of $\mathcal{G}^{[3]}$ then the possibility (ii) in (7.1.2) takes place. Therefore there is no way we can get down to the possibility (i) looking at the amalgam $\mathcal{G}$ only and some further subgroups of G should be brought into play.

7.2   Tracking $2^{11} : M_{24}$

In Sections 3.7 and 3.8 we have seen that $G^{[1]}$ is a semidirect product of $Q^{[m]} \cong 2^{11}$ and $A^{[1]}L^{[1]} \cong 2^4 : L_4(2)$. The relevant action is isomorphic to the action of the octad stabilizer in M_{24} on the irreducible Todd module $\bar{\mathcal{C}}_{11}$. In Section 7.4 we will prove the following.

Proposition 7.2.1 *Let* $G^{[m]}$ *be the subgroup in* G *generated by the normalisers of* $Q^{[m]}$ *in* $G^{[1]}$, $G^{[2]}$, *and* $G^{[3]}$. *Then*

$$G^{[m]} \cong 2^{11} : M_{24},$$

more specifically $G^{[m]}$ is the semidirect product of $Q^{[m]} \cong \bar{C}_{11}$ and M_{24} with respect to the natural action.

For $i = 1$, 2 and 3 put $G^{[mi]} = N_{G^{[i]}}(Q^{[m]})$, $\mathcal{G}^{[m]} = \{G^{[mi]} \mid 1 \leq i \leq 3\}$ and $\bar{\mathcal{G}}^{[m]} = \{G^{[mi]}/Q^{[m]} \mid 1 \leq i \leq 3\}$, so that $\bar{\mathcal{G}}^{[m]}$ is the quotient of $\mathcal{G}^{[m]}$ over $Q^{[m]}$. Notice that $G^{[m1]} = G^{[m]}$.

Lemma 7.2.2 *The following assertions hold:*

(i) $G^{[m1]}/Q^{[m]} \cong 2^4 : L_4(2)$
(ii) $G^{[m2]}/Q^{[m]} \cong 2^6 : (L_3(2) \times \mathrm{Sym}_3)$;
(iii) *either*
    (a) *(7.1.2 (i)) takes place and* $G^{[m3]}/Q^{[m]} \cong 2^6 : 3 \cdot \mathrm{Sym}_6$, *or*
    (b) *(7.1.2 (ii)) takes place and* $G^{[m3]}/Q^{[m]} \cong (2^6 \times 3) : 3 \cdot \mathrm{Sym}_6$.

Proof Statement (i) follows is directly from (3.8.1). In order to establish (ii) we locate $Q^{[m]}$ inside $G^{[2]}$. It is clear that $Q^{[m]}$ is contained in $G^{[2]}$ (for instance because $[G^{[1]} : G^{[12]}] = 15$ is odd and $Q^{[m]}$ is a normal 2-subgroup in $G^{[1]}$). By (4.2.7) $|N^{[2]} \cap Q^{[m]}| = 2^9$ and the image of $Q^{[m]}$ in $G^{[2]}/N^{[2]} \cong \mathrm{Sym}_5$ is an elementary abelian subgroup of order 4, which stabilizes an edge of $\Gamma^{[2]}$ as a whole but not vertexwisely. This means that $Q^{[m]}N^{[2]}/N^{[2]}$ is contained in the commutator subgroup of $G^{[2]}/N^{[2]}$, isomorphic to Alt_5.

Let S_7 be a Sylow 7-subgroup in $G^{[2]}$, $C \cong \mathrm{Sym}_5$ be the complement to S_7 in $C_{G^{[2]}}(S_7)$ (compare (4.9.1)) and R be the elementary abelian subgroup of order 4 in C, such that $RN^{[2]}/N^{[2]} = Q^{[m]}N^{[2]}/N^{[2]}$. We claim that R is contained in $Q^{[m]}$. In fact, by (3.7.1) $G^{[1]}/Q^{[m]} \cong 2^4 : L_4(2)$ and since $[G^{[1]} : G^{[12]}] = 15$ is not divisible by 7, $Q^{[m]}$ is normalized by a Sylow 7-subgroup in $G^{[2]}$. By Sylow's theorem without loss we assume that this subgroup is S_7. By (4.9.1) (ii) $C_{Q^{[2]}}(S_7) = 1$. Therefore

$$C_{Q^{[m]}}(S_7)N^{[2]}/N^{[2]} = Q^{[m]}N^{[2]}/N^{[2]}$$

and the claim follows. Next we claim that $Q^{[m]} = RC_{Q^{[2]}}(R)$. Since $R \leq Q^{[m]}$ and $Q^{[m]}$ is abelian, $Q^{[m]}$ is obviously in the centralizer of R in $Q^{[2]}$ and hence we only have to show that $|C_{Q^{[2]}}(R)|$ is at most 2^9. The subgroup $C_{G^{[2]}}(R)$ contains S_7, therefore $C_{Q^{[2]}}(R)$ is normalized by S_7. Clearly $C_{Q^{[2]}}(R)$ contains $Z^{[2]}$ and therefore every dent of $Q^{[2]}$ is either completely contained in $C_{Q^{[2]}}(R)$ or intersects $C_{Q^{[2]}}(R)$ in $Z^{[2]}$. In addition, since R commutes with S_7 and every dent is the direct sum of two non-isomorphic S_7-modules, whenever R normalizes a dent, it necessarily centralizes it. Now it only remains to recall that by (4.7.4) C acts on the set of dents as it acts on the edge-set of the Petersen graph $\Gamma^{[2]}$. Finally, R stabilizes exactly three edges of $\Gamma^{[2]}$ (these edges form the antipodal triple containing $\{x, y\}$).

By the above paragraph the number of conjugates of $Q^{[m]}$ in $G^{[2]}$ is equal to the number of conjugates of R in C (which is five). Since

$$N_C(R) \cong \mathrm{Sym}_4 \cong 2^2 : \mathrm{Sym}_3,$$

(ii) follows.

The stabilizer in G of an edge $e = \{u, v\}$ of Γ is a conjugate of $G^{[1]}$ and by (3.7.1) this stabilizer contains a unique normal elementary abelian subgroup Q_e of order 2^{11}, which is of course a conjugate of $Q^{[m]}$. By the above paragraph whenever two edges e and f are contained in a common geometric cubic subgraph $\Sigma \cong \Gamma^{[2]}$ and are antipodal in the line graph of Σ, the equality

$$Q_e = Q_f$$

holds (notice that there are 15 edges in Σ and only 5 different conjugates of $Q^{[m]}$ in $G^{[2]}$). Let $\Phi = \Phi_\Gamma$ be the local antipodality graph of Γ, so that Φ is a graph on the edge-set of Γ in which two edges are adjacent if they are contained in a common geometric cubic Petersen subgraph Σ and are antipodal in the line graph of Σ. Then $Q_e = Q_f$ whenever e and f are in the same connected component of Φ.

Let us turn to (iii). It is clear that $Q^{[m]} \leq G^{[3]}$. On the other hand, since $Q^{[3]} = O_2(G^{[3]}) \cong 2^{1+12}_+$ is extraspecial, while $Q^{[m]}$ is elementary abelian, $|Q^{[m]} \cap Q^{[3]}| \leq 2^7$ by (1.6.7). Let $\Psi = \Phi_{\Gamma^{[3]}}$ be local antipodality graph of $\Gamma^{[3]}$ and Ψ^c be the connected component of Ψ containing $\{x, y\}$. Since $\Gamma^{[3]}$ is either the octet graph or the Ivanov–Ivanov–Faradjev graph, by (11.4.4) and the paragraph after that lemma Ψ^c contains 15 or 45 edges of $\Gamma^{[3]}$ depending on whether we are in case (a) or (b). By the above paragraph the stabilizer S of Ψ^c in $G^{[3]}$ is contained in $G^{[m3]}$. Furthermore, S contains $N^{[3]}$ and $S/N^{[3]}$ is $2^4 : \mathrm{Sym}_6$ and $(2^4 \times 3) \cdot \mathrm{Sym}_6$ in the respective cases (a) and (b). Since $Q^{[m]} N^{[3]}/N^{[3]} = O_2(S/N^{[3]})$, using the well-known fact that $K_h = N_K(O_2(K_h))$ for the stabilizer $K_h \cong 2^4 : \mathrm{Sym}_6$ of a hexad in $K \cong \mathrm{Aut}\,(M_{22})$, we conclude that S is the whole of $G^{[m3]}$, which completes the proof of (iii). ∎

Lemma 7.2.3 *Suppose that* (7.1.2 *(i))* *takes place. Then*

(i) *the coset geometry corresponding to the embedding into* $G^{[m]}/Q^{[m]}$ *of the amalgam*

$$\{G^{[mi]}/Q^{[m]} \mid 1 \leq i \leq 3\}$$

is described by the locally truncated diagram

$$C_4^t : \quad \underset{2}{\circ}=\!\!=\!\!=\underset{2}{\circ}-\!\!-\!\!-\underset{2}{\circ}-\!\!-\!\!-\square ;$$

(ii) $G^{[m]}/Q^{[m]} \cong M_{24};$
(iii) $G^{[m]}$ *splits over* $Q^{[m]}$.

Proof First notice that the assertion (7.2.2 (iii) (a)) holds. Calculating the intersections of the $G^{[mi]}$'s we obtain (i). Now (ii) is by (i) and (11.2.1), while (iii) is by (ii), (3.7.1) and Gaschütz's theorem. ∎

7.3 *P*-geometry of $G^{[4]}$

In this section for a subsequence α of 0123 we denote the subgroup $G^{[\alpha 4]}$ by $F^{[\alpha]}$. This convention also applies when α is empty, so that $F = G^{[4]}$. Let $\mathcal{F} = \{F^{[0]}, F^{[1]}\}$ be the corresponding subamalgam in F, let $\Xi = \Lambda^{[4]} = \Lambda(\mathcal{G}^{[4]}, \varphi^{[4]}, F)$ be the coset graph associated with the completion

$$\varphi^{[4]} : \mathcal{G}^{[4]} \to F$$

(which is the restriction of φ to $\mathcal{G}^{[4]}$). At this stage we do not know yet that F is the whole of G, but at any event the action of F on Ξ is faithful (since $N^{[4]}$ is trivial by (5.4.1)) and locally projective of type $(4, 2)$. Let $\{u, v\}$ be the edge of Ξ such that

$$\mathcal{F} = \{F(u), F\{u, v\}\},$$

where $\mathcal{F}$ is identified with its image in F under $\varphi^{[4]}$. For $i = 2$ and 3 let $\Xi^{[i]}$ be the geometric subgraph in Ξ induced by the images of u under $F^{[i]}$ and let $I^{[i]}$ be the vertexwise stabilizer of $\Xi^{[i]}$ in F.

Lemma 7.3.1 *The following assertions hold:*

(i) $F^{[0]} = G^{[04]} \cong 2^{4+4} : 2^6 : L_4(2)$;

(ii) $F^{[1]} = \langle G^{[014]}, t_1 \rangle \cong 2^{6+5+6} \cdot (L_3(2) \times 2) \cong 2^{11} : 2^{1+6}_+ : L_3(2)$;

(iii) $F^{[2]} \cong 2^{3+12} \cdot (\mathrm{Sym}_4 \times \mathrm{Sym}_5)$, $\Xi^{[2]}$ *is the Petersen graph and* $I^{[2]} \cong 2^{3+12} \times \mathrm{Sym}_4$;

(iv) $F^{[3]} \cong 2^{1+12}_+ \cdot 3 \cdot \mathrm{Aut}\,(M_{22})$, $\Xi^{[3]}$ *is the Ivanov–Ivanov–Faradjev graph and* $I^{[3]} \cong 2^{1+12}_+$.

Proof Since $F^{[0]}$ and $F^{[1]}$ are the stabilizers of U_1 in $G^{[0]}$ and $G^{[1]}$, respectively (i) and (ii) are quite clear. In terms of Section 3.8 $F^{[1]}$ is a semidirect product of $Q^{[m]} \cong 2^{11}$ and the stabilizer of U_1 in $A^{[1]}L^{[1]} \cong 2^4 : L_4(2)$. The latter stabilizer coincides with the centralizer of a central involution in $L^{[0]} \cong L_5(2)$, isomorphic to $2^{1+6}_+ : L_3(2)$. We know that $F^{[2]}$ is the subgroup in $G^{[2]}$ generated by $G^{[024]}$ and $G^{[124]}$. The set $\mathcal{P}$ of geometric subgraphs of valency 7 in Γ containing $\Gamma^{[2]}$ is of size 7 (of course $\Gamma^{[3]} \in \mathcal{P}$). The action (isomorphic to $L_3(2)$) of $N^{[2]}$ on $\mathcal{P}$ induces a structure of the projective plane of order 2. Then $F^{[2]}$ is the stabilizer in $G^{[2]}$ of a line in that projective plane structure which gives (iii).

The subgroup $F^{[3]}$ is generated by $G^{[034]}$ and $G^{[134]}$. Since $Q^{[3]} = O_2(N^{[3]}) = N^{[3]} \cap G^{[014]}$ we immediately conclude that

$$I^{[3]} \geq Q^{[3]} \cong 2^{1+12}_+.$$

On the other hand, the whole of $N^{[3]}$ could not be in $I^{[3]}$ since it is not even in $G^{[014]}$. The action of $F^{[3]}$ on $\Xi^{[3]}$ is locally projective of type $(3, 2)$ and by (ii) $\Xi^{[2]}$ is a geometric cubic subgraph in $\Xi^{[3]}$ isomorphic to the Petersen graph. By (11.4.3) this implies that $\Xi^{[3]}$ is either the octet graph $\Gamma(M_{22})$ of the

Ivanov–Ivanov–Faradjev graph $\Gamma(3 \cdot M_{22})$. Let

$$\chi : G^{[3]} \to \mathrm{Out}\, Q^{[3]}$$

be the natural homomorphism. By (5.2.4) the image of χ is isomorphic to $3 \cdot \mathrm{Aut}\,(M_{22})$. For $\alpha = 0$ and 1 the subgroups $G^{[\alpha 34]}$ and $N^{[3]} \cong 2_+^{1+12} : 3$ factorize $G^{[\alpha 3]}$ and hence by (5.2.1 (i), (ii)) we have

$$\chi(F^{[03]}) \cong 2^3 : L_3(2) \times 2, \quad \chi(F^{[13]}) \cong (2_+^{1+4} : \mathrm{Sym}_3 \times 2) \cdot 2.$$

Since $\chi(G^{[3]})$ does not split over $O_3(\chi(G^{[3]}))$,

$$3 \cdot \mathrm{Aut}\,(M_{22}) \cong \chi(G^{[3]}) = \chi(F^{[3]}) = \langle \chi(F^{[03]}), \chi(F^{[13]}) \rangle$$

and therefore $F^{[3]}/I^{[3]}$ possesses a homomorphism onto $3 \cdot \mathrm{Aut}\,(M_{22})$. Thus (iv) follows. ∎

Let $\mathcal{G}(G^{[4]})$ be the geometry, whose elements of type 1 are the vertices of Ξ, the elements of type 2 are the edges of Ξ, the elements of type 3 are the geometric cubic subgraphs in Ξ and the elements of type 4 are the geometric subgraphs of valency 7 in Ξ; the incidence relation is via inclusion. As a direct consequence of (7.3.1) we obtain the following

Proposition 7.3.2 *The geometry $\mathcal{G}(G^{[4]})$ is a P-geometry of rank 4 with the diagram*

$$P_4 : \quad \underset{1}{\circ} \overset{\text{P}}{\underset{}{\rule{2cm}{0.4pt}}} \underset{2}{\circ} \rule{2cm}{0.4pt} \underset{2}{\circ} \rule{2cm}{0.4pt} \underset{2}{\circ}.$$

The group $G^{[4]}$ acts on $\mathcal{G}(G^{[4]})$ faithfully and flag-transitively. The residue in $\mathcal{G}(G^{[4]})$ of an element of type 4 is isomorphic to the geometry $\mathcal{G}(3 \cdot M_{22})$.

In Section 7.4 we will show that $G^{[4]}$ is the whole of G and by the Main Theorem the latter is J_4. Therefore the geometry in (7.3.2) is the P-geometry $\mathcal{G}(J_4)$ of J_4 first constructed in (Ivanov 1987).

For $1 \leq i \leq 3$ put $F^{[mi]} = F^{[i]} \cap Q^{[m]}$ and $F^{[m]} = \langle F^{[mi]} \mid 1 \leq i \leq 3 \rangle$.

Lemma 7.3.3 *The following assertions hold:*

(i) $F^{[m1]}/Q^{[m]} \cong 2_+^{1+6} : L_3(2)$ *and* $F^{[m1]}$ *splits over* $Q^{[m]}$;

(ii) $F^{[m2]}/Q^{[m]} \cong 2^6 : (\mathrm{Sym}_4 \times \mathrm{Sym}_3)$;

(iii) $F^{[m3]}/Q^{[m]} \cong 2^6 : 3 \cdot \mathrm{Sym}_6$;

(iv) *the coset geometry $\mathcal{M}$ corresponding to the embedding of the amalgam $\{F^{[mi]}/Q^{[m]} \mid 1 \leq i \leq 3\}$ into $F^{[m]}/Q^{[m]}$ is described by the tilde diagram*

$$T_3 : \quad \underset{2}{\circ} \overset{\sim}{\rule{2cm}{0.4pt}} \underset{2}{\circ} \rule{2cm}{0.4pt} \underset{2}{\circ};$$

(v) $F^{[m]}/Q^{[m]} \cong M_{24}$ *and* $\mathcal{M} \cong \mathcal{G}(M_{24})$;

(vi) $Q^{[m]}$ *is the irreducible Todd module* $\bar{\mathcal{C}}_{11}$;

(vii) $F^{[m]}$ *splits over* $Q^{[m]}$.

Proof A mere comparison of (7.2.2) and (7.3.1) gives (i) to (iii). The diagram of $\mathcal{M}$ can be recovered by direct calculating the intersection of the $F^{[mi]}$'s. Alternatively one can employ the following combinatorial realization of $\mathcal{M}$. Let $\Upsilon = \Phi_{\Xi}$ be the local antipodality graph of Ξ. Then, arguing as in the proof of (7.2.2), one can see that $F^{[m]}$ coincides with the stabilizer in F of the connected component Υ^c of Υ containing $\{u, v\}$. The elements of $\mathcal{M}$ are the vertices of Υ^c (which are edges of Ξ), the intersections of the vertex set of Υ^c with edge-sets of geometric subgraphs of valency 3 and 7. Then by (7.3.2) and the paragraph after (11.4.4) we obtain the desired diagram.

By (11.2.2) the assertions (i) to (iv) imply that $F^{[m]}/Q^{[m]}$ is either M_{24} or He. Since $Q^{[m]}$ is a non-trivial module in which $F^{[m3]}$ stabilizes the 1-dimensional subspace $Z^{[3]}$ the latter possibility is excluded, since the index of $2^6 : 3 \cdot \mathrm{Sym}_6$ in He is $29,155$ (cf. Conway *et al.* (1985) and Section 11.2) and hence (v) follows. The subgroup $F^{[m2]}$ stabilizes in $Q^{[m]}$ the 2-dimensional subspace $Z^{[2]}$ which contains the 1-dimensional subspace $Z^{[3]}$ stabilized by $F^{[m3]}$. In terms of Ivanov and Shpectorov (2002) this means that $Q^{[m]}$ is a quotient of the universal representation group of $\mathcal{M} \cong \mathcal{G}(M_{24})$, so that (vi) follows from Proposition 4.3.1 in Ivanov and Shpectorov (2002). Finally (vii) follows from (i) in view of Gaschütz's theorem. $\blacksquare$

It is worth mentioning that the proof of (7.3.3 (v)) is the only place in the present volume where we essentially make use of a result (which is (11.2.2)) whose proof relies on computer-aided calculations.

7.4 $G^{[4]} = G$

First we show that $G^{[m4]} = G^{[m]}$ (recall that $G^{[m4]} \cong 2^{11} : M_{24}$ by (7.3.3 (v), (vi), (vii)).

Lemma 7.4.1 *Suppose that $G^{[m4]} \cong 2^{11} : M_{24}$ is a proper subgroup in $G^{[m]}$. Then the coset geometry $\mathcal{N}$ corresponding to the embedding into $G^{[m]}/Q^{[m]}$ of the amalgam*

$$\{G^{[mi]}/Q^{[m]} \mid 1 \leq i \leq 4\}$$

is described by the rank 4 tilde diagram

$$T_4 : \quad \overset{\sim}{\underset{2}{\circ}\!\!=\!\!\!=\!\!\underset{2}{\circ}}\!\!-\!\!\underset{2}{\circ}\!\!-\!\!\underset{2}{\circ}.$$

Proof We claim that under the hypothesis (7.1.2 (ii)) takes place. In fact, otherwise

$$G^{[m]} \cong G^{[m4]} \cong 2^{11} : M_{24}$$

by (7.2.3), (7.3.3 (v), (vi), (vii)) and the order comparison. Then the structure of the $G^{[mi]}$'s can be read from (7.2 (i), (ii), (iii) (b)), and (7.3 (v), (vi), (vii)). Calculating the intersections we get the diagram. $\blacksquare$

Lemma 7.4.2 $G^{[m4]} = G^{[m]}$.

Proof If the claim fails then by (7.4.1) $G^{[m]}/Q^{[m]}$ acts flag-transitively on a rank 4 tilde geometry $\mathcal{N}$. In terms of Ivanov and Shpectorov (2002) this geometry is of truncated M_{24}-type and it does not exist by Proposition 12.4.6 and 12.5.1 in Ivanov and Shpectorov (2002). ∎

Proof of Proposition 7.2.1 The result is now immediate by (7.3.3) and (7.4.2). ∎

Lemma 7.4.3 *The possibilities* (7.1.2 (*i*)) *and* (7.2.2 (*iii*) (*a*)) *take place, so that*

$$G^{[3]} \cong 2^{1+12}_+ \cdot 3 \cdot \text{Aut } (M_{22}) \text{ and } G^{[3m]}/Q^{[m]} \cong 2^6 : 3 \cdot \text{Sym}_6.$$

Proof By (7.2.1) $G^{[m]}/Q^{[m]} \cong M_{24}$ and the latter group just does not contain subgroups as in (7.2.2 (iii) (b)) already by Lagrange theorem. ∎

We are ready to prove the main result of the section.

Proposition 7.4.4 $G^{[4]} = G$.

Proof By (7.4.1) $G^{[m]} = G^{[m4]} \leq G^{[4]}$. Also $G^{[1]} \leq G^{[4]}$ since $G^{[1]} \leq G^{[m]}$ (as remarked before (7.2.2)). But $G^{[0]}$ is generated by $G^{[01]}$ and $G^{[04]}$, and so also $G^{[0]} \leq G^{[4]}$. This clearly implies $G^{[4]} = G$. ∎

We refer the reader to sections 9.5, 9.6 in Ivanov (1999) for general discussion about the existence/non-existence of geometric subgraphs.

Lemma 7.4.5 *Let* Y *be a Sylow 3-subgroup in* $O_{2,3}(G^{[3]})$. *Then*

(i) $C_{G^{[3]}}(Y) \cong 6 \cdot M_{22}$ *is a non-split central extension of a cyclic group of order 6 by* M_{22};
(ii) $G^{[3]}$ *does not split over* $Q^{[3]} = O_2(G^{[3]}) \cong 2^{1+12}_+$.

Proof Since Y is a Sylow 3-subgroup of $N^{[2]}$ the result is by (4.9.1 (iii)). ∎

By (7.4.5) the Schur multiplier of M_{22} possesses the cyclic group of order 6 as a factor-group. In 1976, when (Janko 1976) was published this cyclic group was believed to be the whole Schur multiplier of M_{22}. In (Mazet 1979) the multiplier of M_{22} was proved to be the cyclic group of order 12.

7.5 Maximal parabolic geometry $\mathcal{D}$

We start this section by summarizing the information about the action of G on Γ we have obtained so far.

Proposition 7.5.1 *Let G be a completion of the amalgam $\mathcal{G}$ which is constrained at level 2 and let Γ be the coset graph associated with this completion. Then*

(i) *Γ is connected of valency 31 and the action of G on Γ is locally projective of type $(5, 2)$;*

(ii) *$G(x) = G^{[0]} \cong 2^{10} : L_5(2)$;*

(iii) *$G\{x, y\} = G^{[1]} \cong 2^{6+4+4} \cdot (L_4(2) \times 2) \cong 2^{11} : 2^4 : L_4(2)$;*

(iv) *the geometric cubic subgraph $\Gamma^{[2]}$ is isomorphic to the Petersen graph and*

$$G\{\Gamma^{[2]}\} = G^{[2]} \cong 2^{3+12} \cdot (L_3(2) \times \mathrm{Sym}_5);$$

(v) *the geometric subgraph $\Gamma^{[3]}$ of valency 7 is isomorphic to octet graph and*

$$G\{\Gamma^{[3]}\} = G^{[3]} \cong 2^{1+12}_+ \cdot 3 \cdot \mathrm{Aut}\,(M_{22});$$

(v) *there are no geometric subgraphs of valency 15 and $G^{[4]} := \langle G^{[04]}, G^{[14]} \rangle$ is the whole of G;*

(vi) *if $\Phi = \Phi_\Gamma$ is the local antipodality graph of Γ and Φ^c is the connected component of Φ containing $\{x, y\}$ then Φ^c is isomorphic to the octad graph $\Gamma(M_{24})$ and*

$$G\{\Phi^c\} = G^{[m]} \cong 2^{11} : M_{24}.$$

Proof (i) and (ii) are already in (4.1.1), (iii) is by (7.1.1), (iv) is by (7.4.3), (v) is by (7.4.4). Finally (vi) is by (7.2.3) since (7.1.2 (i)) takes place by (7.4.3). ∎

Let $\mathcal{F}(G)$ be a geometry such that

(0) the elements of type 0 are the vertices of Γ;
(1) the elements of type 1 are the edges of Γ;
(2) the elements of type 2 are the geometric cubic subgraphs;
(3) the elements of type 3 are the geometric cubic subgraphs of valency 7;

(i) the incidence relation is via inclusion.

Then it is immediate from (7.5.1) that $\mathcal{F}(G)$ belongs to the locally truncated Petersen diagram

$$P_5^t : \quad \overset{\text{P}}{\underset{1}{\circ}\!\!-\!\!\!-\!\!\!-\!\!\!-\!\!\!-\!\!\underset{2}{\circ}\!\!-\!\!\!-\!\!\!-\!\!\underset{2}{\circ}\!\!-\!\!\!-\!\!\!-\!\!\underset{2}{\circ}\!\!-\!\!\!-\!\!\!-\!\!\square}.$$

By the Main Theorem $G \cong J_4$ so $\mathcal{F}(G)$ is another geometry for J_4 constructed in (Ivanov 1987).

More fruitful for our current purposes is the geometry $\mathcal{D} = \mathcal{D}(G)$ whose elements are as in $\mathcal{F}(G)$, only instead of the elements of type 1 (which are the edges of Γ) we take elements of type m which are the connected components of the local antipodality graph Φ of Γ. The incidence relation between the elements of type 0, 2, and 3 is as in $\mathcal{F}(G)$. A connected component of Φ (an element of type m) is adjacent to an element $f \in \mathcal{F}(G)$ if f is incident in $\mathcal{F}(G)$ to an edge of Γ contained in that connected component.

Since G is generated by $G^{[0]}$ and $G^{[1]}$ it is a standard result that both $\mathcal{F}(G)$ and $\mathcal{D}(G)$ are connected.

7.6   Residues in $\mathcal{D}$

Let $\mathcal{D} = \mathcal{D}(G)$ be the geometry defined in Section 7.5. Recall that the set of types of $\mathcal{D}$ is $\{m, 0, 2, 3\}$. For $i \in \{m, 0, 2, 3\}$ the set of elements of type i on $\mathcal{D}$ will be denoted by $\mathcal{D}^{[i]}$. Often we will write the type of an element above its name, for instance we write $\overset{3}{d}$ for an element d of type 3. The stabilizer in G of this element will be denoted by $G_d^{[3]}$. The residue in $\mathcal{D}$ of an element a (whose type will be clear from the context) will be denoted by $\mathcal{D}_a$.

Recall that a *path* in $\mathcal{D}$ is a sequence $\pi = (a_0, a_1, a_2, a_3, \ldots, a_s)$ of its elements such that a_i is incident to a_{i+1} but neither equal nor incident to a_{i+2} for every $1 \leq i \leq s - 1$. In this case s is the length of π.

With every element $\overset{i}{a} \in \mathcal{D}$ we associate a certain combinatorial/geometrical structure (whose isomorphism type depends on i only). Then the residue $\mathcal{D}_a$ of a in $\mathcal{D}$ and the stabilizer $G_a^{[i]}$ of a in G possess natural descriptions in terms of this structure. This works in the following way:

Type m: If $\overset{m}{a}$ is an element of type m then there is a Witt design $W_a^{[24]}$ of type $S(5, 8, 24)$. If $\mathcal{B}_a$, $\mathcal{T}_a$ and $\mathcal{S}_a$ are the octads, trios and sextets of $W_a^{[24]}$, then

$$\mathcal{D}_a^{[0]} = \mathcal{B}_a \times GF(2), \quad \mathcal{D}_a^{[2]} = \mathcal{T}_a, \quad \mathcal{D}_a^{[3]} = \mathcal{S}_a.$$

The incidence relation in $\mathcal{D}_a$ is via the refinement relation on the corresponding partitions of the element set of $W_a^{[24]}$. For instance suppose that $(B, \alpha) \in \mathcal{D}_a^{[0]}$ and $S \in \mathcal{D}_a^{[3]}$, where B is an octad of $W_a^{[24]}$ (identified with the partition of the set of 24 elements into the octad B and its complement), $\alpha \in GF(2)$ and S is a sextet. Then (B, α) and S are incident if and only if B is the union of two tetrads from S. In particular, α does not effect the incidence. The stabilizer $G_a^{[m]}$ is the semidirect product of the automorphism group $M_a^{[m]} \cong M_{24}$ of $W_a^{[24]}$ and the irreducible Todd module $Q_a^{[m]} \cong \bar{C}_{11}$. The module $Q_a^{[m]}$ is considered as a section of the $GF(2)$-permutation module of $M_a^{[m]}$ on the set of elements of $W_a^{[24]}$. In particular $M_a^{[m]}$ has two orbits on the set of non-zero vectors in $Q_a^{[m]}$; the elements in one of the orbits are indexed by pairs of elements of $W_a^{[24]}$, while those from the other orbit are indexed by the sextets from $\mathcal{S}_a$. Dually, the hyperplanes in $Q_a^{[m]}$ are indexed by the octads from $\mathcal{B}_a$ and by the complementary pairs of dodecads. The subgroup $Q_a^{[m]} = O_2(G_a^{[m]})$ is the kernel of the action of $G_a^{[m]}$ on $\mathcal{D}_a^{[2]} \cup \mathcal{D}_a^{[3]}$. Every orbit of $Q_a^{[m]}$ on $\mathcal{D}_a^{[0]}$ is of the form $\{(B, 0), (B, 1)\}$, where $B \in \mathcal{B}_a$. An element $q \in Q_a^{[m]}$ fixes this orbit elementwise if and only if q is in the hyperplane corresponding to B. For every $\alpha \in GF(2)$ the complement $M_a^{[m]}$ stabilizes $\{(B, \alpha) \mid B \in \mathcal{B}_a\}$ as a whole and acts on it as it acts on $\mathcal{B}_a$.

Type 0: If $\overset{0}{b}$ is an element of type 0 then there is a 5-dimensional vector space V_b over $GF(2)$ such that

$$\mathcal{D}_b^{[3]} = \begin{bmatrix} V_b \\ 2 \end{bmatrix}, \quad \mathcal{D}_b^{[2]} = \begin{bmatrix} V_b \\ 3 \end{bmatrix}, \quad \mathcal{D}_b^{[m]} = \begin{bmatrix} V_b \\ 4 \end{bmatrix},$$

(where $\begin{bmatrix} V_b \\ i \end{bmatrix}$ stands for the set of i-dimensional subspaces in V_b). The incidence relation in $\mathcal{D}_b$ is by inclusion. The subspace corresponding to an element x in $\mathcal{D}_b$ will be denoted by $V_b(x)$. The stabilizer $G_b^{[0]}$ is the semidirect product with respect to the natural action of the general linear group $L_b^{[0]} \cong L_5(2)$ and the exterior square $Q_b^{[0]} \cong 2^{10}$. The latter is the kernel of the action of $G_b^{[0]}$ on $\mathcal{D}_b$, while $G_b^{[0]}/Q_b^{[0]} \cong L_b^{[0]}$ acts in the natural way. If b is the vertex x of Γ as in (7.5.1 (ii)) then $V_b = U_5$, $G_b^{[0]} = G^{[0]}$ etc.

Type 2: If $\overset{2}{c}$ is an element of type 2 then there is a Petersen graph Θ_c and a 3-dimensional $GF(2)$-vector space $Z_c^{[2]}$ such that

$$\mathcal{D}_c^{[3]} = \begin{bmatrix} Z_c^{[2]} \\ 1 \end{bmatrix}, \quad \mathcal{D}_c^{[0]} = V(\Theta_c),$$

and $\mathcal{D}_c^{[m]}$ is the set of antipodal triples of edges of Θ_c (considered also as 6-element subsets of $V(\Theta_c)$). Every element from $\mathcal{D}_c^{[3]}$ is incident to every element from $\mathcal{D}_c^{[0]} \cup \mathcal{D}_c^{[m]}$ while the incidence between the elements from $\mathcal{D}_c^{[0]}$ and the elements from $\mathcal{D}_c^{[m]}$ is via inclusion. The stabilizer

$$G_c^{[2]} \cong 2^{3+12} \cdot (L_3(2) \times \mathrm{Sym}_5)$$

is isomorphic to the pentad group. Furthermore, $Q_c^{[2]} = O_2(G_c^{[2]})$ is the kernel of the action of $G_c^{[2]}$ on $\mathcal{D}_c$; $N_c^{[2]} \cong 2^{3+12} : L_3(2)$ is the kernel of the action of $G_c^{[2]}$ on Θ_c. Let $L_c^{[2]} \cong L_3(2)$ be a complement to $Q_c^{[2]}$ in $N_c^{[2]}$ and let $S_c^{[2]} \cong \mathrm{Sym}_5$ be a complement to $N_c^{[2]}$ in $G_c^{[2]}$ (recall that $Q_c^{[2]}$ is not complemented in $G_c^{[2]}$). If c is the geometric cubic subgraph $\Gamma^{[2]}$ in Γ as in (7.5.1 (iv)), then $G_c^{[2]} = G^{[2]}$, $Q_c^{[2]} = Q^{[2]}$ etc.

Type 3: If $\overset{3}{d}$ is an element of type 3 then there is a Witt design $W_d^{[22]}$ of type $S(3, 6, 22)$ associated with d. If $\mathcal{O}_d$ is the set of octets, $\mathcal{H}_d$ is the set of hexads and $\mathcal{P}_d$ is the set of pairs in $W_d^{[22]}$ then

$$\mathcal{D}_d^{[0]} = \mathcal{O}_d, \quad \mathcal{D}_d^{[m]} = \mathcal{H}_d, \quad \mathcal{D}_d^{[2]} = \mathcal{P}_d$$

with the incidence relation as in the geometry $\mathcal{H}(M_{22})$. The stabilizer $G_d^{[3]}$ is of the form

$$G_d^{[3]} \cong 2_+^{1+12} \cdot 3 \cdot \mathrm{Aut}\,(M_{22})$$

and $N_d^{[3]} = O_{2,3}(G_d^{[3]})$ is the kernel of the action of $G_d^{[3]}$ on the residue $\mathcal{D}_d$. Let $M_d^{[3]} \cong 6 \cdot \mathrm{Aut}\,(M_{22})$ be the normalizer in $G_d^{[3]}$ of a Sylow 3-subgroup $Y_d^{[3]}$ in $N_d^{[3]}$ (compare (7.4.5)). Then $M_d^{[3]} \cap Q_d^{[3]} = Z_d^{[3]}$ and $Q_d^{[3]} M_d^{[3]} = G_d^{[3]}$ (where $Q_d^{[3]} = O_2(G_d^{[3]})$ and $Z_d^{[3]} = Z(Q_d^{[3]})$). If d is the geometric subgraph $\Gamma^{[3]}$ of valency 7 in Γ as in (7.5.1 (v)) then $G_d^{[3]} = G^{[3]}$, $N_d^{[3]} = N^{[3]}$ etc.

It is immediate from the above that $\mathcal{D}(G)$ belongs to the following diagram (cf. Section 10.4 for the definitions of the relevant rank 2 residues). Instead of types next to every node we indicate the structure of the corresponding stabilizer in G.

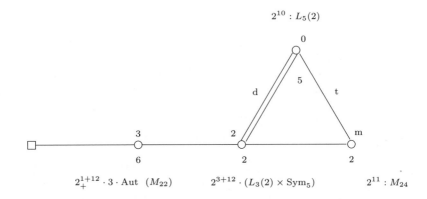

7.7 Intersections of maximal parabolics

Suppose that $\overset{i}{x}$ and $\overset{j}{y}$ are incident elements in $\mathcal{D}$. We require a clear understanding of the structure of the intersection $G_x^{[i]} \cap G_y^{[j]}$ in terms of the chief factors of $G_x^{[i]}$. This information, as summarized in lemmas below, is not so difficult to deduce, keeping in mind that

$$\mathcal{M} = \{G^{[m]}, G^{[0]}, G^{[2]}, G^{[3]}\}$$

is the amalgam of maximal parabolic subgroups associated with the action of G on $\mathcal{D}$.

The action of $G_a^{[m]}$ on $\mathcal{D}_a$ follows from the results in Sections 7.2 and 7.3, particularly from (7.2.2) and (7.3.3).

Lemma 7.7.1 *Let* $a \in \mathcal{D}^{[m]}$ *and let* $G_a^{[m]} \cong 2^{11} : M_{24}$ *be the stabilizer of* a *in* G. *Then*

    (i) *if* $b \in \mathcal{D}_a^{[0]}$ *then*
        (1) $b = (B, \alpha)$, *where* B *is an octad from* $\mathcal{B}_a$ *and* $\alpha \in \{0, 1\}$;

(2) *the subgroup* $M_a^{[m]}(b) \cong 2^4 : L_4(2)$ *stabilizes a unique hyperplane* $P_a(B)$ *in* $Q_a^{[m]}$;

(3) $G_a^{[m]} \cap G_b^{[0]} = P_a(B) : M_b^{[m]}(b)$;

(4) $Q_b^{[0]} = C_{P_a(B)}(O_2(M_a^{[m]}(b)))O_2(M_a^{[m]}(b)) \cong 2^{10}$;

(ii) *if* $c \in \mathcal{D}_a^{[2]}$ *then*

(5) c *is a trio from* $\mathcal{T}_a$;

(6) *the subgroup* $M_a^{[m]}(c) \cong 2^6 : (L_3(2) \times \mathrm{Sym}_3)$ *stabilises in* $Q_a^{[m]}$ *a unique subgroup* $R_a(c)$ *of index 4*;

(7) $G_a^{[m]} \cap G_c^{[2]} = Q_a^{[m]} : M_a^{[m]}(c)$;

(8) $Q_c^{[2]} = R_a(c)O_2(M_a^{[m]}(c)) \cong 2^{3+12}$.

(iii) *if* $d \in \mathcal{D}_a^{[3]}$ *then*

(9) d *is a sextet from* $\mathcal{S}_a$;

(10) *if* Y *is a Sylow 3-subgroup of* $O_{2,3}(M_a^{[m]}(d))$, *where* $M_a^{[m]}(d) \cong 2^6 : 3 \cdot \mathrm{Sym}_6$, *then* $Q_a^{[m]}/[Q_a^{[m]}, Y] \cong 2^5$;

(11) $G_a^{[m]} \cap G_b^{[3]} = Q_a : M_a^{[m]}(d)$;

(12) $Q_d^{[3]} = [Q_a^{[m]}, Y]O_2(M_a^{[m]}(d)) \cong 2_+^{1+12}$;

(13) Y *is a Sylow 3-subgroup of* $O_{2,3}(G_d^{[3]})$.

An element b of type 0 in $\mathcal{D}(G)$ is a vertex of the locally projective graph Γ. The edges containing b are in the natural bijection with the elements of type m incident to b in $\mathcal{D}(G)$. Therefore the action of $G_b^{[0]}$ on $\mathcal{D}_b$ is isomorphic to the action of $H^{[0]}$ on the corresponding residue in the dual polar space $\mathcal{O}^+(10, 2)$ (cf. (2.1.2), (2.1.3)).

Lemma 7.7.2 *Let* $b \in \mathcal{D}^{[0]}$ *and* $G_b^{[0]} \cong 2^{10} : L_5(2)$ *be the stabilizer of* b *in* G.

(i) *If* $x \in \mathcal{D}_a^{[i]}$ *for* $i = m$, 2, *or* 3 *then*

(1) x *is a subspace in* V_b *of dimension 4, 3, or 2, respectively;*

(2) $G_b^{[0]} \cap G_x^{[i]} = Q_b^{[0]} : L_b^{[0]}(x)$, *where* $L_b^{[0]}(x)$ *is isomorphic to*

$$2^4 : L_4(2), \quad 2^6 : (L_3(2) \times \mathrm{Sym}_3) \text{ and } 2^6 : (L_3(2) \times \mathrm{Sym}_3)$$

in the respective three cases;

(3) $Q_x^{[i]} \cap G_b^{[0]} = [Q_b^{[0]}, O_2(L_b^{[0]}(x))]O_2(L_b^{[0]}(x))$;

(4) *if* x *is of type* m *then* $O_2(G_x^{[m]})$ *intersects* $G_b^{[0]}$ *in a subgroup of index 2 in* $O_2(G_x^{[m]})$ *and* $Q_b^{[0]}/[Q_b^{[0]}, O_2(L_b^{[0]}(x))] \cong 2^4$;

(5) *the subgroup* $O_2(G_x^{[i]})$ *is contained in* $G_b^{[0]}$ *for* $i = 2$ *and* 3, *while* $Q_b^{[0]}/[Q_b^{[0]}, O_2(L_b^{[0]}(x))]$ *is isomorphic to 2 and* 2^3 *in the respective cases.*

The next result follows from the properties of the pentad group established in Sections 4.8 and 4.9.

Lemma 7.7.3 *Let* $c \in \mathcal{D}^{[2]}$ *and let* $G_c^{[2]} \cong 2^{3+12} \cdot (L_3(2) \cdot \mathrm{Sym}_5)$ *be the stabilizer of* c *in* G. *Then*

(i) *if* $a \in \mathcal{D}_c^{[m]}$ *then*
   (1) a *is an antipodal triple in the Petersen graph* Θ_c;
   (2) $S_c^{[2]}(a) \cong \mathrm{Sym}_4$;
   (3) $G_c^{[2]} \cap G_a^{[m]} = N_c^{[2]} S_c^{[2]}(a)$;
   (4) $Q_a^{[m]} = C_{Q_c^{[2]}}(O_2(S_c^{[2]}(a)))O_2(S_c^{[2]}(a))$;
   (5) $Z_c^{[2]} \le Q_a^{[m]}$;

(ii) *if* $b \in \mathcal{D}_c^{[0]}$ *then*
   (6) b *is a vertex of* Θ_c;
   (7) $S_c^{[2]}(b) \cong \mathrm{Sym}_3 \times 2$;
   (8) $G_c^{[2]} \cap G_b^{[0]} = N_c^{[2]} S_c^{[2]}(b)$;
   (9) $Q_b^{[0]} = C_{Q_c^{[2]}}(O_2(S_c^{[2]}(b)))O_2(S_c^{[2]}(b))$;

(iii) *if* $d \in \mathcal{D}_c^{[3]}$ *then*
   (10) d *is a 1-dimensional subspace in* $Z_c^{[2]}$;
   (11) $L_c^{[2]}(d) \cong \mathrm{Sym}_4$;
   (12) $G_c^{[2]} \cap G_d^{[3]} = Q_c^{[2]} S_c^{[2]} L_c^{[2]}(d)$;
   (13) $Q_d^{[3]} = C_{Q_c^{[2]}}(O_2(L_c^{[2]}(d)))O_2(L_c^{[2]}(d))$.

The structure of $G_d^{[3]}$ and its action on $\mathcal{D}_d$ follows from results in Sections 5.2, 7.3, and 7.4.

Lemma 7.7.4 *Let* $d \in \mathcal{D}^{[3]}$ *and let* $G_d^{[3]} \cong 2_+^{1+12} \cdot 3 \cdot \mathrm{Aut}\,(M_{22})$ *be the stabilizer of* d *in* G. *Then*

(i) *if* $a \in \mathcal{D}_d^{[m]}$ *then*
   (1) a *is a hexad from* $\mathcal{H}_d$;
   (2) $M_d^{[3]}(a) \cong 2^5 : 3 \cdot \mathrm{Sym}_6$;
   (3) $G_d^{[3]} \cap G_a^{[m]} = Q_d^{[3]} M_d^{[3]}(a)$;
   (4) $Q_a^{[m]} = C_{Q_d^{[3]}}(O_2(M_d^{[3]}(a)))O_2(M_d^{[3]}(a))$.

(ii) *if* $b \in \mathcal{D}_d^{[0]}$ *then*
   (5) b *is an octet from* $\mathcal{O}_d$;
   (6) $M_d^{[3]}(b) \cong 2 \times \mathrm{Sym}_3 \times 2^3 : L_3(2)$;
   (8) $G_d^{[3]} \cap G_b^{[0]} = Q_d^{[3]} M_d^{[3]}(b)$;
   (9) $Q_b^{[0]} = C_{Q_d^{[3]}}(O_2(M_d^{[3]}(b)))O_2(M_d^{[3]}(b)) \cong 2^{10}$;

(iii) *if* $c \in \mathcal{D}_d^{[2]}$ *then*
   (10) c *is a pair from* $\mathcal{P}_d$;
   (11) $M_d^{[3]}(c) \cong 2^6 : 3 : \mathrm{Sym}_5$;

(12) $G_d^{[3]} \cap G_c^{[2]} = Q_d^{[3]} M_d^{[3]}(c)$;

(13) $Q_c^{[2]} = [O_2(G_d^{[3]}), O_2(M_d^{[3]}(c))]O_2(M_d^{[3]}(c)) \cong 2^{3+12}$.

Exercises

1. Let $\mathcal{J} = \{G^{[0]}, G^{[1]}, G^{[2]}\}$ be the amalgam defined in Section 7.1. Show that the actions of $N_{G^{[1]}}(Q^{[m]})$ and $N_{G^{[2]}}(Q^{[m]})$ on $Q^{[m]}$ generate the Mathieu group M_{24}.

2. Show directly that the amalgam $\{F^{[m1]}/Q^{[m]}, F^{[m2]}/Q^{[m]}, F^{[m3]}/Q^{[m]}\}$ as in (7.3.3) is isomorphic to $\mathcal{A}(M_{24})$.

3. Give a computer-free proof of the simple connectedness of the rank 3 tilde geometry $\mathcal{G}(M_{24})$.

8

173,067,389-VERTEX GRAPH Δ

In this chapter we study the graph Δ on the set of elements of type m in $\mathcal{D}(G)$ (with stabilizers $2^{11} : M_{24}$) in which two vertices are adjacent whenever in $\mathcal{D}(G)$ they are incident to a common element of type 2 (whose stabilizer is the pentad group $2^{3+12} \cdot (L_3(2) \times \mathrm{Sym}_5)$). The ultimate goal is to show that the number of vertices in Δ is as given in the title of the chapter. The number of vertices gives the order of G; already half way through we establish the simplicity of G. Thus the chapter completes the proof of the Main Theorem. We follow section 8 in (Ivanov and Meierfrankenfeld 1999) which is mainly due to U. Meierfrankenfeld.

8.1 Defining the graph

The graph Δ we will study in this chapter is defined as follows: the vertices are the elements of type m in the geometry $\mathcal{D}(G)$; two such distinct elements are adjacent whenever in $\mathcal{D}(G)$ they are incident to a common element of type 2. Here G is an *arbitrary* faithful completion group of the rank 2 amalgam $\mathcal{G} = \{G^{[0]}, G^{[1]}\}$ which is constrained at level 2 (equivalently G is a faithful completion of the rank 3 amalgam $\mathcal{J} = \{G^{[0]}, G^{[1]}, G^{[2]}\}$); $\mathcal{D}(G)$ is the coset geometry of G defined as in Section 7.5.

We can define Δ in terms of the locally projective graph Γ of valency 31 as in (7.5.1). Let Σ be a (geometric) Petersen subgraph in Γ. Let e and f be two distinct edges of Σ which are *not* antipodal in the line graph of Σ. Then the connected components Φ_e and Φ_f of the local antipodality graph Φ of Γ (these components are elements of type m in $\mathcal{D}(G)$ and therefore they are vertices of Δ) are adjacent in Δ. Unfortunately our knowledge of the structure of Γ is not sufficient already for accomplishing the next obvious step: calculating the valency of Δ. We have to show that Σ is the only geometric Petersen subgraph in Γ whose edge-set intersects both Φ_e and Φ_f. Within our current knowledge this appears not at all obvious. This forces us to give up on Γ (at least for a time being) and make use of the structure of maximal parabolics in G as reviewed in Section 7.7. Notice that Δ is connected since Γ is such.

If X is a subset of elements of $\mathcal{D}$ consisting, say of $a, b, \ldots$ then the vertexwise stabilizer of X in G will be denoted by $G_{ab\ldots}$, while the set of elements of type i in $\mathcal{D}$ incident to every element in X will be denoted by $\mathcal{D}^{[i]}_{ab\ldots}$.

Proposition 8.1.1 *Let $a \in V(\Delta)$ and $e \in \Delta(a)$. Then*

(i) *the set $\mathcal{D}^{[2]}_{ae}$ contains a unique element, say c;*

(ii) *$N^{[2]}_c \le G_{ae}$ and $G_{ae}/N^{[2]}_c \cong \mathrm{Sym}_3$;*

(iii) $G_{ac} = Q_a^{[m]} G_{ace}$;

(iv) $Q_a^{[m]} \cap Q_e^{[m]} = Z_c^{[2]}$;

(v) $Q_a^{[m]} Q_e^{[m]} = Q_a^{[m]} O_2(M_a^{[m]}(c))$;

(vi) $Q_a^{[m]}$ acts on $\Delta(a)$ with orbits of length 4 indexed by the trios in $\mathcal{T}_a$;

(vii) the valency of Δ is

$$\Delta(a) = |\mathcal{T}_a| \cdot 4 = 3795 \cdot 4 = 15{,}180 = 2^2 \cdot 3 \cdot 5 \cdot 11 \cdot 23.$$

Proof Since a and e are adjacent vertices of Δ, by the definition there is at least one $c \in \mathcal{D}_{ae}^{[2]}$. The set $\mathcal{D}_c^{[m]}$ consists of five vertices (including a and e). The group $G_c^{[2]}$ induces on $\mathcal{D}_c^{[m]}$ the natural action of Sym_5 with kernel $N_c^{[2]}$. Since $Q_a^{[m]}$ stabilizes c, it stabilizes $\mathcal{D}_c^{[m]}$ as a whole. By (7.7.3 (i) (4)) $Q_a^{[m]}$ induces on $\mathcal{D}_c^{[m]}$ an elementary abelian group of order 4 which permutes regularly the vertices in $\mathcal{D}_c^{[m]} \backslash \{a\}$ with kernel $R_a(c)$. Thus $Q_a^{[m]}$ acts on $\Delta(a)$ with orbits of length 4 and $R_a(c)$ is the kernel at one of the orbits. The subgroup $R_a(c)$ is normalized by

$$M_a^{[m]}(c) \cong 2^6 : (L_3(2) \times \mathrm{Sym}_3)$$

and the latter is a maximal subgroup in $M_a^{[m]} \cong M_{24}$ by (11.2.3 (i)). Therefore $G_{ac} = Q_a^{[m]} M_a^{[m]}(c)$ is the normalizer of $R_a(c)$ in $G_a^{[m]}$. Since $Q_a^{[m]} \cap G_e^{[m]} = Q_a^{[m]} \cap Q_c^{[2]}$ we have

$$G_{ae} \leq N_{G_a^{[m]}}(R_a(c)) = G_{ac}.$$

Therefore G_{ae} stabilizes c. Since $G_c^{[2]}$ acts on $\mathcal{D}_c^{[m]}$ doubly transitively, the action of G_{ae} on $\mathcal{D}_a^{[2]} \cap \mathcal{D}_e^{[2]}$ is transitive. Hence c is unique as claimed in (i). Statements (ii) and (iii) follow from the above discussions.

The subgroup $Q_a^{[m]} \cap Q_e^{[m]}$ is normalized by both $Q_a^{[m]}$ and $Q_e^{[m]}$. On the other hand, the actions of $Q_a^{[m]}$ and $Q_e^{[m]}$ on $\mathcal{D}_c^{[m]}$ generate Alt_5. It follows from the structure of the pentad group (4.7.4) that Alt_5 permutes transitively the dents in $Q_c^{[2]}$, hence the subgroup $Q_a^{[m]} \cap Q_e^{[m]}$ is contained in $Z_c^{[2]}$. By (7.7.3 (i) (5)) $Z_c^{[2]}$ is contained in $Q_c^{[2]} \cap Q_x^{[m]}$ for $x = a$ and e. Therefore (iv) follows and immediately implies (v). Now (vi) and (vii) are rather straightforward. ∎

The next easy lemma describes the subgraph in Δ induced by the vertices incident to a given element of $\mathcal{D}$.

Lemma 8.1.2 *Let c, b, and d be elements in $\mathcal{D}$ of type 2, 0, and 3, respectively. Then*

(i) *the subgraph in Δ induced by $\mathcal{D}_c^{[m]}$ is complete on 5 vertices;*

(ii) *the subgraph in Δ induced by $\mathcal{D}_b^{[m]}$ is complete on 31 vertices;*

(iii) *the subgraph in Δ induced by $\mathcal{D}_d^{[m]}$ is isomorphic to the hexad graph on 77 vertices.*

Proof Statement (i) follows directly from the definition. The element b is a vertex of Γ. It is incident to 31 edges. Any pair of these edges is contained in a geometric Petersen subgraph. Therefore the connected components of the local antipodality graph containing these edges as vertices are pairwise adjacent as vertices of Δ which gives (ii).

By (7.7.4 (ii)) the set $\mathcal{D}_d^{[m]}$ can be identified with the vertex-set of the hexad graph Ξ as defined before (11.4.5). The group $G_d^{[3]}$ induces the natural action of

$$\text{Aut}\,(M_{22}) \cong G_d^{[3]}/N_d^{[3]} \cong M_d^{[3]}/O_{2,3}(M_d^{[3]})$$

on Ξ. It is immediate from the diagram of $\mathcal{D}(G)$ that, under the above identification, two vertices from $\mathcal{D}_d^{[m]}$ are adjacent in Δ whenever they are adjacent in Ξ. Thus it only remains to show that two vertices from $\mathcal{D}_d^{[m]}$ are not adjacent in Δ if they are at distance 2 in Ξ. Let $a \in \mathcal{D}_d^{[m]}$. Then by (7.7.4 (2)) $M_d^{[3]}(a) \cong 2^5 : 3 \cdot \text{Sym}_6$ induces on $\mathcal{D}_d^{[m]}$ an action isomorphic to $2^4 : \text{Sym}_6$. By (11.4.5) $O_2(M_d^{[3]}(a))$ permutes transitively the 16 vertices in $\Xi_2(a)$. By (7.7.4 (4)) $Q_a^{[m]}N_d^{[3]} = N_d^{[3]}O_2(M_d^{[3]}(a))$. Therefore $Q_a^{[m]}$ acts transitively on $\Xi_2(a)$ of size 16. By (8.1.1 (vi)) this implies that $\Xi_2(a) \not\subseteq \Delta(a)$ and (iii) follows. ∎

In terms of the proof of (8.1.2 (iii)) $\Xi_2(a) \subseteq \Delta_2(a)$ and as a byproduct we obtain the following.

Proposition 8.1.3 *Let $\Delta_2^1(a)$ be the set of vertices in Δ which are incident with a to a common element of type 3 but not to a common element of type 2 and let $h \in \Delta_2^1(a)$. Then*

(i) $\Delta_2^1(a)$ *is a $G_a^{[m]}$-orbit on the set of vertices at distance 2 in Δ from the vertex a;*

(ii) *the set $\mathcal{D}_{ah}^{[3]}$ contains a unique element, say d;*

(iii) $Q_a^{[m]}$ *acts on $\Delta_2^1(a)$ with orbits of length 16 indexed by the sextets in $\mathcal{S}_a$;*

(iv) $|\Delta_2^1(a)| = |\mathcal{S}_a| \cdot 16 = 1771 \cdot 16 = 28,336 = 2^4 \cdot 7 \cdot 11 \cdot 23$;

(v) $G_{ah} \cong 2_+^{1+12} . 3 \cdot \text{Sym}_6$;

(vi) $Q_a^{[m]} \cap Q_h^{[m]} = Z(G_{ah}) \cong 2$.

Proof Statement (i) follows directly from (8.1.2 (iii)). Since $M_a^{[m]}(d) \cong 2^6 : 3 \cdot \text{Sym}_6$ is maximal in $M_a^{[m]} \cong M_{24}$ by (11.2.3 (i)) and $M_d^{[m]} \cong 6 \cdot \text{Aut}\,(M_{22})$ acts transitively on the set of ordered pairs of non-adjacent vertices in the hexad graph Ξ, one can argue as in the proof of (8.1.1 (i)) to establish (ii). Since $Q_a^{[m]}$ stabilizes d, (iii) follows from (7.7.4 (4)). Now (iv) and (v) are direct consequences of the previous assertions. In order to establish (vi) we apply (7.7.1 (12)) and (7.7.4 (4)). ∎

8.2   The local graph of Δ

In this section we identify the subgraph in Δ induced by the set $\Delta(a)$ of vertices adjacent to a given vertex a. We start by describing the action of $G_a^{[m]} \cong 2^{11}$: M_{24} on $\Delta(a)$.

Lemma 8.2.1 *Let $c \in \mathcal{D}_a^{[2]}$. Then for some $e \in \mathcal{D}_c^{[m]} \backslash \{a\}$ we have the equality*

$$G_{ae} = R_a(c)M_a^{[m]}(c)$$

(here $R_a(c)$ is the only subgroup of index 4 in $Q_a^{[m]}$ normalized by $M_a^{[m]}(c) \cong 2^6$: $(L_3(2) \times Sym_3)$).

Proof By (7.7.1 (7)) $G_{ac} = Q_a^{[m]}M_a^{[m]}(c)$. It is easy to deduce from the shape of the chief factors of G_{ac} (cf. (11.3.1 (ii))) that $R_a(c)M_a^{[m]}(c)$ represents the only conjugacy class of subgroups of index 4 in G_{ac}. Since the latter group permutes transitively the four vertices in $\mathcal{D}_c^{[m]} \backslash \{a\}$, the assertion follows.   ∎

Let $\Lambda(a)$ be the set of pairs (T, β), where $T = \{B_1, B_2, B_3\}$ is a trio from $\mathcal{T}_a$ and β is $GF(2)$-valued function on T, whose support is even. Notice that if T is treated as the set of non-zero vectors of a non-singular symplectic 2-space, then β is an associated quadratic form.

Define an action of $G_a^{[m]}$ on $\Lambda(a)$ by the following rule. If $g \in M_a^{[m]}$ and $T^g = \{B_1^g, B_2^g, B_3^g\}$ is the image of T under g then

$$g : (T, \beta) \mapsto (T^g, \beta^g),$$

where $\beta^g(B_i^g) = \beta(B_i)$ for $1 \leq i \leq 3$. If $q \in Q_a^{[m]}$ then

$$q : (T, \beta) \mapsto (T, \beta^q),$$

where $\beta^q(B_i) = \beta(B_i)$ if and only if $q \in P_a(B_i)$ (where $P_a(B_i)$ is the only hyperplane in $Q_a^{[m]}$ stabilized by $M_a^{[m]}(B_i) \cong 2^4 : L_4(2)$). Since

$$R_a(T) = P_a(B_1) \cap P_a(B_2) \cap P_a(B_3),$$

it is straightforward to check that in this way we obtain a faithful transitive action of $G_a^{[m]}$ on $\Lambda(a)$.

Lemma 8.2.2 *The actions of $G_a^{[m]}$ on $\Delta(a)$ and $\Lambda(a)$ are permutation isomorphic.*

Proof It is clear that $|\Lambda(a)|$ is the number of trios in $\mathcal{T}_a$ times the number of quadratic forms in a 2-space associated with a given (non-singular) symplectic form, which is 4. By (8.1.1 (vii)) this means that

$$|\Delta(a)| = |\Lambda(a)|.$$

Thus it only remains to show that the subgroup G_{ae} as in (8.2.1) stabilizes an element from $\Lambda(a)$. Let c be the unique element in $\mathcal{D}_a^{[2]} \cap \mathcal{D}_e^{[2]}$ treated as a trio from $\mathcal{T}_a$ and let β_1 be the function taking value 1 on every octad from c. Then it is straightforward that (c, β_1) is an element of $\Lambda(a)$ stabilized by G_{ae}. ∎

In what follows $\lambda_a : \Delta(a) \to \Lambda(a)$ denotes the bijection commuting with the action of $G_a^{[m]}$ whose existence is guaranteed by (8.2.2).

Lemma 8.2.3 *Let* $a \in \mathcal{D}^{[m]}$ *and let* B *be an octad from* $\mathcal{B}_a$. *Let* $b^{(\varepsilon)} = (B, \varepsilon)$, $\varepsilon = 0$, 1 *be the corresponding elements from* $\mathcal{D}_a^{[2]}$. *Then (up to transposing* $b^{(0)}$ *and* $b^{(1)}$) *the following holds:*

$$\mathcal{D}_{b^{(\varepsilon)}}^{[m]} = \{a\} \cup \{\lambda_a^{-1}(T, \beta) \mid B \in T, \beta(B) = \varepsilon\}.$$

Proof Let $e \in \mathcal{D}_{b^{(\varepsilon)}}^{[m]}$ for $\varepsilon = 0$ or 1, let c be the unique element from $\mathcal{D}_a^{[2]} \cap \mathcal{D}_e^{[2]}$ and let $\lambda_a(e) = (T, \beta)$. Then $B \in T$, $P_a(B)$ contains $R_a(c)$ with index 2 and hence $P_a(B)$ acts on $\mathcal{D}_c^{[m]} \setminus \{a\}$ with two orbits of length 2 and these orbits are distinguished by the value of $\beta(B)$. Let $L \cong L_4(2)$ be a complement to $O_2(M_a^{[m]}(B))$ in $M_a^{[m]}(B) \cong 2^4 : L_4(2)$. Then L permutes transitively the trios containing B. On the other hand, if $l \in L$ then $\beta^l(B^l) = \beta(B)$ since $L \leq M_a^{[m]}$. The subgroup $P_a(B)L$ permutes transitively the vertices in $\mathcal{D}_{b^{(\varepsilon)}}^{[m]} \setminus \{a\}$ and therefore the result follows. ∎

Lemma 8.2.4 *Let* $h \in \mathcal{D}_{b^{(0)}}^{[m]} \setminus \{a\}$, $k \in \mathcal{D}_{b^{(1)}}^{[m]} \setminus \{a\}$. *Suppose that* h *and* k *are adjacent in* Δ. *Then there is a trio* T *in* $\mathcal{T}_a$ *such that*

$$\lambda_a(h) = (T, \beta^{(h)}), \quad \lambda_a(k) = (T, \beta^{(k)}),$$

(this is equivalent to the claim that the unique element from $\mathcal{D}_{hk}^{[2]}$ *is incident to* a).

Proof Let $\{c\} = \mathcal{D}_{ah}^{[2]}$ and let $\{f\} = \mathcal{D}_{ak}^{[2]}$. Clearly h and k are adjacent whenever $c = f$, so suppose that $c \neq f$. In this case c and f are distinct trios in $\mathcal{T}_a$ sharing the octad B. Therefore these trios are refined by a common sextet which means that there is $d \in \mathcal{D}_c^{[3]} \cap \mathcal{D}_f^{[3]}$. By (11.4.6) and (8.1.2 (iii)) h is adjacent to only two vertices from $\mathcal{D}_f^{[m]} \setminus \{a\}$ and these two vertices are in $\mathcal{D}_{b^{(0)}}^{[m]}$, so k is not among them. ∎

It appears useful to interpret the situation in terms of the locally projective graph Γ of valency 31. The element a is a connected component of the local antipodality graph Φ on the edge-set of Γ. The edges of Γ in this connected component are indexed by the octads from $\mathcal{B}_a$. Thus B is an edge in Γ, say $\{x, y\}$. Then (up to renaming) we have $b^{(0)} = x$ and $b^{(1)} = y$, while $\mathcal{D}_{b^{(0)}}^{[m]}$ and $\mathcal{D}_{b^{(1)}}^{[m]}$ are the connected components of the local antipodality graph containing the edges incident to x and y, respectively. Therefore h is the connected component of the local antipodality graph containing $\{x, u\}$ for some $u \in \Gamma(x) \setminus \{y\}$ and k is a similar component containing $\{y, v\}$ for some $v \in \Gamma(y) \setminus \{x\}$. If h and k are adjacent in Δ then there is a Petersen subgraph Σ whose edge-set intersects both h and k. If Σ contain $\{x, u\}$ and $\{y, v\}$ then (because Γ does not contain triangles) Σ is forced to contain $\{x, y\}$ as on the figure below.

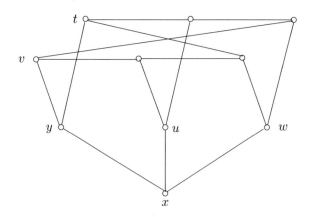

By (8.2.4) whenever Σ intersects h and k it necessarily contains $\{x, u\}$ and $\{y, v\}$, which is not at all obvious at first glance.

From now on we start using the notation for paths in $\mathcal{D}$ as introduced in Section 7.6. Consider a path

$$\overset{m}{(e}, \overset{2}{c}, \overset{m}{a}, \overset{2}{f}, \overset{m}{h})$$

in $\mathcal{D}$. Then c and f are distinct elements from $\mathcal{D}_a^{[2]}$, $e \in \mathcal{D}_c^{[m]}\backslash\{a\}$, and $h \in \mathcal{D}_f^{[m]}\backslash\{a\}$. Let us explain the way we are going to use the diagrams for $\mathcal{H}(M_{24})$ and $\mathcal{H}(M_{22})$. The elements c and f are trios in $\mathcal{T}_a$. By the diagram $D_t(M_{24})$ we see that the subgroup

$$M_a^{[m]}(c) \cong 2^6 : (L_3(2) \times \mathrm{Sym}_3)$$

has four orbits on $\mathcal{T}_a\backslash\{c\}$ with lengths

$$2 \cdot 21, \quad 2^2 \cdot 14, \quad 2^4 \cdot 63, \quad \text{and} \quad 2^6 \cdot 42.$$

Since the orbit lengths are pairwise different we can and will refer to an orbit via its length. When the orbit length is given in the form $2^l \cdot k$, the subgroup $O_2(M_a^{[m]}(c)) \cong 2^6$ acts on this orbit with k orbits of length 2^l each. Depending on the (length of the) $M_a^{[m]}(c)$-orbit containing f we write that

$$\angle caf = 2 \cdot 21, \quad 2^2 \cdot 14, \quad 2^4 \cdot 63, \quad \text{and} \quad 2^6 \cdot 42$$

in the respective cases.

Lemma 8.2.5 *Let* $\overset{m}{(e}, \overset{2}{c}, \overset{m}{a}, \overset{2}{f}, \overset{m}{h})$ *be a path in* $\mathcal{D}$. *Then* $Q_a^{[m]}(c) = R_a(c)$ *acts transitively on* $\mathcal{D}_f^{[m]}\backslash\{a\}$ *unless* $\angle caf = 2 \cdot 21$, *in which case there are two orbits* *of length* 2. *Furthermore, if* $\lambda_a(e) = (T^{(e)}, \beta^{(e)})$ *and* $\lambda_a(h) = (T^{(h)}, \beta^{(h)})$ *then the following assertions hold:*

(i) *if* $\angle caf = 2^4 \cdot 63$ *or* $2^6 \cdot 42$ *then* e *and* h *are not adjacent in* Δ;

(ii) *if* $\angle caf = 2 \cdot 21$ *then* e *and* h *are adjacent in* Δ *if and only if* $\beta^{(e)}(B) = \beta^{(h)}(B)$, *where* $B = T^{(e)} \cap T^{(h)}$;

(iii) *if* $\angle caf = 2^2 \cdot 14$ *then* e *and* h *are adjacent in* Δ.

Proof Since $N_{G_a^{[m]}}(R_a(c)) = G_a^{[m]}(c)$ (because of the maximality of $M_a^{[m]}(c) \cong$ $2^6 : (L_3(2) \times \mathrm{Sym}_3)$ in $M_a^{[m]} \cong M_{24}$ by (11.2.3 (i))), we have $R_a(c) \neq R_a(f)$. Therefore the action of $R_a(c)$ on $\mathcal{D}_f^{[m]} \setminus \{a\}$ is non-trivial. The three hyperplanes in $Q_a^{[m]}$ containing $R_a(c)$ are $P_a(B_1)$, $P_a(B_2)$ and $P_a(B_3)$ where $c = \{B_1, B_2, B_3\}$ (a trio). The subgroup $R_a(f)$ is contained in one of these hyperplanes if and only if the trios c and f intersect. Since the latter takes place only if $\angle caf = 2 \cdot 21$ (compare the diagram $D_t(M_{24})$) the first assertion follows.

By (8.1.1 (vi)) h is adjacent to e in Δ only if the orbit of h under $Q_e^{[m]}$ has length 4. On the other hand, $Q_a^{[m]} Q_e^{[m]} = Q_a^{[m]} O_2(M_a^{[m]}(c))$ by (7.7.1 (7)). Since f is the only element in $\mathcal{D}_{ah}^{[2]}$, the orbit of h under $Q_e^{[m]}$ is at least as long as the orbit of f under $Q_e^{[m]}$. By $D_t(M_{24})$ the orbit of f under $O_2(M_a^{[m]}(c))$ has length 2, 2^2, 2^4, and 2^6 when $\angle caf = 2 \cdot 21$, $2^2 \cdot 14$, $2^4 \cdot 63$, and $2^6 \cdot 42$, respectively. This immediately implies (i).

If $\angle caf = 2 \cdot 21$ then the trios $T^{(e)}$ and $T^{(h)}$ share an octad B, say. Now (ii) is immediate from (8.2.3).

If $\angle caf = 2^2 \cdot 14$ then $T^{(e)}$ and $T^{(h)}$ are disjoint but contained in a common quad in the octad graph on $\mathcal{O}_a$ (equivalently there is a sextet which refines both $T^{(e)}$ and $T^{(h)}$). Therefore there is an element $d \in \mathcal{D}^{[3]}$ incident to both c and f. On the diagram of $\mathcal{D}$ the nodes 3 and m are disjoint, therefore e and h are also contained in $\mathcal{D}_d^{[m]}$. By (8.1.2 (iii)) the latter subgraph is isomorphic to the hexad graph Ξ as in (11.4.5). In terms of Ξ the subgraphs $\mathcal{D}_c^{[m]}$ and $\mathcal{D}_f^{[m]}$ correspond to intersecting pairs in the hexad corresponding to a. Therefore (iii) follows from (11.4.6). ∎

We summarize (8.2.3) and (8.2.5) in the following.

Corollary 8.2.6 *Let* $\Lambda(a)$ *be the set of pairs* (T, β), *where* T *is a trio from* $\mathcal{T}_a$ *and* β *is a* $GF(2)$-*valued function on* T *whose support is even. Let* $G_a^{[m]}$ *act on* $\Lambda(a)$ *as described before (8.2.2). Then there is a bijection* $\lambda_a : \Delta(a) \to \Lambda(a)$ *which commutes with the action of* $G_a^{[m]}$. *Let* e *and* h *be distinct vertices from* $\Delta(a)$, *let* $\lambda_a(e) = (T^{(e)}, \beta^{(e)})$ *and let* $\lambda_a(h) = (T^{(h)}, \beta^{(h)})$. *Then* e *and* h *are adjacent in* Δ *if and only if one of the following holds:*

(i) $T^{(e)} = T^{(h)}$;

(ii) $T^{(e)}$ *and* $T^{(h)}$ *share an octad* B *and* $\beta^{(e)}(B) = \beta^{(h)}(B)$;

(iii) $T^{(e)}$ *and* $T^{(h)}$ *are disjoint but are refined by a common sextet.*

8.3   Distance two neighbourhood

In this section we study the set of vertices at distance 2 in Δ from a given vertex.

Lemma 8.3.1 *Let* $\pi = (\overset{m}{a}, \overset{2}{c}, \overset{m}{e}, \overset{2}{f}, \overset{m}{h})$ *be a path in* $\mathcal{D}$ *such that* $d_\Delta(a, h) = 2$. *Then the* G-*orbit containing* π *is uniquely determined by* $\angle cef$ *which is* $2 \cdot 21$, $2^4 \cdot 63$, *or* $2^6 \cdot 42$.

Proof The assertion is a direct consequence of (8.2.5).   ∎

Lemma 8.3.2 *Let* $\Delta_2^1(a)$ *be defined as in* (8.1.3). *Let* $\Delta_2^2(a)$ *and* $\Delta_2^3(a)$ *be the sets of vertices* h *in* $\mathcal{D}$ *joint with* a *by paths* $(\overset{m}{a}, \overset{2}{c}, \overset{m}{e}, \overset{2}{f}, \overset{m}{h})$ *with* $\angle cef = 2^4 \cdot 63$ *and* $2^6 \cdot 42$, *respectively. Then*

- (i) $G_a^{[m]}$ *acts transitively on* $\Delta_2^2(a)$ *and on* $\Delta_2^3(a)$;
- (ii) *the orbits of* $Q_a^{[m]}$ *on* $\Delta_2^3(a)$ *are of length* 2^{10} *while those on* $\Delta_2^2(a)$ *are of length at most* 2^8;
- (iii) $\Delta_2^1(a)$, $\Delta_2^2(a)$ *and* $\Delta_2^3(a)$ *are the (pairwise distinct) orbits of* $G_a^{[m]}$ *on* $\Delta_2(a)$.

Proof It follows from (8.1.3), (8.2.5), and (8.2.6) that every vertex at distance 2 from a in Δ is contained in $\Delta_2^i(a)$ for $i = 1$, 2, or 3. By (8.3.1) $G_a^{[m]}$ acts transitively on each of the $\Delta_2^i(a)$'s. We have to show that the three orbits are pairwise different. We accomplish this by estimating the lengths of the orbits of $Q_a^{[m]}$ on each of the $\Delta_2^i(a)$'s. By (8.1.3 (iii)) the orbits of $Q_a^{[m]}$ on $\Delta_2^1(a)$ are of length 16. Let 2^l be the length of the orbit of f under $O_2(M_a^{[m]}(c))$. Then $2^l = 2^4$ and $2^l = 2^6$ for $h \in \Delta_2^2(a)$ and $h \in \Delta_2^3(a)$, respectively. By (8.1.1 (v)) the length of the orbit of h under $Q_a^{[m]}$ is at least 2^l and at most

$$2^l \cdot |\mathcal{D}_c^{[m]} \setminus \{a\}| \cdot |\mathcal{D}_f^{[m]} \setminus \{e\}| = 2^l \cdot 4 \cdot 4 = 2^{l+4}.$$

This immediately shows that $\Delta_2^1(a)$ is different from the remaining two orbits while the upper bounds for the lengths are 2^8 and 2^{10} for $\Delta_2^2(a)$ and $\Delta_2^3(a)$. We claim that in the latter case the upper bound is attained (later we will see that in the former case the upper bound is not attained (8.3.15 (ii))).

By (8.1.1 (iv)) $Q_a^{[m]} \cap Q_e^{[m]} = Z_c^{[2]}$. We are going to show that the latter group acts transitively on $\mathcal{D}_f^{[m]} \setminus \{e\}$ provided that $\angle cef = 2^6 \cdot 42$. Let B_1, B_2, and B_3 be the trios constituting f (here f is considered as a trio from $\mathcal{T}_e$). We have to show that $Z_c^{[2]} \not\leq P_e(B_i)$ for $i = 1$, 2, and 3. Since $Z_c^{[2]}$ is of order 2^3, it is contained in $2^8 - 1$ hyperplanes of $Q_e^{[m]}$. On the other hand, we can see from the diagram $D_t(M_{24})$ that every octad B_i is from an orbit of length $2^5 \cdot 21$ of $M_e^{[m]}(c)$ on $\mathcal{B}_e$. Since $2^5 \cdot 21 > 2^8 - 1$ we obtain the required transitivity. Therefore the second factor 4 in the above upper bound is attained.

By (8.2.6) e is the only vertex in $\mathcal{D}_c^{[m]}$ adjacent to h. Since $Q_a^{[m]}$ acts transitively on $\mathcal{D}_c^{[m]}\backslash\{a\}$, the first factor 4 in the upper bound is also attained and the result follows. ■

The following lemma is a direct consequence of (8.3.1) and (8.3.2).

Lemma 8.3.3 *If $h \in \Delta_2(a)$ then G_{ah} acts transitively on $\Delta(a) \cap \Delta(h)$.*

For $h \in \Delta_2^i(a)$ put $\mu_i(a, h) = \Delta(a) \cap \Delta(h)$ and

$$\mu_i = |\mu_i(a, h)|.$$

Then μ_i is precisely the number of paths π as in (8.3.1) for a fixed $h \in \Delta_2^i(a)$. Thus, as a direct consequence of (8.2.5) and (8.3.2) we obtain the following.

Lemma 8.3.4 *Let $h \in \Delta_2^i(a)$ for $i = 1$, 2 or 3. Let $(\overset{m}{a}, \overset{2}{c}, \overset{m}{e}, \overset{2}{f}, \overset{m}{h})$ be a path in $\mathcal{D}$ and let $\delta = 2$ if $i = 1$ and $\delta = 4$ if $i = 2$ or 3. Then*

$$|\Delta_2^i(a)| = \frac{1}{\mu_i}(|\Delta(a)| \cdot \angle cef \cdot \delta),$$

in particular

$$|\Delta_2^2(a)| = \frac{1}{\mu_2}(2^8 \cdot 3^3 \cdot 5 \cdot 7 \cdot 11 \cdot 23);$$

$$|\Delta_2^3(a)| = \frac{1}{\mu_3}(2^{11} \cdot 3^2 \cdot 5 \cdot 7 \cdot 11 \cdot 23).$$

Proof By (8.2.5) δ is the length of the orbit of h under $R_e(c)$ on $\mathcal{D}_f^{[m]}\backslash\{e\}$. The rest is easy counting. ■

Since $|\Delta_2^1(a)|$ is given by (8.1.3 (iv)), in view of (8.3.4) the value of μ_1 can easily be calculated.

Lemma 8.3.5 *Let $h \in \Delta_2^1(a)$ and let $d \in \mathcal{D}_{ah}^{[3]}$. Then*

(i) $\Delta(a) \cap \Delta(h) \subseteq \mathcal{D}_d^{[m]}$;
(ii) $\mu_1 := |\Delta(a) \cap \Delta(h)| = 45$.

Proof By (8.1.3 (ii)) d is uniquely determined and hence it is stabilized by G_{ah}. Since $\Delta(a) \cap \Delta(h)$ intersects $\mathcal{D}_d^{[m]}$ and by (8.3.3) G_{ah} acts transitively on $\Delta(a) \cap \Delta(h)$, (i) follows. By (8.1.2 (iii)) the assertion (ii) is immediate from the intersection diagram of the hexad graph (cf. the paragraph before (11.4.5)). ■

We will calculate μ_2 and μ_3 by showing that for $i = 2$ and 3 a vertex $h \in \Delta_2^i(a)$ can be reached from a along a path in $\mathcal{D}$, satisfying a uniqueness criteria in the sense that for every h there is a unique path of that type. In order to accomplish this we start by analysing various 2-paths in $\mathcal{D}$.

8.3.1   *Analysing 2-paths*

Let $(\overset{i}{x}, \overset{k}{z}, \overset{j}{y})$ be a 2-path in $\mathcal{D}$. This means that both x and y are incident to z, but x and y are not incident. We would like to analyse the structure of G_{xy}. Since both x and y are in the residue of z in $\mathcal{D}$ the success of the analysis depends on how easily we find our way in the residues of $\mathcal{D}$.

The residue of $\overset{m}{a}$ in $\mathcal{D}$ is closely related to the locally truncated geometry $\mathcal{H}(M_{24})$ of the Mathieu group M_{24}. Only for every octad in $\mathcal{B}_a$ we have got two elements in $\mathcal{D}_a^{[0]}$, which are $(B, 0)$ and $(B, 1)$ and they are transposed by elements $Q_a^{[m]}$. Keeping in mind this remark we can use the diagrams for $\mathcal{H}(M_{24})$ presented in Section 11.5. These diagrams show the orbits of the stabilizer $M_a^{[m]}(x)$ of an element x in $\mathcal{H}(M_{24})$ on the remaining elements of $\mathcal{H}(M_{24})$. The length of every orbit is given in the form $2^l \cdot k$, which means that $O_2(M_a^{[m]}(x))$ has k orbits of length 2^l each (if $l = 0$ the first factor is dropped). In particular the diagram $D_b(M_{24})$ shows the orbits of $M_a^{[m]}(b) \cong 2^4 : L_4(2)$ where $b \in \mathcal{D}_a^{[0]}$. It is immediate from this diagram that the elements in $\mathcal{D}_a$ fixed by $O_2(M_a^{[m]}(b))$ are exactly those which are incident to b. Similarly for $\{i, j\} = \{2, 3\}$ and for $\overset{i}{x} \in \mathcal{D}_a^{[i]}$ the elements in $\mathcal{D}_a^{[j]}$ fixed by $O_2(M_a^{[m]}(x))$ are exactly those incident to x. These observations can be stated in the following two lemmas.

Lemma 8.3.6 *Let* $a \in \mathcal{D}^{[m]}$, *let* $b \in \mathcal{D}_a^{[0]}$ *and let* $\overset{i}{x} \in \mathcal{D}_a^{[i]}$, *where* $i = 2$ *or* 3. *Then* x *is incident to* b *if and only if there is a subgroup* $R \leq O_2(G_{ab})$ *such that* R *stabilizes* x *and* $Q_a^{[m]}R = Q_a^{[m]}O_2(M_a^{[m]}(b))$.

Lemma 8.3.7 *Let* $a \in \mathcal{D}^{[m]}$, *let* $\{i, j\} = \{2, 3\}$ *and let* $\overset{i}{x}, \overset{j}{y} \in \mathcal{D}_a$. *Then* x *is incident to* y *if and only if there is a subgroup* $R \leq O_2(G_{ax})$ *such that* R *stabilizes* y *and* $Q_a^{[m]}R = Q_a^{[m]}O_2(M_a^{[m]}(x))$.

The next lemma supplies further conditions for elements in $\mathcal{D}_a$ to be incident.

Lemma 8.3.8 *Let* $a \in \mathcal{D}^{[m]}$ *and let* $\overset{0}{b}, \overset{2}{c}, \overset{3}{d} \in \mathcal{D}_a$. *Then the following assertions hold:*

(i) *for* $\overset{i}{x} = \overset{2}{c}$ *or* $\overset{3}{d}$ *the elements* x *and* b *are incident*
   (1) *if and only if* $Z_x^{[i]} \leq Q_b^{[0]} \cap Q_a^{[m]}$;
   (2) *if and only if* $Q_x^{[i]} \cap Q_a^{[m]} \leq G_b^{[0]} \cap Q_a^{[m]}$;
(ii) *the elements* c *and* d *are incident*
   (3) *if and only if* $Z_d^{[3]} \leq Z_c^{[2]}$;
   (4) *if and only if* $Q_d^{[3]} \cap Q_a^{[m]} \leq Q_c^{[2]} \cap Q_a^{[m]}$;
(iii) $Z_d^{[3]} \leq Q_c^{[2]}$ *if and only if* $\angle cad \neq 2^6 \cdot 21$.

Proof The *only if* statements in (i) and (ii) follow from (7.7.1), (7.7.2), (7.7.3), and (7.7.4). With x as in (i) suppose that $Z_x^{[i]} \leq Q_b^{[0]} \cap Q_a^{[m]}$. Then $Q_b^{[0]}$ centralizes

$Z_x^{[i]}$. Since G_{ax} is maximal in $M_a^{[m]}$ by (11.2.3 (i)), we have

$$Q_b^{[0]} \leq N_{G_a^{[m]}}(Z_x^{[i]}) = G_{ax}$$

and hence $Q_b^{[0]}$ stabilizes x. By (7.7.1 (7))

$$Q_b^{[0]}Q_a^{[m]} = O_2(G_{ab}Q_a^{[m]})$$

and hence $O_2(G_{ab}Q_a^{[m]})$ stabilizes x. Since

$$O_2(G_{ab})Q_a^{[m]} = Q_a^{[m]}O_2(M_a^{[m]}(b)),$$

by (8.3.6) x and b are incident which proves (1).

Suppose now that $Q_x^{[i]} \cap Q_a^{[m]} \leq G_b^{[0]} \cap Q_a^{[m]}$. Since $[Q_a^{[m]}, Q_x^{[i]}] \leq Q_a^{[m]} \cap Q_x^{[i]}$, we conclude that $Q_x^{[i]}$ normalizes $G_2^{[i]} \cap Q_a^{[m]}$, and so $Q_x^{[i]} \leq G_{ab}Q_a^{[m]}$. As above (8.3.6) forces x to be incident to b which gives (2).

The proof of statements in (ii) are similar to those in (i).

To establish (iii) suppose that $\angle cad \neq 2^6 \cdot 21$. Then by the digram $D_t(M_{24})$ in Section 11.5 there is a path

$$(\overset{2}{c}, \overset{3}{x}, \overset{2}{y}, \overset{3}{d})$$

in $\mathcal{D}_a$. By (i) and (ii) we obtain

$$Z_d^{[3]} \leq Z_y^{[2]} \leq Q_x^{[3]} \cap Q_a^{[m]} \leq Q_c^{[2]} \cap Q_a^{[m]}.$$

This proves the *if* statement in (iii). To establish the *only if* part, observe that $Z_d^{[3]} \leq Q_a^{[m]}$ and that $Q_a^{[m]}$ acts faithfully on $\Delta(a)$. This means that $Z_d^{[3]}$ acts non-trivially on $\mathcal{D}_c^{[m]} \setminus \{a\}$ for some $c \in \mathcal{D}_a^{[2]}$. In this case $Z_d^{[3]} \not\leq Q_c^{[2]}$. By the *if* part we necessarily have $\angle cad = 2^6 \cdot 21$. ∎

Lemma 8.3.9 *If* $(\overset{m}{a}, \overset{3}{d}, \overset{2}{c})$ *is a path in* $\mathcal{D}$ *with* $\angle adc = 2^4 \cdot 6$ *then*

   (i) $Q_a^{[m]} \cap G_c^{[2]} \leq Q_d^{[3]}$;
   (ii) $Q_d^{[3]} = (Q_a^{[m]} \cap Q_d^{[3]})(Q_c^{[2]} \cap Q_d^{[3]})$;
   (iii) $Q_a^{[m]} \cap Z_c^{[2]} = Z_d^{[3]}$;
   (iv) $G_{ac} = G_{adc}$;
   (v) $|(Q_a^{[m]} \cap Q_c^{[2]})Z_c^{[2]}/Z_c^{[2]}| = 2^4$.

Proof By (7.7.4 (4)) $Q_d^{[3]}Q_a^{[m]} = Q_d^{[3]}O_2(M_d^{[3]}(a))$ while by the diagram $D_h(M_{22})$ we observe that $O_2(M_d^{[3]}(a))$ acts fixed-point freely on the $M_d^{[3]}(a)$-orbit of length $2^4 \cdot 6$, which gives (i). By the above sentence we also see that $Q_a^{[m]}$ does not fix c and therefore $Q_a^{[m]}$ does not normalize $Q_d^{[3]} \cap Q_c^{[2]}$. Hence $Q_a^{[m]} \cap Q_d^{[3]} \not\leq Q_d^{[3]} \cap Q_c^{[2]}$. Since $N_d^{[3]} \cong 2_+^{1+12}$. 3 acts irreducibly on $Q_d^{[3]}/(Q_d^{[3]} \cap Q_c^{[2]}) \cong 2^2$, we obtain (ii).

In particular

$$|Q_a^{[m]} \cap Q_c^{[2]}| = \frac{|Q_a^{[m]} \cap Q_d^{[3]}|}{|Q_d^{[3]}/(Q_d^{[3]} \cap Q_c^{[2]})|} = \frac{2^7}{2^2} = 2^5.$$

Suppose that $Q_a^{[m]} \cap Z_c^{[2]} \neq Z_d^{[3]}$. Since $N_d^{[3]}$ acts irreducibly on $Z_c^{[2]}/Z_d^{[3]} \cong 2^2$, we conclude that $Z_c^{[2]} \leq Q_a^{[m]}$. But in that case $Q_a^{[m]}$ centralizes $Z_c^{[2]}$ and $Q_a^{[m]} \leq G_c^{[2]}$, contrary to (i). This gives (iii) and the latter immediately implies (iv) and (v). ∎

Lemma 8.3.10 *The group G permutes transitively the paths $(\overset{m}{a}, \overset{2}{c}, \overset{0}{b})$ in $\mathcal{D}$. For such a path the following assertions hold:*

(i) $G_{acb}/Q_c^{[2]} \cong L_3(2) \times \mathrm{Sym}_3$;

(ii) $Q_a^{[m]} \cap Q_b^{[0]} = Z_c^{[2]}$;

(iii) $Q_c^{[2]} = (Q_c^{[2]} \cap Q_a^{[m]})(Q_c^{[2]} \cap Q_b^{[0]})$;

(iv) $G_{ab} = G_{acb}$;

(v) $G_{acb}Q_a^{[m]} = G_{ac}$ *and* $G_{acb}Q_b^{[0]} = G_{cb}$;

(vi) $Q_a^{[m]} \cap G_b^{[0]} = Q_a^{[m]} \cap Q_c^{[2]}$ *and* $Q_b^{[0]} \cap G_a^{[m]} = Q_b^{[0]} \cap Q_c^{[2]}$;

(vii) $(Q_a^{[m]} \cap G_b^{[0]})Q_b^{[0]} = O_2(G_{cb})$ *and* $(Q_b^{[0]} \cap G_a^{[m]})Q_a^{[m]} = O_2(G_{ac})$.

Proof If Θ_c is the Petersen subgraph associated with c then b is a vertex of Θ_c disjoint from the antipodal triple a of edges. It is clear that $S_c^{[2]} \cong \mathrm{Sym}_5$ acts transitively on the set of such pairs (a, b) with stabilizer isomorphic to Sym_3. This gives the transitivity assertion along with (i). In terms of the figure after the proof of (8.2.4) $b = u$, $c = \Sigma$, and a is the connected component of the local antipodality graph containing the edge $\{x, y\}$.

Let us turn to (ii). From (7.7.3 (1), (9)) we conclude that both $Q_a^{[m]}$ and $Q_b^{[0]}$ contain $Z_c^{[2]}$ and also that $Q_a^{[m]}$ and $Q_b^{[0]}$ induce on $\mathcal{D}_c^{[2]}$, respectively, an elementary abelian group of order 4 and a group generated by a transposition which does not normalize the action of $Q_a^{[m]}$. The latter implies that

$$\langle Q_a^{[m]}, Q_b^{[0]} \rangle Q_c^{[2]}/Q_c^{[2]} \cong \mathrm{Sym}_5.$$

By (4.7.4) the latter group permutes transitively the dents in $Q_c^{[2]}$. On the other hand, since both $Q_a^{[m]}$ and $Q_b^{[0]}$ are abelian, they centralize their intersection, which gives (ii). Now (iii) is immediate by the order consideration.

By (ii) we obtain

$$G_{ab} \leq N_{G_a^{[m]}}(Q_a^{[m]} \cap Q_b^{[0]}) = N_{G_a^{[m]}}(Z_c^{[2]}) = G_{ac},$$

which gives (iv). It is immediate from the elementary properties of the Petersen graph that $Q_a^{[m]}$ permutes transitively the four elements in $\mathcal{D}_c^{[0]} \backslash \mathcal{D}_a^{[0]}$, while $Q_b^{[0]}$ permutes transitively the two elements in $\mathcal{D}_c^{[m]} \backslash \mathcal{D}_b^{[m]}$, we obtain (v). The remaining statements (vi) and (vii) are now easy to verify. ∎

Recall that if $b \in \mathcal{D}^{[0]}$ then V_b is a 5-dimensional $GF(2)$-space associated with b and for $x \in \mathcal{D}_b$ by $V_b(x)$ we denote the subspace in V_b associated with x.

Lemma 8.3.11 *Let $(\overset{i}{x}, \overset{0}{b}, \overset{2}{c})$ be a path in $\mathcal{D}$, where $i = 2$ or 3. Then $Z_x^{[i]} \not\leq Q_c^{[2]}$ if and only if $\langle V_b(x), V_b(c) \rangle = V_b$.*

Proof By (7.7.2) we observe that $Q_b^{[0]} \not\leq Q_c^{[2]}$, that

$$Q_b^{[0]} = \langle Z_y^{[i]} \mid y \in \mathcal{D}_b^{[i]} \rangle,$$

and that G_{bc} acts transitively on the set

$$\{y \in \mathcal{D}_b^{[i]} \mid \langle V_b(y), V_b(c) \rangle = V_b\}.$$

Therefore, it is sufficient to show that $Z_y^{[i]} \leq Q_c^{[2]}$ whenever $y \in \mathcal{D}_b^{[i]}$ and $\langle V_b(y), V_b(c) \rangle \neq V_b$. Take $a \in \mathcal{D}^{[m]}$ such that $\langle V_b(y), V_b(c) \rangle \geq V_b(a)$. In this case $a \in \mathcal{D}_{ybc}^{[m]}$. Then by (8.3.8 (i)) we have $Z_y^{[i]} \leq Q_a^{[m]} \cap Q_b^{[0]} \leq Q_c^{[2]} \cap Q_b^{[0]}$. ∎

Lemma 8.3.12 *Let $(\overset{3}{e}, \overset{2}{c}, \overset{3}{d})$ be a path in $\mathcal{D}$. Then*

(i) $N_e^{[3]}$ *acts transitively on $\mathcal{D}_c^{[3]} \backslash \{e\}$;*

(ii) $G_{ecd} N_d^{[3]} = G_{cd}$;

(iii) $Q_e^{[3]} \cap Q_d^{[3]} \leq Q_c^{[2]}$;

(iv) $|(Q_e^{[3]} \cap Q_d^{[3]})/Z_c^{[2]}| = 2^4$;

(v) $Q_c^{[2]} = (Q_e^{[3]} \cap Q_c^{[2]})(Q_d^{[3]} \cap Q_c^{[2]})$;

(vi) $(Q_e^{[3]} \cap G_d^{[3]}) N_d^{[3]} / N_d^{[3]} = O_2(G_{cd}/N_d^{[3]})$,

(vii) $G_{ecd} = C_{G_e}(Z_d^{[3]}) = G_{ed}$.

Proof Since $Z_e^{[3]} \leq Z_c^{[2]} \leq Q_e^{[3]}$ and $N_e^{[3]}$ acts fixed-point freely on $Q_e^{[3]}/Z_e^{[3]}$, we conclude that $N_e^{[3]}$ acts transitively on $Z_c^{[2]}/Z_e^{[3]}$. Now $[Z_c^{[2]}, Q_e^{[3]}] = Z_e^{[3]}$, and so $N_e^{[3]}$ acts transitively on $Z_c^{[2]} \backslash Z_e^{[3]}$ and also on $\mathcal{D}_c^{[3]} \backslash \{e\}$. So (i) is established and implies the equality $G_{ecd} N_e^{[3]} = G_{ec}$ which by symmetry gives (ii). Since

$$|(Q_e^{[3]} \cap Q_c^{[2]})/Z_c^{[2]}| = \frac{2^{13}}{2^{2+3}} = 2^8$$

and $|Q_c^{[2]}/Z_c^{[2]}| = 2^{12}$, we have the following lower and upper bounds:

$$2^4 \leq |(Q_e^{[3]} \cap Q_c^{[2]} \cap Q_d^{[3]})/Z_c^{[2]}| \leq 2^8$$

The lower bound is attained if and only if (v) holds. Since G_{ec} is a maximal subgroup in $G_c^{[2]}$, we have $Q_e^{[3]} \cap Q_c^{[2]} \neq Q_d^{[3]} \cap Q_c^{[2]}$, therefore the upper bound is not attained. Moreover, by the properties of the pentad group (4.9.1 (i)) we know that the elements of order 5 in G_{ecd} act fixed-point freely on $Q_c^{[2]}/Z_c^{[2]}$ and hence also on $(Q_e^{[3]} \cap Q_c^{[2]} \cap Q_d^{[3]})/Z_c^{[2]}$. This means that the lower bound is attained and (v) holds. By the order consideration we also obtain (iii) and (iv). By (7.7.3 (13)) $Q_c^{[2]}Q_d^{[3]} = O_2(G_{cd})$ and therefore $(Q_e^{[3]} \cap Q_c^{[2]})Q_d^{[3]} = O_2(G_{cd})$. It is clear that $(Q_e^{[3]} \cap G_d^{[3]})N_d^{[3]}/N_d^{[3]} \leq O_2(G_{cd}/N_d^{[3]})$ and hence (vi) follows. Since $Q_e^{[3]}$ is extraspecial,

$$[Z_c^{[2]}, Q_e^{[3]} \cap G_d^{[3]}] = [Z_c^{[2]}, C_{Q_e^{[3]}}(Z_d^{[3]})] = Z_e^{[3]} \not\leq Z_d^{[3]}$$

and so $Q_e^{[3]} \cap G_d$ inverts $N_d^{[3]}/Q_d^{[3]}$. Since the conjugates of $Z_d^{[3]}$ under $N_e^{[3]}$ generate $Z_c^{[2]}$, we have $C_{G_e}(Z_d^{[3]}) \leq N_{G_e}(Z_c^{[2]}) = G_{ec}$ and hence

$$G_{ed} \leq C_{G_e}(Z_d^{[3]}) \leq C_{G_{ec}}(Z_d^{[3]}) = G_{ecd},$$

proving (vii). $\blacksquare$

8.3.2   *Calculating* μ_2

Recall that our strategy is to reach $h \in \Delta_2^2(a)$ from a by a path in $\mathcal{D}$ which satisfies a uniqueness condition.

Lemma 8.3.13 *Let* Φ *be the set of paths*

$$\varphi = (\overset{m}{a}, \overset{2}{c}, \overset{0}{b}, \overset{2}{e}, \overset{m}{h})$$

in $\mathcal{D}$ *subject to the condition that* $\langle V_b(c), V_b(e)\rangle = V_b$. *Then the action of* G *on* Φ *is transitive and the following assertions hold for a path* φ *from* Φ:

   (i) $|Z_c^{[2]} \cap G_h^{[m]}| = 2^2$;
   (ii) $|G_{acbeh}| = 2^{14} \cdot 3^2$;
   (iii) $G_{cbe} = G_{acbeh}Q_b^{[0]}$;
   (iv) $G_{acbeh}Q_a^{[m]}/Q_a^{[m]} = N_{G_{ac}}(Z_c^{[2]} \cap G_h^{[m]})/Q_a^{[m]}$ *is of order* $2^{10} \cdot 3^2$;
   (v) $G_{ah} = G_{acbeh}$;
   (vi) $|G_{ah}| = 2^{14} \cdot 3^2$ *and* $|G_{ah} \cap Q_a^{[m]}| = 2^4$;
   (vii) φ *is the unique path in* Φ *which joins* a *with* h;
   (viii) $G_{ah}Q_b^{[0]}/Q_b^{[0]} \cong \mathrm{Sym}_4 \times \mathrm{Sym}_4$;
   (ix) $Q_a^{[m]} \cap Q_h^{[m]} = 1$.

Proof By (8.3.10 (v)) $G_{acb}Q_b^{[0]} = G_{cb}$ and so there exists a unique G-orbit on paths $(\overset{m}{a}, \overset{2}{c}, \overset{0}{b}, \overset{2}{e})$ with $\langle V_b(c), V_b(e)\rangle = V_b$. By (8.3.11) $Z_c^{[2]} \not\leq Q_e^{[2]}$ and so $Z_c^{[2]}$ permutes transitively the pair of elements in $\mathcal{D}_e^{[m]} \backslash \mathcal{D}_b^{[m]}$. Thus the uniqueness assertion is proved along with (i). Since

$$|G_{acbeh}| = \frac{|G_{acb}|}{\angle cbe \cdot \angle beh} = \frac{2^{19} \cdot 3^2 \cdot 7}{2^4 \cdot 7 \cdot 2} = 2^{14} \cdot 3^2,$$

(ii) holds. Moreover, by (8.3.11) $Z_c^{[2]}$ is not contained in $Q_e^{[2]}$. We claim that $Z_c^{[2]}$ is not contained in $N_e^{[2]}$. In fact, otherwise $Z_c^{[2]}$ would act non-trivially on $\mathcal{D}_b^{[m]} \backslash \mathcal{D}_e^{[m]}$ contrary to the inclusion $Z_c^{[2]} \leq Q_b^{[0]}$. Hence $Z_c^{[2]}$ permutes the pair of vertices in $\mathcal{D}_e^{[m]} \backslash \mathcal{D}_b^{[0]}$. This gives (iii) and also proves the equality $G_{acbeh} Z_c^{[2]} = G_{acbe}$. By (8.3.10 (vii)) $(Q_a^{[m]} \cap G_b^{[0]}) Q_b^{[0]} = O_2(G_{cb})$ and therefore $(Q_a^{[m]} \cap G_b^{[0]})$ acts transitively on the set of 3-dimensional subspaces U in V_b such that $U \cap V_b(c) = V_b(e) \cap V_b(c)$ (there are exactly 16 such subspaces and each of them can be taken as $V_b(e)$ for a suitable path in Φ). From this conclusion it is easy to deduce (iii). Furthermore, the established transitivity implies the equalities

$$|(Q_a^{[m]} \cap G_b^{[0]})/(Q_a^{[m]} \cap G_{be})| = 2^4, \quad |Q_a^{[m]} \cap G_{be}| = 2^9/2^4 = 2^5$$

and

$$|Q_a^{[m]} \cap G_{beh}| = 2^4.$$

Since $Q_a^{[m]}$ fixes a and c, easy counting gives $|G_{acbeh} Q_a/Q_a| = 2^{10} \cdot 3^2$. Since

$$|G_{ac}/Q_a^{[m]}| = |2^6 : (L_3(2) \times \mathrm{Sym}_3)| = 2^{10} \cdot 3^2 \cdot 7$$

and G_{ab} acts transitively on the set of 7 subgroups of order 4 in $Z_c^{[2]}$, we conclude that $|N_{G_{ac}}(Z_c^{[2]} \cap G_e^{[m]})|$ is also $2^{10} \cdot 3^2$ and (v) follows. Notice that $N_{G_{ac}}(Z_c^{[2]} \cap G_e^{[m]}) Q_a^{[m]}/Q_a^{[m]}$ is a conjugate of the subgroup $N_{M_t}(R)$ of $M_{24} \cong G_a^{[m]}/Q_a^{[m]}$ as in (11.2.3 (iii)).

Since $\langle V_b(c), V_b(e) \rangle = V_b$, $\mathcal{D}_{cbe}^{[3]} = \emptyset$ and hence $Z_c^{[2]} \cap Z_e^{[2]} = 1$. By (8.3.10 (ii)) $Q_b^{[0]} \cap Q_h^{[m]} = Z_e^{[2]}$, and so $Z_c^{[2]} \cap Q_h^{[m]} = Z_c^{[2]} \cap Z_e^{[2]} = 1$. In particular, $Q_a^{[m]} \cap G_h^{[m]} \not\leq Q_h^{[m]}$, and since G_{ah} normalizes $Q_a^{[m]} \cap G_h^{[m]}$, we conclude that $G_{ah} Q_h^{[m]} \neq G_h^{[m]}$. By symmetry $G_{ah} Q_a^{[m]} \neq G_a^{[m]}$. By (11.2.3 (iii)) the only group between $N_{G_{ac}}(Z_c^{[2]} \cap G_h^{[m]})$ and $G_a^{[m]}$ is G_{ac}. Hence $G_{ah} \leq G_c^{[2]}$. By symmetry $G_{ah} \leq G_e^{[2]}$. Since $Z_c^{[2]} Q_e^{[2]} = Q_b^{[0]} Q_e^{[2]}$, we have

$$G_{ce} \leq N_{G_e^{[2]}}(Q_b^{[0]} Q_e^{[2]}) = G_{be},$$

which gives $G_{ce} = G_{cbe}$ and also $G_{ah} = G_{acbeh}$, which is (vi). Comparing (ii), (v), and (vi) we obtain (vii). The transitivity of the action of G on Φ and (vii) imply the uniqueness condition (viii). Hence

$$G_{ah} Q_b^{[0]}/Q_b^{[0]} = G_{acbeh} Q_b^{[0]}/Q_b^{[0]} = G_{cbe}/Q_b^{[0]} \cong \mathrm{Sym}_4 \times \mathrm{Sym}_4$$

and (ix) follows. It only remains to prove (x). As $\langle V_b(c), V_b(e) \rangle = V_b$, we have $Q_c^{[2]} \cap Q_e^{[2]} \leq Q_b^{[0]}$. By (8.3.10 (vi)) $Q_a^{[m]} \cap G_b^{[0]} \leq Q_c^{[2]}$ and so

$$Q_a^{[m]} \cap Q_h^{[m]} \leq Q_a^{[m]} \cap Q_c^{[2]} \cap Q_e^{[2]} \cap Q_h^{[m]}$$
$$\leq (Q_a^{[m]} \cap Q_b^{[0]}) \cap (Q_h^{[m]} \cap Q_b^{[0]}) = Z_c^{[2]} \cap Z_e^{[2]} = 1.$$

∎

Lemma 8.3.14 *If $h \in \Delta_2^2(a)$ then there is a path $\pi = (\overset{m}{a}, \overset{2}{p}, \overset{0}{q}, \overset{2}{r}, \overset{m}{h})$ in $\mathcal{D}$, such that $\langle V_q(p), V_q(r)\rangle = V_q$.*

Proof Let $(\overset{m}{a}, \overset{2}{c}, \overset{m}{e}, \overset{2}{f}, \overset{m}{h})$ be a path joining a and h as in (8.3.4) such that $\angle cef = 2^4 \cdot 63$. We are going to 'shift' the path π to obtain a path from the orbit Φ as in (8.3.13) which still joins a with h. The shifting process can be seen on the following diagram where edges indicate incidence in $\mathcal{D}$.

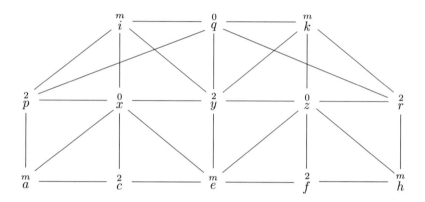

By the diagram $D_t(M_{24})$ in the residue $\mathcal{D}_e$ there is a path $(\overset{2}{c}, \overset{0}{x}, \overset{2}{y}, \overset{0}{z}, \overset{2}{f})$ such that

$$\angle cex = 3, \quad \angle cey = 2 \cdot 21, \quad \angle cez = 2^2 \cdot 21.$$

Recall that for every $B \in \mathcal{B}_e$ there are two elements in $\mathcal{D}_e^{[0]}$ forming an orbit under $Q_e^{[m]}$. In particular the elements x and z are contained in different $Q_e^{[m]}$-orbits on $\mathcal{D}_e^{[0]}$. Therefore we can independently choose x and z in their $Q_e^{[m]}$-orbits so that a is incident to x and h is incident to z.

Let us take a closer look at the residue of y. The elements x and z are non-adjacent vertices of the Petersen graph Θ_y, while e is an antipodal triple of edges, containing both x and z. Since the diameter of Θ_y is two, the vertices x and z, being non-adjacent, are joint by a (unique) 2-path (x, q, z) in Θ_y. Therefore there is a path $(\overset{0}{x}, \overset{m}{i}, \overset{0}{q}, \overset{m}{k}, \overset{0}{z})$ in $\mathcal{D}_y$, where i and k are the connected components of the local antipodality graph of Γ containing the edges $\{x, q\}$ and $\{q, z\}$ respectively (now treated as edges of Γ).

Treating x as a vertex of Γ the elements a and i are the connected components of two distinct edges incident to x in the local antipodality graph. Let p be the element of type 2 corresponding to the unique geometric Petersen subgraph containing these edges. Let r be a similar element defined with respect to z, h

and k. Then

$$\{p\} = \mathcal{D}^{[2]}_{axi}, \quad \{r\} = \mathcal{D}^{[2]}_{hzk}.$$

Since q is adjacent in Γ to both x and z, both p and r are incident to q. Considering the Petersen graph Θ_p we observe that a and q are not incident in $\mathcal{D}$. Similarly in Θ_r we observe that h and q are not incident. Therefore

$$(\overset{m}{a}, \overset{2}{p}, \overset{0}{q}, \overset{2}{r}, \overset{m}{h})$$

is a path in $\mathcal{D}$.

Now it only remains to show that $\langle V_q(p), V_q(r)\rangle = V_q$. Suppose the contrary. Then $V_q(p) \cap V_q(r)$ contains a 2-dimensional subspace $V_q(w)$ for an element $w \in \mathcal{D}^{[3]}_{pqr}$. Since the nodes 3 and m are disjoint on the diagram of $\mathcal{D}$, the element w is incident to i and k. In the residue of q we observe that w is also incident to y. Again directly from the diagram of $\mathcal{D}$ we see that w further is incident to x and e. Now a and w are both incident to p and x and hence they are incident to each other. Next we observe that

$$V_x(c) = V_x(a) \cap V_x(c) \geq V_x(w)$$

and c is incident to w. By symmetry w is incident to f. Then $w \in \mathcal{D}^{[3]}_{cef}$ contrary to the assumption that $\angle cef = 2^4 \cdot 63$ and the diagram $D_t(M_{24})$. Hence no such w exists, $(\overset{m}{a}, \overset{2}{p}, \overset{0}{q}, \overset{2}{r}, \overset{m}{h})$ belongs to the orbit Φ as in (8.3.13) and the result follows. ∎

Proposition 8.3.15 *The following assertions hold:*

  (i) $|\Delta_2^2(a)| = 2^7 \cdot 3 \cdot 5 \cdot 7 \cdot 11 \cdot 23 = 3,400,320;$
 (ii) $Q_a^{[m]}$ *acts on* $\Delta_2^2(a)$ *with orbits of length* $2^7;$
(iii) *if* $h \in \Delta_2^2(a)$ *then* $G_{ah}Q_a^{[m]} = N_{M_t}(R)Q_a^{[m]}$ *where* $N_{M_t}(R) \cong 2^6 : (\mathrm{Sym}_4 \times \mathrm{Sym}_3)$ *is the stabilizer in* $M_a^{[m]} \cong M_{24}$ *of an element of the intermediate type from the rank 3 tilde geometry* $\mathcal{G}(M_{24});$
 (iv) $\mu_2 = 18.$

Proof Since $\Delta_2^2(a)$ is a $G_a^{[m]}$-orbit, $|\Delta_2^2(a)| = [G_a^{[m]} : G_{ah}]$ for $h \in \Delta_2^2(a)$, therefore (8.3.13 (vii)) and (8.3.14) give (i) and (ii). Statement (iii) is by (8.3.13 (v), (vi)) and the remark made at the end of the first paragraph of the proof of (8.3.13). Finally (iv) is by (i) and (8.3.4). ∎

8.3.3   Calculating μ_3

Similarly to what has been accomplished in the previous section we will reach $h \in \Delta_2^3(a)$ along a path in $\mathcal{D}$ satisfying a uniqueness condition.

We need an extra piece of notation. If a, b, c, d are elements of $\mathcal{D}$ such that $a, d \in \mathcal{D}_{bc}$ (which means each of a and d is incident to both b and c) then $\angle a\overset{b}{\underset{c}{}}d$ denotes the length of the orbit of d under G_{abc}.

Lemma 8.3.16 *The set of paths* $(\overset{m}{a}, \overset{3}{b}, \overset{2}{c}, \overset{3}{d})$ *in* $\mathcal{D}$ *with* $\angle abc = 2^4 \cdot 6$ *is non-empty and* G *acts on this set transitively. For such a path the following assertions hold:*

  (i) $|G_{abcd}| = 2^{15} \cdot 3 \cdot 5$;
  (ii) $G_{abcd} N_d^{[3]} = G_{cd}$;
  (iii) $Z_d^{[3]} Q_a^{[m]} / Q_a^{[m]}$ *is generated by a non-2-central involution of* $G_a^{[m]}/Q_a^{[m]} \cong M_{24}$;
  (iv) $G_{abcd} Q_a^{[m]}/Q_a^{[m]}$ *is the centralizer of* $Z_d^{[3]} Q_a^{[m]}/Q_a^{[m]}$ *in* $G_a^{[m]}/Q_a^{[m]}$;
  (v) $G_{abcd} = C_{G_a^{[m]}}(Z_d^{[3]}) = G_{ad}$.

Proof Let $(\overset{m}{a}, \overset{3}{b}, \overset{2}{c})$ be a path in $\mathcal{D}$ such that $\angle abc = 2^4 \cdot 6$. Then by (8.3.12 (i)) $N_b^{[3]}$ acts transitively on $\mathcal{D}_c^{[3]} \setminus \{b\}$. Therefore the existence and the transitivity statements hold with

$$|G_{abcd}| = \frac{|G_{ab}|}{\angle abc \times \angle bcd} = \frac{2^{21} \cdot 3^3 \cdot 5}{2^4 \cdot 6 \cdot 6} = 2^{15} \cdot 3 \cdot 5,$$

which gives (i). By (8.3.12 (vi)) $(Q_b^{[3]} \cap G_d^{[3]}) N_d^{[3]} / N_d^{[3]} = O_2(G_{dc}/N_d^{[3]})$ and by $D_p(M_{22})$ the group $Q_b^{[3]} Q_c^{[2]} / Q_b^{[3]}$ acts regularly on the set

$$\{a \mid a \in \mathcal{D}_b^{[m]}, \angle abc = 2^4 \cdot 6\}.$$

Hence (ii) holds. Note that $Z_d^{[3]} \leq Z_c^{[2]}$ and that $Z_d^{[3]} \neq Z_b^{[3]}$. By (8.3.9 (iii)) $Q_a^{[m]} \cap Z_c^{[2]} = Z_b^{[3]}$. Thus $Z_d^{[3]} \nleq Q_a^{[m]}$ and d is not incident to a. Since G_{abcd} centralizes $Z_d^{[3]}$ and has order divisible by 5, we obtain (iii) from the information about centralizers of involutions in M_{24}. Therefore the centralizer of $Z_d^{[3]} Q_a^{[m]}/Q_a^{[m]}$ in $G_a^{[m]}/Q_a^{[m]}$ has order $2^9 \cdot 3 \cdot 5$. Since $Q_a^{[m]} \cap G_c^{[2]} \leq Q_b^{[3]}$ by (8.3.9 (i)) and $C_{G_b^{[3]}}(Z_d^{[3]}) = G_{bcd}$ by (8.3.12 (vii)), we obtain

$$C_{Q_a^{[m]}}(Z_d^{[3]}) = Q_a^{[m]} \cap G_{bcd} = C_{Q_a^{[m]} \cap Q_b^{[3]}}(Z_d^{[3]}).$$

Furthermore, since $Q_b^{[3]}$ is extraspecial and $Q_a^{[m]} \cap Q_b^{[m]}$ is a maximal subgroup of $Q_b^{[3]}$, we have $|Q_a^{[m]} \cap G_{bcd}| = |Q_a^{[m]} \cap Q_b^{[3]}|/2 = 2^6$. Thus

$$|G_{abcd} Q_a^{[m]}/Q_a^{[m]}| = \frac{|G_{abcd}|}{2^6} = 2^9 \cdot 3 \cdot 5,$$

which gives (iv) and (v). ∎

Lemma 8.3.17 *The group* G *acts transitively on the set of paths* $(\overset{m}{a}, \overset{3}{b}, \overset{2}{c}, \overset{3}{d}, \overset{m}{e})$ *with* $\angle abc = \angle edc = 2^4 \cdot 6$. *For such a path the following assertions hold:*

  (i) $|G_{abcde}| = 2^{10} \cdot 3 \cdot 5$;
  (ii) $Z_c^{[2]} \leq Q_b^{[3]} \cap Q_d^{[3]}$ *and* $|(Q_b^{[3]} \cap Q_d^{[3]})/Z_c^{[2]}| = 2^4$;

(iii) $G_{ae} = G_{abcde}$;

(iv) $Q_a^{[m]} \cap Q_e^{[m]} = 1$;

(v) $G_{ae}/(Q_b^{[3]} \cap Q_d^{[3]}) \cong \mathrm{Sym}_5$.

Proof The transitivity assertion and (i) follow from (8.3.16 (i), (ii)). By (8.3.12 (iv)) $|(Q_b^{[3]} \cap Q_d^{[3]})/Z_c^{[2]}| = 2^4$ which is (ii). By (8.3.9 (v)) we have $|(Q_a^{[m]} \cap Q_c^{[2]})Z_c^{[2]}/Z_c^{[2]}| = 2^4$. By (8.3.12 (vi)) $(Q_d^{[3]} \cap G_b^{[3]})N_b^{[3]}/N_b^{[3]} = O_2(G_{bc}/N_b^{[3]})$. Also by the diagram $D_p(M_{22})$ the group $Q_d^{[3]} \cap G_b^{[3]}$ acts transitively on the set of 32 elements $x \in \mathcal{D}_b^{[m]}$ with $\angle xbc = 2^4 \cdot 6$. Thus $G_{bc} = G_{abc}(Q_d^{[3]} \cap G_b^{[3]})$. Suppose that $(Q_a^{[m]} \cap Q_c^{[2]})Z_c^{[2]} = Q_b^{[3]} \cap Q_d^{[3]}$. Then $G_{bc} = G_{abc}(Q_d^{[3]} \cap G_b^{[3]})$ normalizes $(Q_a^{[m]} \cap Q_c^{[2]})Z_c^{[2]}/Z_c^{[2]}$. But this is impossible, since by the structure of the pentad group (4.9.1) it follows that

$$G_{bc} \cong 2^{3+12} \cdot (\mathrm{Sym}_4 \times \mathrm{Sym}_5)$$

does not normalize subgroups of order 2^4 in $Q_c^{[2]}/Z_c^{[2]}$. So $(Q_a^{[m]} \cap Q_c^{[2]})Z_c^{[2]} \neq Q_b^{[3]} \cap Q_d^{[3]}$. Since 5 divides $|G_{abcd}|$, we have

$$Q_a^{[m]} \cap Q_c^{[2]} \cap Q_d^{[3]} \leq Z_c^{[2]}.$$

Since $O_2(G_{cd}/N_d^{[3]}) \cap G_{de}/N_d^{[3]} = 1$, $Q_a^{[m]} \cap G_{cde} \leq Q_d^{[d]}$. Similarly $Q_a^{[m]} \cap G_c^{[2]} \leq Q_b^{[3]}$. By (8.3.12 (iii)) $Q_b^{[3]} \cap Q_d^{[3]} \leq Q_c^{[2]}$. Hence

$$Q_a^{[m]} \cap G_{cde} \leq Q_a^{[m]} \cap Q_c^{[2]} \cap Q_d^{[3]} \leq Z_c^{[2]}.$$

By (8.3.9 (iii)) we have $Q_a^{[m]} \cap Z_c^{[2]} \leq Z_b^{[3]}$ and thus $Q_a^{[m]} \cap G_{cde} = Z_b^{[3]}$. By symmetry $Q_e^{[m]} \cap G_{abc} = Z_d^{[3]}$. By (8.3.16 (v)) $G_{abcd} = G_{ad}$, and so $Q_e^{[m]} \cap G_a^{[m]} = Q_e^{[m]} \cap G_{abc} = Z_d^{[3]}$. Thus $G_{ae} = C_{G_e^{[m]}}(Z_d^{[3]}) = G_{ed}$ and $G_{ae} = G_{abcde}$ which is (iii). By symmetry $Q_a^{[m]} \cap G_e = Z_b^{[3]}$ and so $Q_a^{[m]} \cap Q_e^{[m]} = 1$, which is (iv). Finally since $G_{ae}/O_2(G_{ae}) \cong \mathrm{Sym}_5$ and $O_2(G_{ae}) \leq Q_b^{[3]} \cap Q_d^{[3]}$ we have $G_{ae}/(Q_b^{[3]} \cap Q_d^{[3]}) \cong \mathrm{Sym}_5$, completing the proof of (v). ∎

Lemma 8.3.18 *Let $h \in \Delta_2^3(a)$ and let $(\overset{m}{a}, \overset{2}{c}, \overset{m}{e}, \overset{2}{f}, \overset{m}{h})$ be a path as in* (8.3.4) *joining a and h, such that $\angle cef = 2^6 \cdot 42$. Then there is a path $(\overset{m}{a}, \overset{3}{l}, \overset{2}{g}, \overset{3}{k}, \overset{m}{h})$ in* $\mathcal{D}$, *such that $l, g, k \in \mathcal{D}_e$ and $\angle alg = \angle hkg = 2^4 \cdot 6$.*

Proof Since $\angle cef = 2^6 \cdot 42$ by $D_t(M_{24})$ there is a path $(\overset{2}{c}, \overset{3}{l}, \overset{2}{g}, \overset{3}{k}, \overset{2}{f})$ in $\mathcal{D}_e$ with $\angle c\overset{l}{e}g = 8 = \angle c\overset{k}{e}g$. Since the nodes m and 3 are disjoint on the diagram of $\mathcal{D}$, a is incident to l and h is incident to k. Since $\angle f\overset{l}{e}g = 8$ we conclude from the diagram $D_h(M_{22})$ that $\angle alg = 2^4 \cdot 6$, and by symmetry $\angle hkg = 2^4 \cdot 6$. Therefore the result follows. ∎

Proposition 8.3.19 *The following assertions hold:*

(i) $|\Delta_2^3(a)| = 2^{11} \cdot 3^2 \cdot 7 \cdot 11 \cdot 23 = 32,643,072$;

(ii) *if* $h \in \Delta_2^3(a)$ *then the subgroup* $G_h^{[m]} \cap Q_a^{[m]}$ *is of order 2 and it is a conjugate in* $G_a^{[m]}$ *of* $Z_c^{[3]}$ *for some* $c \in \mathcal{D}_a^{[3]}$;

(iii) *let* $M_s \cong 2^6 : 3 \cdot \mathrm{Sym}_6$ *be the stabilizer in* $M_a^{[m]} \cong M_{24}$ *of a sextet* S *from* $\mathcal{S}_a$, *let* F *be a subgroup in* M_s *such that*

    (1) $F \cap O_{2,3}(M_s) = O_2(M_s)$;

    (2) $F/O_2(M_s) \cong \mathrm{Sym}_5$;

    (3) F *acts on the set of 15 trios from* $\mathcal{T}_a$ *refined by the sextet* S *with two orbits of lengths 5 and 10;*

then $Q_a^{[m]}G_{ah}$ *and* $Q_a^{[m]}F$ *are conjugate in* $G_a^{[m]}$;

(iv) $\mu_3 = 5$.

Proof Because of the transitivity of the action of $G_a^{[m]}$ on $\Delta_2^3(a)$, since the order of $G_a^{[m]}$ is known (8.3.18) and (8.3.17 (i)) imply (i). By (8.3.2 (ii)) the orbit of h under $Q_a^{[m]}$ has length 2^{10}. Therefore $Q_a^{[m]} \cap G_h^{[m]}$ is of order 2 and clearly $G_{ah} \leq C_{G^{[m]}}(Q_a^{[m]} \cap G_h^{[m]})$. The group $Q_a^{[m]}$ is a Todd module for $M_a^{[m]} \cong G_a^{[m]} \cong M_{24}$ and the latter has two orbits on the set of subgroups of order 2 in $Q_a^{[m]}$ with lengths 1771 and 276. The corresponding stabilisers are

$$2^6 : 3 \cdot \mathrm{Sym}_6 \text{ of order } 2^{10} \cdot 3^2 \cdot 5$$

and

$$\mathrm{Aut}\,(M_{22}) \text{ of order } 2^8 \cdot 3 \cdot 5 \cdot 7 \cdot 11.$$

By the order consideration $G_h^{[m]} \cap Q_a^{[m]}$ is from the former orbit which gives (ii) and (iii). Finally (iv) is by (i) and (8.3.4). ∎

8.3.4   *The μ-subgraphs*

Let a be a vertex of Δ and let $h \in \Delta_2(a)$. In this section we describe the subgraph in the local graph $\Delta(a)$ as in (8.2.6) induced by the set $\mu(a, h) = \Delta(a) \cap \Delta(h)$. This subgraph is commonly known as a μ-subgraph in Δ. If $h \in \Delta_2^\alpha(a)$ then by (8.3.5), (8.3.15), and (8.3.19) the number of vertices in the μ-subgraph $\mu(a, h)$ is 45, 18, and 5 for $\alpha = 1$, 2, and 3, respectively. Therefore we have established the following fragment of the suborbit diagram of Δ:

In order to deal with μ-subgraphs we first introduce some further terminology concerning the locally truncated geometry $\mathcal{H}(M_{24})$ and the tilde geometry $\mathcal{G}(M_{24})$ of the Mathieu group M_{24} (cf. Section 11.2). Let Σ be a sextet, let $\mathcal{B}_\Sigma$ and $\mathcal{T}_\Sigma$ be respectively the octads and trios refined by Σ. By the definition $\mathcal{B}_\Sigma$ and $\mathcal{T}_\Sigma$ are the octads and trios incident to Σ in $\mathcal{H}(M_{24})$. Clearly $\mathcal{B}_\Sigma$ corresponds to the pairs of the tetrads in Σ while $\mathcal{T}_\Sigma$ corresponds to the partitions of the tetrads into three disjoint pairs. Thus by Section 1.8 there is a non-singular 4-dimensional symplectic space V_Σ such that $\mathcal{T}_\Sigma$ are the non-zero vectors in V_Σ; $\mathcal{B}_\Sigma$ corresponds to the totally singular 2-dimensional subspaces; the incidence (in $\mathcal{H}(M_{24})$) is via inclusion. The group $\bar{M}_s = M_s/O_{2,3}(M_s) \cong \mathrm{Sym}_6$ (where $M_s \cong 2^6 : 3 \cdot \mathrm{Sym}_6$ is the stabilizer of Σ in $M \cong M_{24}$) induces the full symplectic group of the space V_Σ.

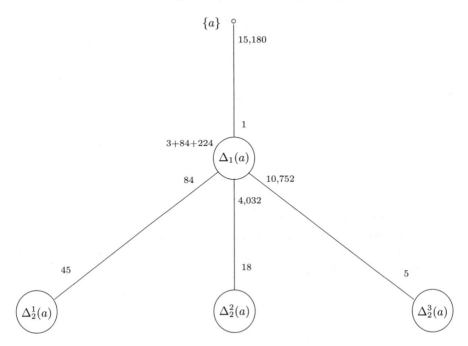

By (1.8.5) $\bar{M}_s$ contains up to conjugation two Sym_5-subgroups. One of them acts transitively on $\mathcal{T}_\Sigma$ while the other one stabilizes an associated quadratic form of minus type on V_Σ and has two orbits on $\mathcal{T}_\Sigma$ with lengths 5 and 10 (isotropic and non-isotropic vectors, respectively).

Let T be a trio and let $\mathcal{S}_T$ be the set of sextets which refine T. Then $|\mathcal{S}_T| = 7$ and $M_t \cong 2^6 : (L_3(2) \times \mathrm{Sym}_3)$ (the stabilizer of T in M) induces on $\mathcal{S}_T$ the natural action of $L_3(2)$. This action preserves on $\mathcal{S}_T$ a unique structure π of the projective plane of order 2. A line

$$l = \{\Sigma_1, \Sigma_2, \Sigma_3\}$$

of π is called a *sextet-line* . The stabilizer of l in M is contained in M_t,

$$M_t(l) \cong 2^6 : (\mathrm{Sym}_4 \times \mathrm{Sym}_3)$$

and $M_t(l)$ is the stabilizer of an element from $\mathcal{G}(M_{24})$ of the intermediate type.

Lemma 8.3.20 *Let $h \in \Delta_2^1(a)$. Then there is a unique $\Sigma \in \mathcal{S}_a$ such that*

(i) *if $T \in \mathcal{T}_a$, then $(T, \beta) = \lambda_a(e)$ for some $e \in \mu(a, h)$ and for a GF(2)-valued function β on T if and only if $T \in \mathcal{T}_\Sigma$ (equivalently if T is refined by Σ);*

(ii) *if $T \in \mathcal{T}_\Sigma$ then there are exactly three functions β on T such that $\lambda_a^{-1}(T, \beta) \in \mu(a, h)$.*

Proof Let d be the unique element from $\mathcal{D}_{ah}^{[3]}$. Then d is a sextet from $\mathcal{S}_a$ and the elements $b \in \mathcal{D}_a^{[2]}$ such that $\mathcal{D}_b^{[m]} \subseteq \mathcal{D}_d^{[m]}$ are the trios in $\mathcal{T}_a$ refined by d. In view of (8.3.3) the result now follows from (8.3.5). $\blacksquare$

Lemma 8.3.21 *Let $h \in \Delta_2^2(a)$. Then there is a unique trio $T^h \in \mathcal{T}_a$ and a sextet-line $l = \{\Sigma_1, \Sigma_2, \Sigma_3\} \subseteq \mathcal{S}_{T^h}$ such that*

(i) *if $T \in \mathcal{T}_a$, then $(T, \beta) = \lambda_a(e)$ for some $e \in \mu(a, h)$ and for a GF(2)-valued function β on T if and only if $T \in \mathcal{T}_{\Sigma_i}$ for $i = 1$, 2, or 3 and there is a unique octad in $\mathcal{B}_a$ which is contained in both T and T^h;*

(ii) *if $(T, \beta) = \lambda_a(e)$ for $e \in \mu(a, h)$ then β is uniquely determined by T.*

Proof By (8.3.15) $G_{ah}Q_a^{[m]} = N_{M_t}(R)Q_a^{[m]}$, where $N_{M_t}(R)$ is the stabilizer of an element of the intermediate type from the tilde geometry $\mathcal{G}(M_{24})$. We claim that $N_{M_t}(R)$ has a unique orbit X on $\mathcal{T}_a$ of length 18 and no orbits of length 9. In fact, $N_{M_t}(R)$ contains $O_2(M_t)$, therefore by the diagram $D_t(M_{24})$ an orbit of length 18 or 9 must be contained in the orbit Y of M_t on $\mathcal{T}_a$ of length $42 = 2 \cdot 21$. Notice that if T^h is the trio stabilized by M_t then Y are the trios intersecting T^h in one octad. It is now elementary to check that $N_{M_t}(R)$ acts on Y with two orbits of lengths 18 and 24. This proves the claim. Now the result is rather easy to deduce. $\blacksquare$

Lemma 8.3.22 *Let $h \in \Delta_2^3(a)$. There there is a unique $\Sigma \in \mathcal{S}_a$ and a unique associated quadratic form q of minus type on the symplectic space V_Σ such that*

(i) *if $T \in \mathcal{T}_a$, then $(T, \beta) = \lambda_a(e)$ for some $e \in \mu(a, h)$ and for a GF(2)-valued function β on T if and only if $T \in \mathcal{T}_\Sigma$ (equivalently if T is refined by Σ) and $q(T) = 0$;*

(ii) *if $T \in \mathcal{T}_\Sigma$ and $q(T) = 0$ then there is a unique function β on T such that $\lambda_a^{-1}(T, \beta) \in \mu(a, h)$.*

Proof The assertion is a reformulation of (8.3.19 (iii)). $\blacksquare$

8.4   Earthing up $\Delta_3^1(a)$

Lemma 8.4.1 *If $(\overset{m}{a}, \overset{3}{b}, \overset{m}{c}, \overset{2}{d}, \overset{m}{e})$ is an arc in $\mathcal{D}$ with $d_\Delta(a, e) = 3$ then $\angle abc = 2^4$ and $\angle bcd = 2^6 \cdot 45$.*

Proof Clearly $d_\Delta(a, c) = 2$, and so $\angle abc = 2^4$. If d is incident to b, then b is incident to e in which case the diagram $D_h(M_{22})$ shows that $d_\Delta(a, e) = 2$. This is a contradiction.

Let $x \in \mathcal{D}_{bc}^{[2]}$ and suppose that $\angle xcd = 2^2 \cdot 14$. Then by (8.2.5 (iii)) the set $\mathcal{D}_x^{[m]}$ is contained in $\{e\} \cup \Delta(e)$. On the other hand, by $\mathcal{D}_t(M_{24})$ we observe that $\Delta(a) \cap \mathcal{D}_x^{[m]} \neq \emptyset$. Therefore $d_\Delta(e, a) \leq 2$, which is again a contradiction.

Thus $\angle xcd \neq 2^2 \cdot 14$ for all $x \in \mathcal{D}_{bc}^{[2]}$. Suppose that $\angle bcd = 2^3 \cdot 45$. Then by $D_s(M_{24})$ there exists a path $(\overset{3}{b}, \overset{2}{x}, \overset{3}{y}, \overset{2}{d})$ in the residue $\mathcal{D}_c$ with $\angle x \overset{c}{y} d = 8$. Thus by $D_t(M_{24})$ we observe that $\angle xcd$ must be $2^2 \cdot 14$, which is a contradiction.

Suppose that $\angle bcd = 2^2 \cdot 45$. Then the residue $\mathcal{D}_c$ contains a path $(\overset{3}{b}, \overset{0}{x}, \overset{2}{d})$. Since $\angle abc = 2^4$ we can choose x in its $Q_c^{[m]}$-orbit in such a way that $\angle abx = 2^2 \cdot 15$ (compare the diagram $D_h(M_{22})$), in which case there exists $y \in \mathcal{D}_{abx}^{[2]}$. Thus we have found a path

$$(\overset{m}{a}, \overset{2}{y}, \overset{0}{x}, \overset{2}{d}, \overset{m}{e}).$$

Since $d_\Delta(a, e) = 3$, $\mathcal{D}_{yxd}^{[m]} = \emptyset$ and hence $\langle V_x(y), V_x(d) \rangle = V_x$. By (8.3.13) and (8.3.14) once again we have $d_\Delta(a, e) = 2$, which is a contradiction.

Thus $\angle bcd = 2^6 \cdot 45$ and the lemma is proved. ∎

Lemma 8.4.2 *There exists a unique G-orbit on the set of paths*

$$(\overset{m}{a}, \overset{3}{b}, \overset{m}{c}, \overset{2}{d}, \overset{m}{e})$$

in $\mathcal{D}$ with $\angle abc = 2^4$ and $\angle bcd = 2^6 \cdot 45$. Moreover for such a path there is a path

$$(\overset{m}{a}, \overset{2}{h}, \overset{0}{f}, \overset{3}{g}, \overset{m}{e})$$

with $\langle V_f(h), V_f(g) \rangle = V_f$ and $\angle fge = 2^3 \cdot 7$.

Proof By the properties of the hexad graph (11.4.5) we have $G_{ac}Q_c^{[m]} = G_{bc}$ and so there exists a unique G-orbit on the set of paths

$$(\overset{m}{a}, \overset{3}{b}, \overset{m}{c}, \overset{2}{d})$$

with $\angle abc = 2^4$ and $\angle bcd = 2^6 \cdot 45$. Moreover, $\mathcal{D}_{bcd}^{[0]} = \emptyset$ and so by (8.2.5)

$$(Q_b^{[3]} \cap Q_c^{[m]})(Q_d^{[2]} \cap Q_c^{[m]}) = Q_c^{[m]}.$$

Since $Q_c^{[m]}$ acts transitively on $\mathcal{D}_d^{[m]} \setminus \{c\}$, the uniqueness assertion holds.

By the diagram $D_s(M_{24})$ the residue $\mathcal{D}_c$ contains a path $(\overset{3}{b}, \overset{0}{f}, \overset{3}{g}, \overset{2}{d})$. Replacing, if necessary, f by the other element in its $Q_c^{[m]}$-orbit, we assume that $\angle abf = 2^2 \cdot 45$ (compare the diagram $D_h(M_{22})$), in which case there exists $h \in \mathcal{D}_{abf}^{[2]}$. Note that d is incident to e and g, and so e and g are incident because of the diagram of $\mathcal{D}$. Since $\angle bcd = 2^6 \cdot 15$, we see from the diagram $D_s(M_{24})$ that $\mathcal{D}_{bcg}^{[2]}$ is empty and so $\langle V_f(b), V_f(g) \rangle = V_f(c)$. Moreover, since $\angle abc = 2^4$, c and h are not incident. So $V_f(b) \leq V_f(h) \not\leq V_f(c)$ and therefore $\langle V_f(h), V_f(g) \rangle = V_f$.

Since $\angle bcd = 2^6 \cdot 45$, d is not incident to f. On the other hand, both d and f are incident to c, and we see from $D_o(M_{22})$ that $\angle fgd = 2^2 \cdot 21$ and hence $\angle fge = 2^2 \cdot 14$. ∎

Lemma 8.4.3 *There is a unique G-orbit on the set Ψ of paths*

$$\psi = (\overset{m}{a}, \overset{2}{b}, \overset{0}{c}, \overset{3}{d}, \overset{m}{e})$$

with $V_c(b) \cap V_c(d) = 0$ and $\angle cde = 2^3 \cdot 7$. Moreover, for such a path the following assertions hold:

(i) $|G_{abcde}| = 2^{10} \cdot 3^2$;

(ii) *there is a unique path*

$$\chi = (\overset{m}{a}, \overset{3}{i}, \overset{0}{g}, \overset{2}{h}, \overset{m}{e})$$

    with $V_g(h) \cap V_g(i) = 0$ and $\angle gia = 2^2 \cdot 14$;

(iii) $Q_a^{[m]} \cap G_e^{[m]} = 1$;

(iv) $d_\Delta(a, e) = 3$;

(v) $G_{ae} = G_{abcde}$;

(vi) $Q_g^{[0]} \cap G_{ae}$ *is elementary abelian of order 2^6 normal in G_{ae}.*

Proof By (8.3.10 (vii)) we have $(Q_a^{[m]} \cap G_c^{[0]})Q_c^{[0]} = O_2(G_{cb})$ and so $Q_a^{[m]} \cap G_c^{[0]}$ acts transitively on the 64-element set

$$\{x \in \mathcal{D}_c^{[3]} \mid V_c(b) \cap V_c(x) = 0\}$$

Hence there is a unique G-orbit on the set of paths $(\overset{m}{a}, \overset{2}{b}, \overset{0}{c}, \overset{3}{d})$ with $V_c(b) \cap V_c(d) = 0$. Moreover, in $G_c^{[0]}/Q_c^{[0]} \cong L_5(2)$ we observe that $G_{bcd}/Q_c^{[0]} \cong L_3(2) \times \text{Sym}_3$ is a complement of $O_2(G_{cd}/Q_c^{[0]})$ in $G_{cd}/Q_c^{[0]}$. Thus $G_{cd} = G_{bcd}Q_d$. By (8.3.10 (v)) we have $G_{bc} = G_{abc}Q_c^{[0]}$ and so

$$G_{cd} = G_{bcd}Q_d^{[3]} = ((G_{abc}Q_c^{[0]}) \cap G_d^{[3]})Q_d^{[3]} = G_{abcd}Q_c^{[0]}Q_d^{[3]}.$$

We claim that $Z_b(Q_c^{[0]} \cap Q_d^{[3]}) = Q_c^{[0]}$. In fact, let us identify the orbit of length 155 of $G_c^{[0]}/Q_c^{[0]} \cong L_5(2)$ on $Q_c^{[0]}$ with the set of 2-dimensional subspaces in $Q_c^{[0]} \cong \bigwedge^2 V_c$. Then by (7.7.2 (3)) the intersection $Q_c^{[0]} \cap Q_d^{[3]}$ is generated by the 2-dimensional subspaces having non-zero intersection with $V_c(d)$, while $Z_b^{[2]}$ is generated by the 2-dimensional subspaces contained in $V_c(b)$. Since

$$V_c = V_c(d) \oplus V_c(b)$$

the claim follows. Since $Z_b^{[2]} \le G_{abcd}$, we have the equality

$$G_{cd} = G_{abcd}Q_c^{[0]}Q_d^{[3]} = G_{abcd}Q_d^{[3]},$$

which particularly proves the transitivity of the G-action on the set of paths in Ψ.

By the transitivity assertion $|G_{abcde}|$ is the quotient of $|G_a^{[m]}|$ over the number of paths in Ψ starting with a. Therefore

$$|G_{abcde}| = \frac{|G_a^{[m]}|}{|\mathcal{D}_a^{[2]}| \cdot \angle abc \cdot \angle bcd \cdot \angle cde} = 2^{10} \cdot 3^2,$$

which gives (i).

The shift from ψ to χ which constitutes the proof of (ii) can be illustrated by the following diagram:

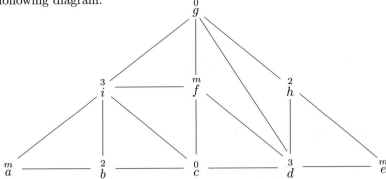

Since $\angle cde = 2^3 \cdot 7$, $\angle edc = 2^4 \cdot 15$, if follows form $D_h(M_{22})$ that there is a unique $f \in \mathcal{D}_{cd}^{[m]}$ with $\angle edf = 2^4$. Moreover, if $\{g, c\}$ is the orbit of c under $Q_f^{[m]}$, then there exists $h \in \mathcal{D}_{gde}^{[2]}$. Since $V_c(d) \leq V_c(f)$, $V_c(b) \not\leq V_c(f)$, there is a unique element $i \in \mathcal{D}_{bcf}^{[3]}$. In fact i is determined by the equality $V_c(i) = V_c(f) \cap V_c(b)$. Then because of the shape of the diagram of $\mathcal{D}$ i is adjacent to a and to g as well. Since $V_c(b) \cap V_c(d) = 0$ and $V_c(i) \leq V_c(b)$, we have $\langle V_c(i), V_c(d) \rangle = V_c(f)$. Conjugation under $Q_f^{[m]}$ yields $\langle V_g(i), V_g(d) \rangle = V_g(f)$. Since $\angle fde = 2^4$, f is not incident to h, and so since $V_g(h) \geq V_g(d)$, the equality $V_g(i) \cap V_g(h) = 0$ holds. Consider the path $(\overset{0}{g}, \overset{m}{f}, \overset{0}{c}, \overset{2}{b}, \overset{m}{a})$ in $\mathcal{D}_i$. By the diagram $D_o(M_{22})$ and since $\{g, c\}$ is a $Q_f^{[m]}$-orbit, we have $\angle gic = 7$ and $\angle gib = 2^2 \cdot 7$. Now it is immediate to observe from the diagram $D_o(M_{22})$ that $\mathcal{D}_{ic}^{[m]}$ and $\mathcal{D}_{ic}^{[2]}$ are the point-set and the line-set of the projective plane of order 2 with the natural incidence relation. In view of this observation and the indexes in $D_o(M_{22})$, every element from

$$\{x \mid x \in \mathcal{D}_{ib}^{[m]}, \angle gix = 2 \cdot 7\}$$

is incident to c. Hence $\angle gia = 2^3 \cdot 7$, which proves (ii).

Since

$$C_{Q_i^{[3]} \cap G_h^{[2]}}(V_g) = C_{Q_i^{[3]} \cap G_h^{[2]}}(V_g(i) \cap V_g(h)) \leq \langle V_g(i), V_g(h) \rangle = V_g,$$

we have

$$Q_i^{[3]} \cap G_h^{[2]} \leq Q_g^{[0]},$$

and so $Z_b^{[2]} \cap G_h^{[2]} \leq Q_c^{[0]} \cap Q_g^{[0]}$. Assuming as above that 2-dimensional subspaces in V_c are considered as elements of $Q_c^{[0]} \cong \bigwedge^2 V_c$ we obtain

$$Z_b^{[2]} \cap Q_c^{[0]} \cap Q_g^{[0]} = Z_i^{[3]}.$$

Hence $Z_b^{[2]} \cap G_h^{[2]} = Z_i^{[3]}$. Since $V_g(i) \cap V_g(h) = 0$, $Z_i^{[3]} \not\leq Q_g^{[0]} \cap Q_h^{[2]}$. Moreover, $Q_g^{[0]} \cap G_e^{[m]} = Q_g^{[0]} \cap Q_e^{[2]}$ by (8.3.10 (vi)), and thus $Z_i^{[3]} \not\leq G_e^{[m]}$.

Put $K = G_{abcde}$. Since each of f, g, h, and i is uniquely determined by (a, b, c, d, e), $K \leq G_{fghi}$, and in particular

$$Z_b^{[2]} \cap G_e^{[m]} = Z_b^{[2]} \cap K \leq Z_i^{[3]} \cap G_e^{[m]} = 1.$$

From the equalities $Q_a^{[m]} \cap G_c^{[0]} \leq Q_b^{[2]}$, $Q_b^{[2]} \cap G_d^{[3]} \leq Q_c^{[0]}$, and $Q_a^{[m]} \cap Q_c^{[0]} = Z_b^{[2]}$ (cf. (8.3.10 (ii)) for the latter equality) we conclude that $Q_a^{[m]} \cap K = 1$. By (i) $|K| = 2^{10} \cdot 3^2$. Since $K \leq G_{abi}$ and

$$|G_{abi}/Q_a^{[m]}| = \frac{|G_{ab}/Q_a^{[m]}|}{7} = 2^{10} \cdot 3^2,$$

we have $KQ_a^{[m]} = G_{abi}$.

Suppose that $Q_a^{[m]} \cap G_e^{[m]} \neq 1$ and let R be a Sylow 2-subgroup of K. Then $C_{Q_a^{[m]} \cap G_e^{[e]}}(R) \neq 1$, and since $C_{Q_a^{[m]}}(R) = C_{Q_a^{[m]}}(RQ_a^{[m]}) = Z_i^{[3]}$, we get $Z_i^{[3]} \leq G_e^{[m]}$, a contradiction. Hence $Q_a^{[m]} \cap G_e^{[m]} = 1$ which gives (iii).

By (8.1.3 (iii)) and (8.3.2 (ii)) we know that for every $y \in \Delta_2(a)$ there is a non-identity element in $Q_a^{[m]}$ which stabilizes y. Therefore by (iii) we conclude that $d_\Delta(a, e) = 3$ which is (iv).

To establish (v) suppose to the contrary that $K \neq G_{ae}$. Then by (11.2.3 (ii)) the subgroup $G_{ae}Q_a^{[m]}$ is one of $G_a^{[m]}$, G_{ab} and G_{ai}. Suppose that $G_{abe} \neq K$. Then $G_{abe}Q_a^{[m]} = G_{ab}$. In particular $|G_{abe}/K| = 7$ and so $G_{abc} = O^{2,3}(G_{abe})K$. From the equality $\angle abc = 4$ we conclude that $O^{2,3}(G_b^{[2]}) \leq G_c^{[0]}$. Now also $K \leq G_c^{[0]}$ and so $G_{abe} \leq G_c^{[0]}$. Note that

$$Q_c^{[0]} \cap G_e^{[m]} = Z_b^{[2]}(Q_c^{[0]} \cap Q_d^{[3]}) \cap G_e^{[m]} = Q_c^{[0]} \cap Q_d^{[3]},$$

and so $G_{ce} \leq N_{G_c^{[0]}}(Q_c^{[0]} \cap Q_d^{[3]}) = G_{cd}$. Thus $G_{ce} = G_{cde}$ and $G_{abe} = G_{abce} = G_{abcde} = K$, a contradiction.

Therefore $G_{abe} = K$ and $G_{ae}Q_a^{[m]} = G_{ai}$. On the other hand, by (8.3.10 (iv)) $G_{ge} \leq G_{ghe}$, and as seen above $Q_i^{[3]} \cap G_h^{[2]} \leq Q_g^{[0]}$. In particular $Q_i^{[3]} \cap G_e^{[m]} \leq Q_i^{[3]} \cap G_h^{[2]} = Q_i^{[3]} \cap Q_g^{[0]}$. Since $Z_i^{[3]}$ permutes transitively the set of two elements in $\mathcal{D}_h^{[m]} \setminus \mathcal{D}_{gh}^{[m]}$, $Q_i^{[3]} \cap Q_g^{[0]} = Z_i^{[3]}(Q_i^{[3]} \cap G_e^{[m]})$. Thus $G_{ae} = G_{aie} \leq N_{G_i^{[3]}}(Q_i^{[3]} \cap Q_g^{[0]}) = G_{ig}$. This is a contradiction, since 3^3 divides $|G_{ae}| = |G_{ai}/Q_a^{[m]}|$ but not $|G_{ig}|$.

Thus $G_{ae} = G_{abcde} = G_{abcde} = G_{abcdefghi}$ completing the proof of (v).

Now it only remains to prove (vi). By (v) and the transitivity assertion ψ is uniquely determined by a and e. By (ii) χ is uniquely determined by ψ.

Therefore G_{ae} stabilizes g and hence normalizes $Q_g^{[0]} \cap G_{ae}$. Let us identify the latter intersection. We identify the 2-dimensional subspaces in V_g with elements in $Q_g^{[0]} \cong \bigwedge^2 V_g$. Since $V_g(h) \cap V_g(i) = 0$ we have $V_g = V_g(h) \oplus V_g(i)$. Therefore $Q_g^{[0]} \leq G_{igh}$,

$$F := G_{igh}/Q_g^{[0]} \cong \mathrm{Sym}_3 \times L_3(2)$$

and as an F-module $Q_g^{[0]}$ is the direct sum of three irreducible submodules W_i, W_h, and W_{ih} of dimensions 3, 1 and 6, respectively. Here W_i is generated by the 2-dimensional subspace of V_g contained in $V_g(i)$, W_h is generated by $V_g(h)$ (which is itself a 2-dimensional subspace) and W_{ih} is generated by the subspaces which intersect both $V_g(i)$ and $V_g(h)$. By (7.7.3 (8)) $Q_g^{[0]}$ induces an action of order 2 on $\mathcal{D}_i^{[m]}$ and by (7.7.4 (9)) $Q_g^{[0]}$ induces an action of order 2^4 on $\mathcal{D}_g^{[m]}$. Clearly $Q_g^{[0]}$ permutes transitively the pair of elements in $\mathcal{D}_h^{[m]} \setminus \mathcal{D}_g^{[m]}$ (notice that a is one of them), while by $D_o(M_{22})$ the orbit of e under $Q_g^{[0]}$ has length 2^3. In the above terms this means that $G_{ae} \cap Q_g^{[0]} = W_{ih}$ completing the proof of (vi).   ■

Define $\Delta_3^1(a)$ to be the set of vertices $e \in \Delta$ such that there is a path $\psi \in \Psi$ as in (8.4.3) which joins a and e.

Lemma 8.4.4 *The following assertions hold:*

(i) $\Delta_3^1(a)$ *is an orbit of* $G_a^{[m]}$ *on* $\Delta_3(a)$;

(ii) $|\Delta_3^1(a)| = 2^{11} \cdot 3 \cdot 5 \cdot 7 \cdot 11 \cdot 23 = 54,405,120$;

(iii) $Q_a^{[m]}$ *acts fixed-point freely on* $\Delta_3^1(a)$;

(iv) *if* $e \in \Delta_3^1(a)$ *then the set* $\Delta(e) \cap \Delta_2^1(a)$ *is of size 6 and the action of* G_{ae} *on this set is transitive.*

Proof Statement (i) follows from the transitivity assertion in (8.4.3) and the definition of $\Delta_3^1(a)$. By (i) $|\Delta_3^1(a)|$ is $|G_a^{[m]}|$ divided by $|G_{ae}|$. Therefore (8.4.3 (i)) and (8.4.3 (v)) give (ii). Statement (iii) is by (8.4.3 (iii)). By (8.4.1), (8.4.2), and (8.4.3 (iv)) for every vertex $c \in \Delta_2^1(a)$ the set $\Delta(c) \cap \Delta_3^1(a)$ is of size $2880 = 2^6 \cdot 45$ and G_{ae} acts transitively on this set. Comparing the size of $\Delta_3^1(a)$ in (ii) with the size of $\Delta_2^1(a)$ in (8.1.3) we obtain (iv).   ■

A direct consequence of (8.4.1), (8.4.2), and (8.4.3) is the following result.

Corollary 8.4.5 *Let* $(\overset{m}{a}, \overset{3}{j}, \overset{m}{k}, \overset{2}{l}, \overset{m}{e})$ *be an arc in* $\mathcal{D}$. *Then either* $d_\Delta(a, e) \leq 2$ *or* $e \in \Delta_3^1(a)$.

Now we are ready to prove the final result of the section.

Lemma 8.4.6 *Let* $(\overset{m}{a}, \overset{3}{j}, \overset{m}{k}, \overset{3}{l}, \overset{m}{e})$ *be a path in* $\mathcal{D}$. *Then either* $d_\Delta(a, e) \leq 2$ *or* $e \in \Delta_3^1(a)$.

Proof First consider an arc $(\overset{m}{a},\overset{0}{b},\overset{3}{c},\overset{0}{d},\overset{m}{e})$. Then by $D_o(M_{22})$ there is an arc $(\overset{0}{b},\overset{2}{f},\overset{m}{g},\overset{0}{d})$ in D_c. Take $i \in D_{abf}^{[3]}$ and $h \in D_{gde}^{[2]}$. Then by the diagram of $\mathcal{D}$ the element i is incident to g while h is incident to e. Thus we have an arc $(\overset{m}{a},\overset{3}{i},\overset{m}{g},\overset{2}{h},\overset{m}{e})$. By (8.4.5) the assertion holds.

Now consider a path $(\overset{m}{a},\overset{3}{j},\overset{m}{k},\overset{3}{l},\overset{m}{e})$ as in the hypothesis of lemma. By the diagram $D_s(M_{24})$ there is a path $(\overset{3}{j},\overset{2}{f},\overset{3}{g},\overset{2}{h},\overset{3}{l})$ in D_k. If $d_\Delta(a,k) \leq 1$ or $d_\Delta(k,e) \leq 1$ then we are done by (8.4.5). So suppose that $\angle ajk = 2^4 = \angle kle$. Then by the diagram $D_h(M_{22})$ applied to the residues of j and l we conclude that

$$\angle ajf \neq 2^4 \cdot 6, \quad \angle elh \neq 2^4 \cdot 6$$

and there exist $b \in \mathcal{D}_{ajf}^{[0]}$ and $d \in \mathcal{D}_{elh}^{[0]}$. By the diagram of $\mathcal{D}$ both b and d are incident to g. Hence we are done by the first paragraph of the proof. ∎

8.5 Earthing up $\Delta_3^2(a)$

Lemma 8.5.1 *Let* $\varphi = (\overset{3}{b},\overset{2}{c},\overset{3}{d},\overset{2}{e},\overset{3}{f})$ *be a path in* $\mathcal{D}$ *such that* $[Z_b^{[3]}, Z_f^{[3]}] \neq 1$. *Then*

(i) $\angle cde = 2^4 \cdot 10$;

(ii) $Q_b^{[3]} \cap Q_f^{[3]} = Z_d^{[3]}$;

(iii) $Q_d^{[3]} \leq (Q_b^{[3]} \cap Q_d^{[3]})Q_e^{[2]}$;

(iv) $Q_b^{[3]} \cap G_e^{[2]}$ *acts transitively on the set of four elements in*

$$\{\alpha \in \mathcal{D}_e^{[3]} \mid [Z_b^{[3]}, Z_\alpha^{[3]}] \neq 1\};$$

(v) $G_{bdf}Q_d^{[3]}/Q_d^{[3]} \cong \mathrm{Sym}_3 \times 2$;

(vi) $(Q_b^{[3]} \cap Q_d^{[3]} \cap G_f^{[3]})N_f^{[3]}/N_f^{[3]} = (Q_d^{[3]} \cap G_f^{[3]})N_f^{[3]}/N_f^{[3]} = O_2(G_{ef}/N_f^{[3]})$.

Proof By the diagram $D_p(M_{22})$ if $\angle cde \neq 2^4 \cdot 10$ then there exists $x \in \mathcal{D}_{cde}^{[m]}$, in which case $Z_c^{[2]}Z_e^{[2]} \leq Q_c^{[m]}$ and $[Z_c^{[2]}, Z_e^{[2]}] = 1$, which contradicts the hypothesis, since $Z_b^{[3]} \leq Z_c^{[2]}$ and $Z_f^{[3]} \leq Z_e^{[2]}$. Hence (i) follows.

We are going to construct some bypasses of the path φ as shown on the following diagram:

Since $\angle cde = 2^4 \cdot 10$, by the diagram $D_p(M_{22})$ there exits a unique path $(\overset{2}{c},\overset{0}{g},\overset{m}{h},\overset{0}{i},\overset{2}{e})$ in $\mathcal{D}_d$ such that $\{i,g\}$ is a $Q_h^{[m]}$-orbit. Then h is not incident to c. On the other hand, by the diagram of $\mathcal{D}$ the element b is incident to g, while the element f is incident to i.

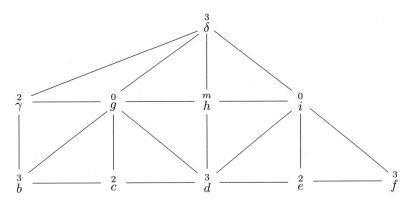

Notice that $\langle V_g(b), V_g(d) \rangle = V_g(c)$, that $V_g(c) \not\leq V_g(h)$ and that $V_g(h) \geq V_g(d)$. Thus $V_g(b) \not\leq V_g(h)$ and so h is not incident to b. By symmetry h is not incident to f. Hence

$$Z_f^{[3]} \not\leq Q_h^{[m]} \cap Q_i^{[0]} = Q_h^{[m]} \cap Q_g^{[0]}.$$

Put $R = Q_g^{[0]} \cap Q_h^{[m]}$. Then $R \cap Q_b^{[3]} = (Q_b^{[3]} \cap Q_g^{[0]}) \cap (Q_h^{[m]} \cap Q_g^{[0]})$. On the other hand, identifying as usual the 2-dimensional subspaces in V_g with the corresponding elements in $Q_g^{[0]}$ we observe the following. The intersection $Q_b^{[3]} \cap Q_g^{[0]}$ is generated by the 2-dimensional subspaces having non-zero intersection with $V_g(b)$, while $Q_h^{[m]} \cap Q_g^{[0]}$ is generated by the 2-dimensional subspaces with non-zero intersection with $V_g(h)$. Since $V_g(b)$ is itself 2-dimensional and $V_g(b)$ intersects $V_g(h)$ in a 1-dimensional subspace, we conclude that the subgroups of order 2 in $R \cap Q_b^{[3]}$ are all of the form $Z_\delta^{[3]}$ for $\delta \in \mathcal{D}_{gh}^{[3]} = \mathcal{D}_{ghi}^{[3]}$ with

$$V_g(b) \cap V_g(h) \leq V_g(\delta) \leq V_g(h).$$

Notice that for any such δ there exists $\gamma \in \mathcal{D}_{bg\delta}^{[2]}$ and so $Z_b^{[3]} \leq Z_\gamma^{[2]} \leq Q_\delta^{[3]}$.

Suppose that $R \cap Q_b^{[3]} \cap Q_f^{[2]} \neq Z_d^{[3]}$ and pick $\delta \in \mathcal{D}_{ghi} \backslash \{d\}$ with $Z_b^{[3]} \leq R \cap Q_b^{[3]} \cap Q_f^{[3]}$. Then $Z_b^{[3]} \leq Q_\delta^{[3]}$, and similarly $Z_f^{[3]} \leq Q_\delta^{[3]}$. Thus both $Z_b^{[3]}$ and $Z_f^{[3]}$ are contained in the elementary abelian group $Q_d^{[3]} \cap Q_\delta^{[3]}$, a contradiction.

Thus $R \cap Q_b^{[3]} \cap Q_f^{[3]} = Z_d^{[3]}$. Since $|R| = 2^6$ and

$$|R \cap Q_b^{[3]}| = 2^3 = |R \cap Q_f^{[3]}|,$$

we get $|R/(R \cap Q_b^{[3]})(R \cap Q_f^{[3]})| = 2$. Since

$$[R \cap Q_b^{[3]}, Q_b^{[3]} \cap Q_f^{[3]}] \leq Z_b^{[3]} \cap R = 1,$$

we have $[(R \cap Q_b^{[3]})(R \cap Q_f^{[3]}), Q_b^{[3]} \cap Q_f^{[3]}] = 1$. Now R is a natural orthogonal module for $G_{ig}/Q_i^{[0]}Q_g^{[0]} \cong \Omega_6^+(2)$, and so no element of G_{ig} acts as a transvection on R. Thus $Q_b^{[3]} \cap Q_f^{[3]} \leq C_{G_{ig}}(R) = Q_i^{[0]}Q_g^{[0]}$. By (7.7.2) we have

$$C_{V_i}(Q_i^{[0]}Q_g^{[0]}) \geq V_i(h), \quad C_{V_i}(Q_f^{[3]}) \geq V_i(f),$$

$$C_{V_i}(Q_i^{[0]}Q_g^{[0]} \cap Q_f^{[3]}) \geq \langle V_i(f), V_i(h) \rangle = V_i,$$

and so $Q_g^{[0]}Q_i^{[0]} \cap Q_f^{[3]} \leq Q_i^{[0]}$. By symmetry $Q_g^{[0]}Q_i^{[0]} \cap Q_b^{[3]} \leq Q_g^{[0]}$ and thus

$$Z_d^{[3]} \leq Q_b^{[3]} \cap Q_f^{[3]} \leq (Q_i^{[0]}Q_g^{[0]} \cap Q_b^{[3]}) \cap (Q_i^{[0]}Q_g^{[0]} \cap Q_f^{[3]})$$
$$= Q_b^{[3]} \cap Q_g^{[0]} \cap Q_i^{[0]} \cap Q_f^{[3]} = Q_b^{[3]} \cap R \cap Q_f^{[3]} = Z_d^{[3]},$$

which gives (ii).

Since $|Q_b^{[3]} \cap Q_d^{[3]}| = 2^7 = |Q_f^{[3]} \cap Q_d^{[3]}|$ and $|Q_d^{[3]}| = 2^{13}$, we conclude that

$$Q_d^{[3]} = (Q_b^{[3]} \cap Q_d^{[3]})(Q_f^{[3]} \cap Q_d^{[3]}) \leq (Q_b^{[3]} \cap Q_d^{[3]})Q_e^{[2]},$$

which is (iii).

By (8.3.12 (vi)) $(Q_b^{[3]} \cap G_d^{[3]})N_d^{[3]}/N_d^{[3]} = O_2(G_{cd}/N_d^{[3]})$ and by (8.3.12 (ii)) $G_{bcd}N_d^{[3]} = G_{cd}$. Thus $G_{bcde}N_d^{[3]} = G_{cde}$. Also

$$(Q_b^{[3]} \cap G_d^{[3]})N_d^{[3]} \cap G_{bcde}N_d^{[3]} = ((Q_b^{[3]} \cap G_d^{[3]}) \cap (G_{bcde}N_d^{[3]}))/N_d^{[3]}$$
$$= (Q_b^{[3]} \cap G_e^{[2]})N_d^{[3]}$$

and

$$(Q_b^{[3]} \cap G_e^{[2]})N_d^{[3]}/N_d^{[3]} = O_2(G_{cd}/N_d^{[3]}) \cap G_{cde}/N_d^{[3]}.$$

It is easy to deduce from the diagram $D_p(M_{22})$ and the structure of the stabilizer $2^5 : \mathrm{Sym}_5$ of a pair in $\mathrm{Aut}\,(M_{22})$ that $G_{bde}Q_d^{[3]}/(Q_b^{[3]} \cap G_e^{[2]})Q_d^{[3]} \cong \mathrm{Sym}_3 \times 2$, while $(Q_b^{[3]} \cap G_e^{[2]})Q_d^{[3]}/Q_d^{[3]}$ has order two and inverts $N_d^{[3]}/Q_d^{[3]} \cong 3$. Since $N_d^{[3]}Q_e^{[2]}/Q_e^{[2]} \cong \mathrm{Alt}_4$ and since by (iii) the equality $Q_d^{[3]}Q_e^{[2]}/Q_e^{[2]} = (Q_b^{[3]} \cap Q_d^{[3]})Q_e^{[2]}/Q_e^{[2]}$ holds, we conclude that $Q_b^{[3]} \cap G_e^{[2]}$ acts as the dihedral group D_8 of order eight on $Z_e^{[2]}$ with $Z_b^{[3]}$ mapping onto the centre of D_8. Hence (iv) follows. Furthermore,

$$[Q_b^{[3]} \cap G_f^{[3]}, Z_e^{[2]}] = [Q_b^{[3]} \cap G_f^{[3]}, C_{Z_e^{[2]}}(Z_b^{[3]})Z_f^{[3]}] \leq Z_d^{[3]},$$

$$Q_b^{[3]} \cap G_f^{[3]} \leq Q_d^{[3]} \quad \text{and} \quad G_{bdf}Q_d^{[3]}/Q_d^{[3]} \cong \mathrm{Sym}_3 \times 2,$$

which is (v). Since by (iii)

$$(Q_b^{[3]} \cap G_f^{[3]})N_f^{[3]} = (Q_b^{[3]} \cap Q_d^{[3]} \cap G_f^{[3]})N_f^{[3]} = (Q_d^{[3]} \cap G_f)N_f^{[3]},$$

(vi) follows from (8.3.12 (vi)). ∎

Lemma 8.5.2 *The group G acts transitively on the set of paths*

$$\rho = (\overset{m}{a}, \overset{3}{b}, \overset{2}{c}, \overset{3}{d}, \overset{2}{e}, \overset{3}{f}, \overset{m}{g})$$

in $\mathcal{D}$ such that $\angle abc = 2^4 \cdot 6 = \angle gfe$ and $[Z_b^{[3]}, Z_f^{[3]}] \neq 1$. Moreover, for such a path ρ the following assertions hold:

(i) $C_{G_{ag}}(Z_d) = G_{adg} = G_{abcdefg}$;

(ii) $G_{adg}Q_d^{[3]}/Q_d^{[3]} \cong \mathrm{Sym}_3 \times 2$;

(iii) $Q_d^{[3]} \cap G_{adg} = Z_d^{[3]}$;

(iv) $|G_{adg}| = 24$;

(v) $Q_a^{[m]} \cap G_g^{[m]} = 1$;

(vi) $d_\Delta(a, g) \geq 3$ and $g \notin \Delta_3^1(a)$;

(vii) *there exists a path*

$$\pi = (\overset{m}{a}, \overset{2}{l}, \overset{0}{k}, \overset{2}{p}, \overset{m}{i}, \overset{2}{n}, \overset{m}{g})$$

in $\mathcal{D}$ such that

(1) $|G_{adg}/(G_d^{[3]} \cap G(\pi))| = 3$;

(2) $i \in \mathcal{D}_d^{[m]}$;

(3) $Z_d^{[3]} \leq Q_i^{[m]} \cap G_d^{[3]}(\pi)$.

Proof By (8.3.16) there is a unique G-orbit of the set of paths $(\overset{m}{a}, \overset{3}{b}, \overset{2}{c}, \overset{3}{d})$ with $\angle abc = 2^4 \cdot 6$. Furthermore, by (8.3.16 (ii), (iv), (v)) we have respectively

$$G_{abcd}N_d^{[3]} = G_{cd},$$

$$G_{abcd}Q_a^{[m]}/Q_a^{[m]} \text{ is the centralizer of } Z_d^{[3]} \text{ in } G_a^{[m]}/Q_a^{[m]} \text{ and}$$

$$G_{abcd} = C_{G_a^{[m]}}(Z_d^{[3]}) = G_{ad}.$$

The latter equality implies that in (i).

From (8.5.1 (i), (ii), (iv)) we conclude that $\angle cde = 2^4 \cdot 10$, $Q_b^{[m]} \cap Q_f^{[3]} = Z_d^{[3]}$ and that $Q_b^{[m]} \cap G_e^{[2]}$ acts transitively on $Z_e^{[2]} \setminus Z_b^{[3]}$. Thus the G-orbit of the subpath in ρ which joins a and f is uniquely determined. Next by (8.5.1 (v)) we have $G_{bdf}Q_d^{[3]}/Q_d^{[3]} \cong \mathrm{Sym}_3 \times 2$. It follows from the diagram $D_p(M_{22})$ that $O_2(G_{ef}/N_f^{[3]})$ acts regularly on $X = \{\alpha \in \mathcal{D}_f^{[m]} \mid \angle ef\alpha = 2^5\}$. Finally by (8.5.1 (vi))

$$O_2(G_{ef}/N_f^{[3]}) = (Q_b^{[3]} \cap Q_d^{[3]} \cap G_f)N_f^{[3]}/N_f^{[3]} = (Q_d^{[3]} \cap G_f^{[3]})N_f^{[3]}/N_f^{[3]},$$

$Q_b^{[3]} \cap Q_d^{[3]} \cap G_f^{[3]}$ acts transitively on X and $Q_d^{[3]} \cap G_f^{[3]} \leq Q_f^{[3]}$. Similarly, $Q_d^{[3]} \cap Q_f^{[3]} \cap G_b^{[3]}$ acts transitively on $Y = \{\beta \in \mathcal{D}_b^{[m]} \mid \angle cb\beta = 2^5\}$, and $Q_d \cap G_{ab} \leq Q_b^{[3]}$. Since

$$(Q_b^{[3]} \cap Q_d^{[3]}) \cap (Q_f^{[3]} \cap Q_d^{[3]}) = Z_d^{[3]}$$

the transitivity assertion follows along with (ii).

Notice that

$$C_{Q_b^{[3]} \cap G_g^{[m]}}(Z_d^{[3]}) \leq Q_b^{[3]} \cap G_{gf} \leq Q_b^{[3]} \cap Q_f^{[3]} = Z_d^{[3]},$$

and so $Q_b^{[3]} \cap G_g^{[m]} = Z_d^{[3]}$. Thus

$$Q_d^{[3]} \cap G_{adg} = Q_b^{[3]} \cap Q_d^{[3]} \cap Q_f^{[3]} = Z_d^{[3]},$$

which is (iii). Furthermore, since

$$|G_{adg}| = \frac{|G_{ab}|}{\angle abc \cdot \angle bcd \cdot \angle cde \cdot \angle def \cdot \angle efg} = \frac{|G_{ab}|}{96 \cdot 6 \cdot 160 \cdot 4 \cdot 32} = 24$$

(which is (iv)) and $Q_a^{[m]} \cap G_c^{[2]} \leq Q_b^{[3]}$, we obtain the equalities

$$Q_a^{[m]} \cap G_{dg} = Q_a^{[m]} \cap Q_b^{[3]} \cap G_g^{[m]} = Q_a^{[m]} \cap Z_d^{[3]} = 1.$$

Hence $C_{Q_a^{[m]} \cap G_g^{[m]}}(Z_d^{[3]}) = 1$ and $Q_a^{[m]} \cap G_g^{[m]} = Q_a^{[m]} \cap G_{ag} = 1$ and (v) follows.

By (v) $d_\Delta(a, g) \geq 3$. Suppose that $g \in \Delta_3^1(a)$. Then by (8.4.3 (vi)) G_{ag} contains a normal elementary abelian subgroup A of order 2^6. If $Z_d^{[3]}$ is in A, then 2^6 divides $|C_{G_{ag}}(Z_d^{[3]})|$, and if $Z_d^{[3]} \notin A$, then 2^3 divides $|C_A(Z_d)|$ and so 2^4 divides $C_A(Z_d)Z_d|$, and in any case we reach a contradiction with (iv) (which is $|G_{adg}| = 24$) and hence (vi) follows.

The proof of (vii) is illustrated by the following diagram:

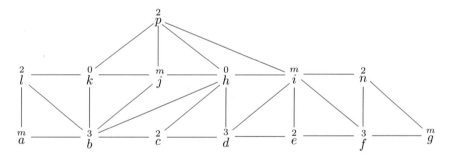

By the diagram $D_p(M_{22})$ in $\mathcal{D}_d$ there are three paths $(\overset{2}{c}, \overset{0}{h}, \overset{m}{i}, \overset{2}{e})$. Since an element of order 3 in $N_d^{[3]}$ acts fixed-point freely on $Q_d^{[3]}/Z_d^{[3]}$, since $Z_b^{[3]} \leq Q_d^{[3]}$ and since $N_d^{[3]} \cap G_b^{[3]} \leq Q_d^{[3]}$, these three paths are transitively permuted by G_{cde}. We choose one of these paths. Then h is incident to b, and since $\angle abc = 2^4 \cdot 6$, it follows from the diagram $D_h(M_{22})$ there is a unique path $(\overset{m}{a}, \overset{2}{l}, \overset{0}{k}, \overset{m}{j}, \overset{0}{h})$ in $\mathcal{D}_b$ such that $\{k, h\}$ is a $Q_j^{[m]}$-orbit. Let p be the unique vertex in $\mathcal{D}_{jhi}^{[2]}$. Since $\{k, h\}$ is a $Q_h^{[m]}$-orbit, p is incident to k. Since $\angle gfe = 2^4 \cdot 6$, i is incident to both e and f. Now $D_h(M_{22})$ shows that there is a unique $n \in \mathcal{D}_{gfi}^{[2]}$. Then by the construction

$\pi = (\overset{m}{a}, \overset{2}{l}, \overset{0}{k}, \overset{2}{p}, \overset{m}{i}, \overset{2}{n}, \overset{m}{g})$ possesses the properties (1) and (2). Since h and i are incident to d, $Z_d^{[3]} \leq G_{adghi} = G_d^{[3]}(\pi)$ and $Z_d^{[3]} \leq Q_i^{[m]}$, which gives (3). ∎

Lemma 8.5.3 *The group G acts transitively on the set of paths*

$$\pi = (\overset{m}{a}, \overset{2}{b}, \overset{0}{c}, \overset{2}{d}, \overset{m}{e}, \overset{2}{f}, \overset{m}{g})$$

satisfying $d_\Delta(a, g) \geq 3$ and $g \notin \Delta_3^1(a)$. Moreover, for such a path π

$$|G(\pi)| = 24 \quad \text{and} \quad G(\pi)/O_2(G(\pi)) \cong \mathrm{Sym}_3.$$

Proof The existence of π has been established in (8.5.2). Suppose there is $x \in \mathcal{D}_{bcd}^{[3]}$. Then by the diagram of $\mathcal{D}$ the element x is incident to both a and e, which contradicts (8.4.6), so $\langle V_c(b), V_c(d) \rangle = V_c$ and the subpath in π joining a and e is as in (8.3.13).

Suppose that $Z_d^{[2]} \cap G_a^{[m]} \cap Q_f^{[2]} \neq 1$. We will reach a contradiction by extending the path π as shown on the following diagram:

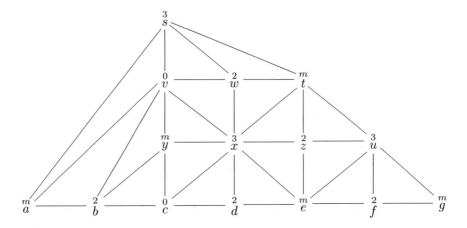

Under the above assumption pick $x \in \mathcal{D}_{dc}^{[3]}$ with $Z_x^{[3]} \leq Z_d^{[2]} \cap G_a^{[m]} \cap Q_f^{[2]}$. Then x is incident to c and $Z_x^{[3]} \leq Q_c^{[0]}$. Since $Q_c^{[0]} \cap G_a^{[m]} \leq Q_b^{[2]}$, we get $Z_x^{[3]} \leq Q_b^{[2]}$. Thus by (8.3.11) $V_c(x) \cap V_c(b) \neq 0$. In particular there exists $y \in \mathcal{D}_{bcx}^{[m]}$. Since $Z_x^{[3]} \leq Q_f^{[2]}$, (8.3.8 (iii)) implies that $\angle fex \neq 2^6 \cdot 21$ and so by $D_s(M_{24})$ there exists a path $(\overset{3}{x}, \overset{2}{z}, \overset{3}{u}, \overset{3}{f})$ in $\mathcal{D}_e$. Then g is incident to u. Let $\{c, v\}$ be the $Q_y^{[m]}$-orbit containing c. Then in the residue of b we observe that a is incident to v, since it is not incident to c. Clearly v is also incident to x. By the diagram $D_p(M_{22})$ there exists a path $(\overset{0}{v}, \overset{2}{w}, \overset{m}{t}, \overset{2}{z})$ in the residue $\mathcal{D}_x$. Then u is incident to t and g. Pick $s \in \mathcal{D}_{avw}^{[3]}$. Then s is incident to t and we have constructed a path $(\overset{m}{a}, \overset{3}{s}, \overset{m}{t}, \overset{3}{u}, \overset{m}{g})$ contrary to (8.5.1).

Hence $Z_d^{[2]} \cap G_a^{[m]} \cap Q_f^{[2]} = 1$. If $\angle def \neq 2^6 \cdot 42$ then the diagram $D_s(M_{24})$ and (8.3.8 (iii)) imply that $|Z_d^{[2]} \cap Q_f^{[2]}| \geq 4$. By (8.3.13 (i)) $|Z_d^{[2]} \cap G_a^{[m]}| = 2^2$. Since $Z_d^{[2]}$ has order 8, the latter means that $Z_d^{[2]} \cap Q_f^{[2]} \cap G_a^{[m]} \neq 1$. So $\angle def = 2^6 \cdot 42$, $|Z_d^{[2]} \cap Q_f^{[2]}| = 2$ and by (8.3.13 (v)) $G_{ae}Q_e^{[m]} = N_{G_{de}}(Z_d^{[2]} \cap G_a^{[m]})$. Put $X = C_{G_{de}}(Z_d)$. Now it is easy to deduce from the diagram $D_t(M_{24})$ that $G_{def}X = N_{G_{de}}(Z_d^{[2]} \cap Q_f^{[2]})$ and $G_{def}/Q_e^{[m]} \cong \mathrm{Sym}_4$. It follows that G_{def} acts transitively on the set

$$\{A \leq Z_d^{[2]} \big| |A| = 4, A \cap Q_f^{[2]} = 1\}.$$

Moreover, $N_{G_{def}}(A)/Q_e^{[m]} \cong \mathrm{Sym}_3$ for any such A. Thus both $N_{G_{de}}(Z_d^{[2]} \cap G_a^{[m]})$ and G_{ae} act transitively on

$$\{f \in \mathcal{D}_e^{[2]} \mid \angle def = 2^6 \cdot 42, \ Z_d^{[2]} \cap G_a^{[m]} \cap Q_f^{[2]} = 1\}.$$

Also the following isomorphisms take place

$$G_{aef}Q_e^{[m]}/Q_e^{[m]} \cong \mathrm{Sym}_3 \cong G_{aef}/O_2(G_{aef}).$$

Moreover, $Q_e^{[m]} = (Z_d^{[2]} \cap G_a^{[m]})(Q_e^{[m]} \cap Q_f^{[2]})$ and so $Z_d^{[2]} \cap G_a^{[m]}$ acts transitively on $\mathcal{D}_f^{[m]} \setminus \{e\}$. Thus the paths as in the hypothesis constitute a single G-orbit and

$$G(\pi)Q_e^{[m]}/Q_e^{[m]} \cong \mathrm{Sym}_3 \cong G(\pi)/O_2(G(\pi)).$$

Finally

$$|G_{abcdefg}| = \frac{|G_{ae}|}{\frac{4}{7} \cdot (2^6 \cdot 42) \cdot 4} = 24$$

and the lemma is proved.   ■

Let g be a vertex of Δ such that the path π as in (8.5.2 (vii)) and (8.5.3) joins a with g. Let $\Delta_3^2(a)$ be the set of images of g under the elements of $G_a^{[m]}$.

Lemma 8.5.4 *The set $\Delta_3^2(a)$ is an orbit of $G_a^{[m]}$ on $\Delta_3(a)$.*

Proof We assume that π is a path which joins a and g as in (8.5.3). In view of the definition we only have to show that $d_\Delta(a, g) = 3$. By (8.5.2 (vi)) $d_\Delta(a, g) \geq 3$. By the first paragraph of the proof of (8.5.3) along with (8.3.13), (8.3.14), and (8.3.15) it follows that $d_\Delta(a, e) = 2$. Clearly $d_\Delta(e, g) = 1$ by the shape of π, the result follows.   ■

Lemma 8.5.5 *Let $g \in \Delta_3^2(a)$. Then*

(i) $|G_{ag}| = 2^3 \cdot 3 \cdot 11 \cdot 23$;

(ii) G_{ag} *has two orbits on $\mathcal{D}_g^{[0]}$ and permutes transitively the $Q_g^{[m]}$-orbits on $\mathcal{D}_g^{[0]}$.*

Proof Let the paths

$$\rho = (\overset{m}{a}, \overset{3}{b}, \overset{2}{c}, \overset{3}{d}, \overset{2}{e}, \overset{3}{f}, \overset{m}{g})$$

and

$$\pi = (\overset{m}{a}, \overset{3}{l}, \overset{0}{k}, \overset{2}{p}, \overset{m}{i}, \overset{2}{n}, \overset{m}{g})$$

be as in (8.5.2). Then by (8.5.2) and (8.5.3) we have the following:

$$|G_{adg}| = 24 = |G(\pi)|, \quad |G_d \cap G(\pi)| = 8,$$

$$C_{G_{ag}}(Z_d^{[3]}) = G_{adg}, \quad G_{ag} \cap Q_g^{[m]} = 1,$$

$$G_{adg}/Z_d^{[3]} \cong \mathrm{Sym}_3 \times 2, \quad G(\pi)/O_2(G(\pi)) \cong \mathrm{Sym}_3.$$

Put $A = Q_i^{[m]}(\pi)$. Then by (8.5.2 (3)) $Z_d^{[3]} \leq A$. Hence A is a non-trivial normal 2-subgroup of $G(\pi)$, and $C_{G(\pi)}(A) \leq G_d^{[3]} \cap G(\pi)$ is a 2-group. Since $|O_2(G(\pi))| = 4$, we get that A is elementary abelian of order 4 and that $G(\pi) \cong \mathrm{Sym}_4$. Thus $G_d(\pi)$ is a dihedral group of order 8, and so $N_{G_{ag}}(G_d^{[3]}(\pi)) \leq C_{M_{ag}}(Z_d^{[3]})$. In particular, $G_d^{[3]}(\pi)$ is a Sylow 2-subgroup of G_{ag}. Moreover, there exists an element $t \in G$ such that

$$(t(a), t(b), t(c), t(d), t(e), t(f), t(g)) = (g, f, e, d, c, b, a).$$

Notice $t \in G_d^{[3]}$, and so t normalizes G_{adg}. Thus we may assume that t normalizes $G_d^{[3]}(\pi)$.

We claim that $A \cap A^t = Z_d^{[3]}$. Clearly, $Z_d^{[3]} \leq A \cap A^t$. By (8.5.2) i is incident to d and since $t \in G_d^{[3]}$, $t(i)$ is also incident to d. Since $d_\Delta(a, g) > 2$, we have $i \neq t(i)$. We claim that $Q_i^{[m]} \cap (Q_i^{[m]})^t \leq G_d^{[3]}$. In fact, if $\mathcal{D}_{idt(i)}^{[2]} = \emptyset$ (i.e. if $\angle idt(i) = 2^4$), then $Q_i^{[m]} \cap (Q_i^{[m]})^t \leq O_2(G_{idt(i)}) \leq Q_d^{[3]}$, and if $\delta \in \mathcal{D}_{idt(i)}^{[2]}$, then by (8.2.5) $Q_i^{[m]} \cap (Q_i^{[m]})^t = Z_\delta^{[2]} \leq Q_d^{[3]}$. By (8.5.2 (iii)) $Q_d^{[3]} \cap G_{ag} = Z_d^{[3]}$, and so

$$A \cap A^t \leq G_{ag} \cap Q_i^{[m]} \cap (Q_i^{[m]})^t \leq G_{ag} \cap Q_d \leq Z_d^{[3]}.$$

In particular, $A \neq A^t$. Put $E = O_2(G_{adg})$. Since $G_{adg}/Z_d^{[3]} \cong \mathrm{Sym}_3 \times 2$, $|E| = 4$. Moreover, t normalizes E, and so $A \neq E \neq A^t$, E is cyclic of order 4 and G_{adg} is a dihedral group of order 24. Let $D = O_3(G_{adg})$ and note that $ED = C_{G_{ag}}(Z_d^{[3]}D)$. Since D centralizes $Z_d^{[3]}$, the information on involution centralizers in M_{24} implies that

$$C_{G_g^{[m]}/Q_g^{[m]}}(D) \cong 3 \times L_3(2).$$

Now a subgroup of $L_3(2)$ with a centralizer of an involution isomorphic to a cyclic group of order four clearly is a cyclic group of order four, and so $C_{G_{ag}}(D) = DE$. In particular, D is a Sylow 3-subgroup of G_{ag}. Note that all involutions in

$G_d^{[3]}(\pi)$ are contained in $A \cup A^t$ and so conjugate into $Z_d^{[3]}$ under $G(\pi)$ and $G(\pi)^t$, respectively. Thus G_{ag} has a unique class of involutions. Let z be any involution in G_{ag} and put

$$C(z) = G_{adg} \cap C_{G_{ag}}(z).$$

If 3 divides $|C(z)|$, $z \in C_{G_{ag}}(D) = DE$ and $z \in Z_d^{[3]}$. Hence exactly one of the following holds: $z \in G_{adg}$, $|C(z)| = 2$, or $C(z) = 1$. Moreover, if $C(z) = \langle y \rangle$ for one of the 12 involutions $y \in G_{adg} \setminus Z_d^{[3]}$, then z is one of the 10 involutions in $C_{G_{ag}}(y) \setminus G_{adg}$. Thus, if r is the number of involutions in G_{ag}, that is, $r = |G_{ag}/G_{adg}|$, then

$$r = 13 + 12 \cdot 10 + 24s = 133 + 24s$$

for some non-negative integer s. On the other hand, since

$$|G_g^{[m]}/Q_g^{[m]}| = 2^{10} \cdot 3^3 \cdot 5 \cdot 7 \cdot 11 \cdot 23,$$

r divides $5 \cdot 7 \cdot 11 \cdot 23$, and we conclude that $r = 11 \cdot 23$ or $r = 5 \cdot 7 \cdot 23$. The latter case is impossible by Burnside's p-complement theorem for $p = 23$, and so $r = 11 \cdot 23$.

Hence $|G_{ag}| = 2^3 \cdot 3 \cdot 11 \cdot 23$. In particular, G_{adg} and $G(\pi)$ are maximal $\{2,3,5,7\}$-subgroups of G_{ag}. Since both G_{fg} and G_{ng} are $\{2,3,5,7\}$-groups, we conclude that $G_{afg} = G_{adg}$ and $G_{ang} = G(\pi)$.

Since $|\mathcal{D}_{ing}^{[0]}| = 3$, we can choose $x \in \mathcal{D}_{nig}^{[0]}$ with $G_d^{[3]}(\pi) \le G_x^{[0]}$. Since the non-trivial elements of odd order in G_{ag} act fixed-point freely on $\mathcal{D}_g^{[2]}$, we conclude that $G_{agx} = G_d^{[3]}(\pi) = G_{agy} = G_{ag}\{x,y\}$, where $\{x,y\}$ is a $Q_g^{[m]}$-orbit. In particular, $|G_{ag}/G_{agx}| = 759$, and the lemma is proved.   ∎

It can be shown using the list maximal subgroups in M_{24} and/or the classification of groups with dihedral Sylow 2-subgroup that in the above lemma G_{ag} is isomorphic to $L_2(23)$.

As a consequence of (8.5.5 (i)) and direct calculations we have the following.

Corollary 8.5.6 $|\Delta_3^2(a)| = 2^{18} \cdot 3^2 \cdot 5 \cdot 7 = 82,575,360.$

We have constructed seven orbits of $G_a^{[m]}$ (including a itself) on the vertex-set of Δ. The information about these orbits from (8.1.1 (vii)), (8.1.3 (iv)), (8.3.15 (i)), (8.3.19 (i)), (8.4.4 (ii)), and (8.5.6) is summarized in Table 3 below. This table shows that we already see the required number of vertices. In the next section we show that all the vertices of Δ are in the accounted orbits.

In the last column of Table 3 the stabilizers G_{ae} are given in the form $(G_{ae} \cap Q_a^{[m]}).(G_{ae}Q_a^{[m]}/Q_a^{[m]})$.

8.6 $|G| = 2^{21} \cdot 3^3 \cdot 5 \cdot 7 \cdot 11^3 \cdot 23 \cdot 29 \cdot 31 \cdot 37 \cdot 43$

Lemma 8.6.1 *Let $g \in V(\Delta)$ and $d_\Delta(a,g) = 3$ then $g \in \Delta_3^1(a) \cup \Delta_3^2(a)$.*

TABLE 3. $G_a^{[m]}$-orbits on Δ

Orbit	Size	Prime decomposition	Stabilizer
$\{a\}$	1	1	$(2^{11}) : (M_{24})$
$\Delta(a)$	15,180	$2^2 \cdot 3 \cdot 5 \cdot 11 \cdot 23$	$(2^9) \cdot (2^6.(L_3(2) \times \mathrm{Sym}_3))$
$\Delta_2^1(a)$	28,336	$2^4 \cdot 7 \cdot 11 \cdot 23$	$(2^7) \cdot (2^6 : 3 \cdot \mathrm{Sym}_6)$
$\Delta_2^2(a)$	3,400,320	$2^7 \cdot 3 \cdot 5 \cdot 7 \cdot 11 \cdot 23$	$(2^4) \cdot (2^6 \cdot (\mathrm{Sym}_4 \times \mathrm{Sym}_3))$
$\Delta_2^3(a)$	32,643,072	$2^{11} \cdot 3^2 \cdot 7 \cdot 11 \cdot 23$	$(2) \cdot (2^6.\mathrm{Sym}_5)$
$\Delta_3^1(a)$	54,405,120	$2^{11} \cdot 3 \cdot 5 \cdot 7 \cdot 11 \cdot 23$	$(1) \cdot (2^6 \cdot (\mathrm{Sym}_4 \times \mathrm{Sym}_3))$
$\Delta_3^2(a)$	82,575,360	$2^{18} \cdot 3^2 \cdot 5 \cdot 7$	$(1) \cdot (L_2(23))$
Total	173,067,389	$11^2 \cdot 29 \cdot 31 \cdot 37 \cdot 43$	

Proof Let e be a vertex of Δ adjacent to g which is at distance 2 from a. If $e \in \Delta_2^1(a)$ then by (8.4.5) $g \in \Delta_3^1(a)$. If $e \in \Delta_2^2(a)$ then $g \in \Delta_3^2(a)$ by (8.5.3), (8.5.4), (8.4.3), and (8.3.13).

So we assume that $e \in \Delta_2^3(a)$. Then by (8.3.18) there exists a path

$$(\overset{m}{a}, \overset{3}{b}, \overset{2}{c}, \overset{3}{d}, \overset{m}{e}, \overset{2}{f}, \overset{m}{g})$$

with $\angle abc = 2^4 \cdot 6 = \angle edc$. We are going to produce an alternative path in $\mathcal{D}$ joining a and g as on the below diagram.

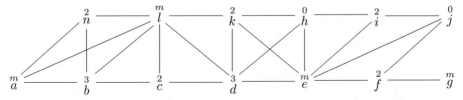

By the diagram $D_s(M_{24})$ there is an arc $(\overset{3}{d}, \overset{0}{h}, \overset{2}{i}, \overset{0}{j}, \overset{2}{f})$ in $\mathcal{D}_e$. Note that $\angle cde = 2^5$ and hence there is $k \in \mathcal{D}_{deh}^{[2]}$ with $\angle cdk = 2^3 \cdot 5$. Thus by the diagram $D_p(M_{22})$ there exists $l \in \mathcal{D}_{cdk}^{[m]}$. Then by the diagram of $\mathcal{D}$ the elements l and b are incident. Next, by the diagram $D_h(M_{22})$ and since $\angle abc = 2^4 \cdot 6$, there exists $n \in \mathcal{D}_{abl}^{[2]}$, and so a and l are incident. Considering the arc $(\overset{2}{k}, \overset{0}{h}, \overset{2}{i}, \overset{0}{j}, \overset{2}{f})$ contained in $\mathcal{D}_e$, we see by the diagram $D_s(M_{24})$ that $\angle kef \neq 2^6 \cdot 42$. Thus by (8.3.18) applied to the arc $(\overset{m}{g}, \overset{2}{f}, \overset{m}{e}, \overset{2}{k}, \overset{m}{l})$ we observe that

$$d_\Delta(l, g) \leq 3 \quad \text{and} \quad l \notin \Delta_2^3(g).$$

Thus by the first paragraph of the proof applied to (g, l, a) in place of (a, e, g), we obtain the inclusion $a \in \Delta_3^1(g) \cup \Delta_3^2(g)$. In view of the obvious symmetry between a and g the result is established. ∎

Proposition 8.6.2 *If z is any vertex of Δ then $d_\Delta(a, z) \leq 3$.*

Proof Suppose the contrary, then by the connectivity of Δ we may assume that $d_\Delta(a, z) = 4$.

We are going to establish the following claim:

$$\text{there does not exist a path } \pi = (\overset{m}{a}, \overset{3}{b}, \overset{m}{c}, \overset{2}{d}, \overset{m}{e}, \overset{2}{f}, \overset{m}{z}).$$

Suppose to the contrary that such a path π exists. Then $d_\Delta(a, e) = 3$, $d_\Delta(c, z) = 2$, by (8.3.2 (iii)) and (8.4.1) this means that $\angle abc = 2^4$, $\angle bcd = 2^6 \cdot 45$ and

$$c \in \Delta_2^\alpha(z) \text{ for } \alpha = 1, \ 2, \text{ or } 3.$$

We are going to consider the three possible values for α successively.

$\alpha = 1$: In this case there exists $\rho \in \mathcal{D}_{cz}^{[3]}$ and we obtain a path

$$(\overset{m}{a}, \overset{3}{b}, \overset{m}{c}, \overset{3}{\rho}, \overset{m}{z}).$$

Hence by (8.4.6) $d_\Delta(a, z) \leq 3$, a contradiction.

$\alpha = 2$: We are going to reduce the path π to a path with $\alpha = 1$. This reduction process is illustrated on the following diagram.

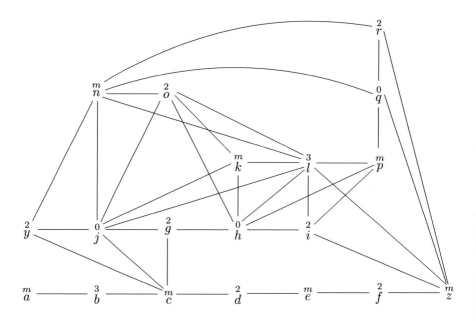

We start by choosing a path $(\overset{m}{c}, \overset{2}{g}, \overset{0}{h}, \overset{2}{i}, \overset{m}{z})$ with $\langle V_h(g), V_h(i) \rangle = V_h$. By the diagram $D_s(M_{24})$ there exists $j \in \mathcal{D}_{cg}^{[0]}$ with $\angle bcj \neq 2^6 \cdot 6$. Replacing, if necessary j by the other vertex from the same $Q_c^{[m]}$-orbit we assume that $\{j, h\}$ is a $Q_k^{[m]}$-orbit for some $k \in \mathcal{D}_{gh}^{[m]}$.

Next take $l \in \mathcal{D}^{[3]}_{khi}$. Then l is incident to j and z. Since $c \in \Delta^2_2(z)$, the set $\mathcal{D}^{[3]}_{cz}$ is empty and hence l is not incident to c. Put

$$\Lambda = \{V_j(c)/V_j(x) \mid x \in \mathcal{D}^{[2]}_{cj}, V_j(x) \geq V_j(c) \cap V_j(l)\}$$

and

$$\Theta = \{V_j(c)/V_j(x) \mid x \in \mathcal{D}^{[2]}_{cj}, \angle bcx \neq 2^6 \cdot 45\}.$$

Then Λ is the set of 1-dimensional factor-spaces of a 4-dimensional space $V_j(c)$. Moreover, since $\angle bij \neq 2^6 \cdot 6$, we deduce from the diagram $D_s(M_{24})$ that Θ is the set of 1-dimensional factor-spaces of $V_j(c)$ over subspaces containing a given 1- or 2-dimensional subspace. Thus

$$|\Theta \cap \Lambda| \geq 3$$

and there exists $y \in \mathcal{D}^{[2]}_{cj}$ with $y \neq g$, $\angle bcy = 2^6 \cdot 45$ and $V_j(y) \geq V_j(c) \cap V_j(l)$. In particular $\langle V_j(y), V_j(l) \rangle = V_j(n)$ for some $n \in \mathcal{D}^{[m]}_{yjl}$. If $n = k$ then $V_j(g) = V_j(k) \cap V_j(c) = V_j(y)$ contrary to the inequality $m \neq g$. Thus $n \neq k$ and there exists a unique $o \in \mathcal{D}^{[2]}_{kjn}$. Since $V_j(o) = V_j(k) \cap V_j(n) \geq V_j(l)$, we have $o \in \mathcal{D}^{[2]}_{kjn}$. Since $V_j(o) = V_j(k) \cap V_j(n) \geq V_j(l)$, the element o is incident to l. Since $\{h, j\}$ is a $Q^{[m]}_k$-orbit, o and i are both incident to l and h. Hence there exists $p \in \mathcal{D}^{[m]}_{ihol}$. Let $\{q, p\}$ be the $Q^{[m]}_h$-orbit containing p. Since n is incident to j and o, we conclude from some elementary properties of the Petersen graph Θ_o, that n is not incident to h. Since $p \in \mathcal{D}^{[m]}_o$, we conclude that n is incident to q. Similarly, since z is not incident to h, calculating in the residue $\mathcal{D}_i$ we observe that z is incident to q. Hence there exists $r \in \mathcal{D}^{[2]}_{nqz}$ and thus we have found a path

$$\begin{pmatrix} m & 3 & m & 2 & m & 2 & m \\ a, & b, & c, & y, & n, & r, & z \end{pmatrix}$$

with $\angle bcy \neq 2^6 \cdot 45$ contrary to the case $\alpha = 1$ of the proof, which is already settled.

$\alpha = 3$: Finally we assume that $c \in \Delta^3_2(z)$ and by (8.3.17) we choose a path $\begin{pmatrix} m & 3 & 2 & 3 & m \\ c, & g, & h, & i, & z \end{pmatrix}$ with $\angle cgh = 2^4 \cdot 6 = \angle zih$. Regard $\mathcal{D}^{[2]}_{cg}$ and $\mathcal{D}^{[0]}_{cg}$ as the 1-dimensional and 2-dimensional isotropic subspaces of a non-singular 4-dimensional symplectic space S over $GF(2)$. Using the diagram $D_s(M_{24})$ we will show that there exits $y \in \mathcal{D}^{[0]}_{cg}$ such that $\angle bcx \neq 2^6 \cdot 42$ for all $x \in \mathcal{D}^{[2]}_{cgy}$. Indeed, if $\angle bcg \neq 2^5 \cdot 45$, we choose y so that x is incident to b. If $\angle bcg = 2^5 \cdot 45$, there exists $u \in \mathcal{D}^{[0]}_{cg}$ such that $v \in \mathcal{D}^{[2]}_{cg}$ is perpendicular to u in S if and only if $\angle bcv \neq 2^6 \cdot 45$. Choose $y = u$ in this case. By the diagram $D_h(M_{22})$ there exists $x \in \mathcal{D}^{[2]}_{cgy}$ with $\angle ghx = 2^3 \cdot 5$. Hence by $D_p(M_{22})$ there exists $n \in \mathcal{D}^{[m]}_{xgh}$. Then n is incident to h and i, and since $\angle zih = 2^4 \cdot 6$, there exists $r \in \mathcal{D}^{[2]}_{inz}$, and again we obtain a path

$$\begin{pmatrix} m & 3 & m & 2 & m & 2 & m \\ a, & b, & c, & x, & n, & r, & z \end{pmatrix}$$

with $\angle bcx \neq 2^6 \cdot 45$ which brings us back to the $\alpha = 1$ case. The path extension we have just performed is illustrated on the following diagram:

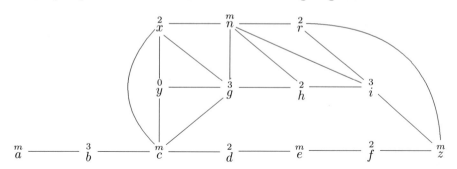

This completes the proof of the claim. We continue to assume that $d_\Delta(a, z) = 4$. Let g be a vertex adjacent to z such that $d_\Delta(g, a) = 3$. By (8.6.1) $g \in \Delta_3^1(a) \cup \Delta_3^2(a)$. By the above claim we have $g \notin \Delta_3^1(a)$ and therefore $g \in \Delta_3^2(a)$. Pick a path $(\overset{m}{e}, \overset{2}{f}, \overset{m}{g}, \overset{3}{h}, \overset{m}{z})$ with $d_\Delta(e, a) = 2$. Let $j \in \mathcal{D}_{ghz}^{[0]}$. Then by (8.5.5 (ii)) and its proof there exists $t \in G_{ag}$ such that $t(j)$ is incident to f, and so $t^{-1}(f)$ is incident to j. Hence, replacing (e, f) by $(t^{-1}(e), t^{-1}(f))$, we may assume that f is incident to j. Since $V_j(g) \geq \langle V_j(f), V_j(h) \rangle$, there exists $k \in \mathcal{D}_{fjh}^{[3]}$. Then k is adjacent to e and z, and we get a contradiction to the claim applied, with the roles of a and z interchanged. $\blacksquare$

By (8.6.1), (8.6.2), and the paragraph after (8.5.6) we deduce that

$$V(\Delta) = \{a\} \cup \Delta(a) \cup \Delta_2^1(a) \cup \Delta_2^2(a) \cup \Delta_2^3(a) \cup \Delta_3^1(a) \cup \Delta_3^2(a).$$

In view of Table 3 this equality gives the following:

Proposition 8.6.3

$$|V(\Delta)| = 173,067,389;$$

$$|G| = |G_a^{[m]}| \cdot |V(\Delta)| = 2^{21} \cdot 3^3 \cdot 5 \cdot 7 \cdot 11^3 \cdot 23 \cdot 29 \cdot 31 \cdot 37 \cdot 43.$$

8.7   The simplicity of G

First we show that the amalgam $\mathcal{J} = \{G^{[0]}, G^{[1]}, G^{[2]}\}$ is *simple* in the sense that it does not possess non-trivial factor-amalgams.

Proposition 8.7.1 *Let* $\bar{\mathcal{J}} = \{\bar{G}^{[0]}, \bar{G}^{[1]}, \bar{G}^{[2]}\}$ *be an amalgam and*

$$\nu : \mathcal{J} \to \bar{\mathcal{J}}$$

be a surjective homomorphism of amalgams. This means that ν *maps (the element-set of)* $\mathcal{J}$ *onto (that of)* $\bar{\mathcal{J}}$ *and for each* $i = 0$, *1, and 2 the restriction* *of ν to $G^{[i]}$ is a homomorphism onto $\bar{G}^{[i]}$. Then unless $\bar{\mathcal{J}}$ contains only one element, ν is an isomorphism.*

Proof Let $K = \{g \in \mathcal{J} \mid \nu(g) = 1\}$ be the kernel of ν. Suppose first that $K \cap G^{[i]}$ is non-identity for some $i = 0$, 1, or 2. Then $K \cap G^{[i]}$ contains the minimal normal subgroup of $G^{[i]}$, which is

$$Q^{[0]} \simeq 2^{10}, \ Z^{[1]} \simeq 2^6, \text{ and } Z^{[2]} \simeq 2^3$$

for $i = 0$, 1, and 2, respectively. Since

$$Z^{[2]} < Z^{[1]} < Q^{[0]},$$

we conclude that K contains $Z^{[2]}$. The normal closure of $Z^{[2]}$ in $G^{[0]}$ is $Q^{[0]}$, the normal closure of $Q^{[0]}$ in $G^{[1]}$ is $Q^{[1]}$ and the normal closure of $Q^{[1]}$ in $G^{[0]}$ is the whole $G^{[0]}$. Finally the normal closure of $G^{[0]} \cap G^{[i]}$ in $G^{[i]}$ is the whole $G^{[i]}$ for $i = 1$ and 2. Therefore in the considered case $K = \mathcal{J}$ and $|\bar{\mathcal{J}}| = 1$.

Next assume that $K \cap G^{[i]} = 1$ for $i = 0$, 1, and 2 (equivalently the restriction of ν to each of the $G^{[i]}$'s is an isomorphism). In this case, unless ν is an isomorphism, there are elements $g \in G^{[i]}$, $h \in G^{[j]}$ for $0 \leq i < j \leq 2$ with $g, h \notin G^{[i]} \cap G^{[j]}$ and

$$\nu(g) = \nu(h).$$

This means that the image of $G^{[i]} \cap G^{[j]}$ under ν is a proper subgroup in $\bar{G}^{[i]} \cap \bar{G}^{[j]}$. On the other hand $G^{[0]} \cap G^{[1]}$ and $G^{[0]} \cap G^{[2]}$ are maximal subgroups in $G^{[0]}$, while $G^{[1]} \cap G^{[2]}$ is maximal in $G^{[1]}$. Since the $G^{[i]}$'s are pairwise non-isomorphic this means that their images cannot have larger intersections than they already have in $\mathcal{J}$. Hence ν is an isomorphism. ∎

As a direct consequence of (8.7.1) we obtain the following.

Lemma 8.7.2 *Let G be a faithful generating completion group of the amalgam $\mathcal{J} = \{G^{[0]}, G^{[1]}, G^{[2]}\}$ and let N be a proper normal subgroup of G. Then (identifying $\mathcal{J}$ with its image in G) the amalgam*

$$\{G^{[0]}N/N, G^{[1]}N/N, G^{[2]}N/N\}$$

is isomorphic to $\mathcal{J}$ and G/N is also a faithful completion of $\mathcal{J}$.

Proposition 8.7.3 *Let G be a faithful generating completion of the amalgam $\mathcal{J}$. Then*

(i) *G is a finite simple group;*
(ii) *G is the universal completion of $\mathcal{J}$;*
(iii) *G is the only faithful generating completion of $\mathcal{J}$.*

Proof Suppose that N is a proper normal subgroup of N and let $\bar{G} = G/N$. Then by (8.7.2) $\bar{G}$ is also a faithful generating completion group of $\mathcal{J}$ and (8.6.3) applies to $\bar{G}$ as good as to G. This is a contradiction and hence (i) is established. By (i) the universal completion is simple and hence there are no further completions which gives (ii) and (iii). ∎

8.8 The involution centralizer

Now we can show that $G^{[3]}$ is the full centralizer of $Z^{[3]}$ in G.

Proposition 8.8.1 *Let z be the generator of $Z^{[3]}$. Then*

$$G^{[3]} \cong 2_{+}^{1+12} \cdot 3 \cdot \text{Aut}\,(M_{22})$$

is the full centralizer of z in G.

Proof Let d be the element from $\mathcal{D}^{[3]}$ such that $G_d^{[3]} = G^{[3]}$. Then z is the generator of $Z_d^{[3]}$. It is clear that $G_d^{[3]}$ is the setwise stabilizer of $\mathcal{D}_d^{[m]}$ (the latter set being of size 77 by (8.1.2 (iii))). On the other hand, $C_G(z)$ clearly stabilizes as a whole the set

$$\Theta(z) = \{x \in V(\Delta) \mid z \in Q_x^{[m]}\}.$$

By (8.3.13 (x)), (8.3.17 (vi)), (8.4.3 (iii)), and (8.5.2)

$$\Theta(z) \subseteq \{a\} \cup \Delta(a) \cup \Delta_2^1(a)$$

Now we apply (8.1.1 (vi)), (8.1.3 (vi)) and perform easy calculations in the Todd module $Q_a^{[m]}$ to show that $|\Theta(z)| = 77$. Thus the sets $\mathcal{D}_d^{[m]}$ and $\Theta(z)$ coincide and the result follows. ∎

Now the proof of the Main Theorem is complete.

Exercises

1. Show that the subgroup G_{ag} in (8.5.5) is isomorphic to $L_2(23)$.
2. Calculate the orbits of $G^{[3]} \cong 2_+^{1+12} \cdot 3 \cdot \text{Aut}\,(M_{22})$ on the vertex-set of Δ.
3. Calculate the automorphism group of the local graph of Δ described in (8.2.6).
4. Suppose that Σ is a graph whose local graph is isomorphic to that of Δ. Is there a geometry $\mathcal{S}$ whose diagram coincides with that of $\mathcal{D}$ such that Σ is the graph on the set of elements of type m in $\mathcal{S}$ in which two of them are adjacent whenever they are incident to a common element of type 2?
5. Classify the graphs whose local graphs are isomorphic to the local graph of Δ.
6. Classify the geometries having the same diagram as $\mathcal{D}$.

7. Calculate the suborbit diagram of the graph Γ of valency 31 associated with J_4. The permutation character χ of J_4 on the vertex-set of Γ (which is the character on the cosets of $G^{[0]} \cong 2^{10} : L_5(2)$) as calculated by Jürger Müller from Aachen is the following:

$$\chi = 1 + 8 + 11 + 14 + 2 \cdot 19 + 2 \cdot 20 + 2 \cdot 21 + 22 + 23 + 24$$
$$+ 29 + 30 + 32 + 36 + 37 + 38 + 39 + 51$$

(we follow the Conway et al. (1985) notation). Thus there are 27 suborbits.

HISTORY AND BEYOND

In this chapter we survey the history of discovery, construction and characterization of J_4. This shows when and how the crucial ingredients of the proof of the Main Theorem emerged.

9.1 Janko's discovery

The first evidence for the existence of the group now known as the fourth Janko group and denoted by J_4 was given by Zvonimir Janko in his remarkable paper (Janko 1976). The main result of Janko (1976) is Theorem A reproduced below (we keep the original notation of Janko, particularly the symmetric group of degree n is denoted by Σ_n unlike Sym_n as in the rest of the book).

Theorem A *Let G be a nonabelian finite simple group which possesses an involution z such that $H = C_G(z)$ satisfies the following conditions.*

  (i) *The subgroup $E = O_2(H)$ is an extraspecial 2-group of order 2^{13} and $C_H(E) \subseteq E$.*

  (ii) *An S_3-subgroup P of $O_{2,3}(H)$ has the order 3 and $C_E(P) = Z(E) = \langle z \rangle$.*

  (iii) *We have $H/O_{2,3}(H) \cong \mathrm{Aut}(M_{22})$, $N_H(P) \neq C_H(P)$, and $P \subseteq (C_H(P))'$.*

Then G has the following properties.

  (1) *The order of G is $2^{21} \cdot 3^3 \cdot 5 \cdot 7 \cdot 11^3 \cdot 23 \cdot 29 \cdot 31 \cdot 43$.*

  (2) *The subgroup $F_0 = C_H(P)$ is isomorphic to the full covering group of M_{22} (i.e. the perfect central extension of a cyclic group of order 6 by M_{22}). Moreover, $F = N_H(P) = N_G(P)$ and the group G has exactly one conjugacy class of elements of order 3.*

  (3) *Let R be an S_3-subgroup of H. Then R is an extraspecial group of order 27 and exponent 3 and we have $N_G(R) = N_H(R) = R(\langle z \rangle \times D)$, where D is a semidihedral group of order 16.*

  (4) *Let T be an S_2-subgroup of G. Then T possesses exactly one elementary abelian subgroup V of order 2^{11}. We have $C_G(V) \subseteq V$ and $N_G(V) = VK$, where K is isomorphic to M_{24}. The orbits of K on $V^{\#}$ have lengths $7 \cdot 11 \cdot 23$ (with representative z') and $4 \cdot 3 \cdot 23$ (with representative t). Here z' is conjugate in G to z and t is not conjugate to z in G. Moreover, we have $C_G(t) = C_G(t) \cap N_G(V)$.*

  (5) *The group G has precisely two conjugacy classes of involutions with representatives z and t. Hence $C_G(t)$ is a splitting extension of an elementary abelian group of order 2^{11} by $\mathrm{Aut}\,(M_{22})$.*

(6) *The group G possesses exactly one conjugacy class of selfcentralizing elementary abelian subgroups of order 2^{10} and if A is one of them, then we have $N_G(A) = AB$, where $B \cong L_5(2)$ and B acts irreducibly on A. The orbits of B on $A^\#$ have lengths $5 \cdot 31$ and $4 \cdot 7 \cdot 31$.*

(7) *Let Q be an S_5-subgroup of H. Then Q is also an S_5-subgroup of G, $C_G(Q) = Q \times J$, where J is isomorphic to the non-splitting extension of an elementary abelian group of order 8 by $L_3(2)$. Also $N_G(Q)$ contains a Frobenius subgroup of order 20. Hence a Sylow 5-normalizer in G has order $2^8 \cdot 3 \cdot 5 \cdot 7$.*

(8) *Let S be an S_7-subgroup of H. Then S is also an S_7-subgroup of G, $C_G(S) = S \times I$, where I is isomorphic with Σ_5 and $|N_G(S)| = 3 \cdot |C_G(S)|$. Hence a Sylow 7-normalizer in G has the order $2^3 \cdot 3^2 \cdot 5 \cdot 7$.*

(9) *The group G possesses a special 2-group L of order 2^{15} with $|Z(L)| = 2^3$, so that $N_G(L)/L \cong \Sigma_5 \times L_3(2)$. Also, $N_G(L)$ does not split over L and nevertheless $N_G(L)$ contains subgroups isomorphic to Σ_5 and to $L_3(2)$. The subgroup $N_G(L)$ contains both a Sylow 5-normalizer in G and a Sylow 7-normalizer in G.*

(10) *A Sylow 11-normalizer in G has order $2^4 \cdot 3 \cdot 5 \cdot 11^3$ and contains a subgroup isomorphic to $GL_2(3)$. An S_{11}-subgroup of G is extraspecial of order 11^3 and exponent 11. The group G has exactly two conjugacy classes of elements of order 11.*

(11) *An S_p-subgroup is self-centralizing in G for $p = 23$, 29, 31, 37, and 43. A Sylow p-normalizer in G has order $23 \cdot 22$, $29 \cdot 28$, $31 \cdot 10$, $37 \cdot 12$, and $43 \cdot 14$, respectively.*

(12) *The group G possesses $PGL_2(23)$ as a subgroup.*

(13) *The group G has exactly 62 conjugacy classes of elements. The character table of G is unique and was computed by J. Conway, S. Norton, J. G. Thompson, and D. Hunt.*

In 1976, Janko deduced almost the whole p-local structure of G from the structure of the involution centralizer. Once in Gaeta in 1990 he mentioned that this was the easier part of the job; the hard part was to eliminate many other possibilities for the structure of centralizer which did not lead to consistent configurations.

At the time of (Janko 1976) the cyclic group $Z(F_0)$ of order 6 in (2) was believed to be the full Schur multiplier of the Mathieu group M_{22} (in (Mazet 1979) the multiplier was proved to be cyclic of order 12).

As pointed out on p. 494 in Kleidman and Wilson (1988) the structure of the Sylow 3-normalizer is slightly mis-stated in (3); the correct structure is $(2 \times 3^{1+2}_+ : 8) : 2$, where the outer involution conjugates the elements of order 8 to its cube times the central involution.

Let us relate the present work with Janko's theorem. The structure of involution centralizer in (i) to (iii) which is the starting point of Janko's work appears in our treatment at the very last stage (8.8.1). Our penultimate result (8.6.3) is the order of G which coincides with that in (1).

The subgroup $N_G(V) = VK$ in (4) is the subgroup $G^{[m]} \cong 2^{11} : M_{24}$ in our terms; the subgroup $N_G(A) = AB$ in (6) is our subgroup $G^{[0]} \cong 2^{10} : L_5(2)$; the subgroup $N_G(L)$ in (9) is our pentad group $G^{[3]} \cong 2^{3+12} \cdot (L_3(2) \times \mathrm{Sym}_5)$. The information on 5- and 7-normalizers in (7) and (8) corresponds to the properties of the pentad group in (4.9.1).

The subgroup $PGL_2(23)$ in (12) is the subgroup $G\{a, g\} = \langle G_{ag}, t \rangle$ in the proof of (8.5.5).

The character table mentioned in (13) can be found in (Conway *et al.* 1985) (under the name J_4 of course). The number 1333 which first appeared in (6.2.1) within our project is the degree of the minimal faithful character of J_4.

Janko left unsettled the questions about existence and uniqueness of the group he had discovered in (Janko 1976).

9.2   Characterizations

Immediately after Janko's discovery the existence of further simple groups with similar 2- and 3-local structure was questioned. A group G is said to be of J_4-type if it satisfies the conditions (i) to (iii) in Janko's Theorem A. Soon after publication of (Janko 1976) it was proved by various authors that each of the properties (2), (4), (5), (6), and (7) is characteristic for groups of J_4-type.

Theorem 9.2.1 (Stroth 1978; Stafford 1979) *Let G be a finite group which is not 3-normal. Furthermore, let x be an element of order 3 in G and $F_0 = C_G(x)$. Then the following assertions hold:*

  (i) *if G is simple and F_0 is isomorphic to $3 \cdot M_{22}$ or $6 \cdot M_{22}$, then G is of J_4-type;*
  (ii) *if G is of characteristic 2-type and F_0 is isomorphic to $6 \cdot M_{22}$, then G of J_4-type.*

Theorem 9.2.2 (Reifart 1978) *Let G be a finite group containing an elementary abelian subgroup V of order 2^{11} such that*

  (i) $C_G(V) = V$;
  (ii) $N_G(V)/V$ *is isomorphic to* M_{24};
  (iii) $O(C_G(z)) = 1$ *for an involution z in the centre of a Sylow 2-subgroup of $N_G(V)$.*

Then one of the following holds:

  (a) V *is normal in G;*
  (b) G *is of J_4-type;*
  (c) G *is simple,* $|G| = |M(24)'|$, *and the centralizer of a 2-central involution of G is isomorphic to the centralizer of a 2-central involution of $M(24)'$;*
  (d) G *is isomorphic to the first Conway group Co_1.*

It is known by now (cf. (Parrott 1981) or theorem 34.1 in Aschbacher 1997) that $M(24)'$ is the only group which satisfies (9.2.2 (c)).

Theorem 9.2.3 (Reifart 1977a, 1977b) *Let G be a finite group containing an involution f such that $V := O_2(C_G(f))$ is elementary abelian of order 2^{11} and $C_G(f)/V \cong \mathrm{Aut}(M_{22})$. Then one of the following holds:*

(i) $G = O(G) : C_G(f)$;
(ii) V is normal in G and $G/V \cong M_{24}$;
(iii) G is of J_4-type.

Theorem 9.2.4 (Lempken 1978a) *Let G be a finite group containing an elementary abelian subgroup A of order 2^{10} such that $N_G(A)/A \cong L_5(2)$ acts irreducibly on A and $O(C_G(z)) = 1$ for a 2-central involution z of $N_G(A)$. Then one of the following holds:*

(i) $G = N_G(A) \cong 2^{10} : L_5(2)$;
(ii) G is isomorphic to $\Omega_{10}^+(2)$ or to $O_{10}^+(2)$;
(iii) G is of J_4-type.

Theorem 9.2.5 (Van Trung 1980) *Let G be a finite simple group of characteristic 2-type and let K be a 2-local subgroup of G satisfying the following:*

(i) $L := O_2(K)$ *is special of order* 2^{15}, $Z := Z(K) \cong 2^3$; $K = N_G(Z)$ *and* $K/R \cong \mathrm{Sym}_5 \times L_3(2)$;
(ii) *if x and y are two commuting elements of K of orders 3 and 7, respectively, then $C_L(x) = Z$ and $C_L(y) = 1$.*

Then G is of J_4-type.

9.3 Ronan–Smith geometry

It was not stated in Janko's Theorem A but can be deduced from its proof that the maximal 2-local subgroups can be chosen to have 'large' pairwise intersections. The following explicit form of this observation led to discovery of the maximal 2-local parabolic geometry of J_4.

Proposition 9.3.1 (Ronan and Smith 1980) *The subgroups*

$$G^{[0]} = N_G(A) \cong 2^{10} : L_5(2), \quad G^{[2]} = N_G(L) \cong 2^{3+12} \cdot (L_3(2) \times \mathrm{Sym}_5),$$

$$G^{[3]} = C_G(z) \cong 2_+^{1+12} \cdot 3 \cdot \mathrm{Aut}\,(M_{22}), \quad G^{[m]} = N_G(V) \cong 2^{11} : M_{24}.$$

in Janko's Theorem A can be chosen in such a way that the coset geometry $\mathcal{D}(G)$ associated with the completion in G of the amalgam $\{G^{[m]}, G^{[0]}, G^{[2]}, G^{[3]}\}$ is

described by the diagram

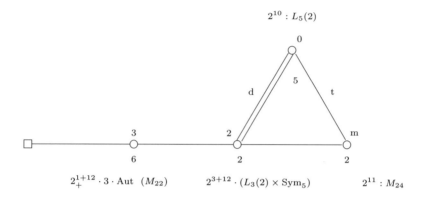

The existence of the geometry $\mathcal{D}(J_4)$ was stated in Ronan and Smith (1980) subject to the existence of the group J_4. In fact these two existences are equivalent and here we have constructed the geometry before the group herself was constructed. The main question about $\mathcal{D}(J_4)$ (as well as about other sporadic geometries) posed in Ronan and Smith (1980) was the question about the simple connectedness. The affirmative answer to this question (originally given in Ivanov (1992)) also follows from our Main Theorem.

9.4   Cambridge five

The group J_4 was constructed in Cambridge in 1980 by D. J. Benson, J. H. Conway, S. P. Norton, R. A. Parker, and J. G. Thackrey as a group of 112×112 matrices over $GF(2)$ and the construction made a heavy use of a computer.

The construction process was very dramatical. This is reflected in the mathematical (Norton 1980) and popular (Conway 1981) summaries. These summaries give the exact date when the last step was accomplished: February 20, 1980.

It would not be wise to reproduce the content of Norton (1980) here 'in my own words'. Although it should be mentioned that a lot of essential information about the action of J_4 on the cosets of $2^{11} : M_{24}$ can be found in Norton (1980) (more details can be found in Benson(1980)). For instance the structure of the 2-point stabilizers from Table 3 at the end of Section 8.5 are already given in Norton (1980).

9.5   *P*-geometry

In Ivanov (1987) it was shown that there exists a graph $\Gamma = \Gamma(J_4)$ of valency 31 on which J_4 acts locally projectively. The graph Γ contains geometric Petersen subgraphs and geometric octet subgraphs. Hence J_4 acts flag-transitively on a locally truncated geometry $\mathcal{F}(J_4)$ with the diagram

$$P_5^t : \underset{1}{\circ}\!\!\overset{\text{P}}{\underline{\hspace{2cm}}}\!\!\underset{2}{\circ}\!\!\underline{\hspace{1.5cm}}\!\!\underset{2}{\circ}\!\!\underline{\hspace{1.5cm}}\!\!\underset{2}{\circ}\!\!\underline{\hspace{1.5cm}}\!\!\square.$$

(cf. Section 7.5). The geometric subgraphs of valency 15 are missing and therefore $\mathcal{F}(J_4)$ is not a truncation of a P-geometry of rank 5. It was noticed by S. V. Shpectorov that the non-existence of the valency 15 geometric subgraphs means that the corresponding geometric subgroup (which still can be defined) is the whole of J_4. Therefore J_4 acts flag-transitively on a rank 4 P-geometry $\mathcal{G}(J_4)$ with the diagram

$$P_4 : \quad \underset{1}{\circ}\!\!\overset{\text{P}}{\underline{\hspace{2cm}}}\!\!\underset{2}{\circ}\!\!\underline{\hspace{1.5cm}}\!\!\underset{2}{\circ}\!\!\underline{\hspace{1.5cm}}\!\!\underset{2}{\circ}.$$

as in (7.3.2).

In Shpectorov (1988) the amalgams of maximal parabolic subgroups associated with flag-transitive actions on P-geometries of rank 4 were classified. It turned out to exists just five such amalgams, one of them being the subamalgam of J_4.

In 1988, G. Stroth and R. Weiss have written down generators and relations which must hold in any group acting locally projectively on a graph of valency 31 and girth 5 with the vertex stabilizer isomorphic to $G^{[0]} \cong 2^{10} : L_5(2)$. This is a presentation for the automorphism group of the universal cover of the geometry $\mathcal{F}(J_4)$. The presentation of the automorphism group of the universal cover of $\mathcal{G}(J_4)$ was also explicitly written down. It was asked in Stroth and Weiss (1988) whether the groups defined by the presentations are isomorphic to J_4. The affirmative answers to these questions are equivalent to the simple connectedness of $\mathcal{F}(J_4)$ and $\mathcal{G}(J_4)$, respectively.

The simple connectedness of $\mathcal{G}(J_4)$, $\mathcal{F}(J_4)$, and $\mathcal{D}(J_4)$ was established in (Ivanov 1992).

Within the theory of P-geometries an important role is played by the so-called *universal abelian representations* of geometries (cf. Ivanov and Shpectorov (2002) for the definition). It was shown in Ivanov and Shpectorov (1990) that $\mathcal{G}(J_4)$ does not possess abelian representations. In a sense this is the reason why $\mathcal{G}(J_4)$ does not appear as a residue in P-geometries of higher rank. In Ivanov and Shpectorov (1997) it was shown that the *universal non-abelian representation group* of $\mathcal{G}(J_4)$ is J_4 itself.

9.6 Uniqueness of J_4

The first uniqueness proof for J_4 was announced in (Norton 1980). It was based on the fact that the amalgam

$$\{G^{[0]}, G^{[1]}\} \cong \{2^{10} : L_5(2), 2^{11} : 2^4 : L_4(2)\}$$

possesses a unique 1333-dimensional representation with a given character (the character being the restriction of a 1333-dimensional character of J_4), see (Thompson 1981). This uniqueness ensures the success of our construction in Chapter 6.

It is a remarkable fact (already mentioned in (Norton 1980)) that the 1333-dimensional representation of a larger amalgam

$$\{G^{[0]}, G^{[m]}\} \cong \{2^{10} : L_5(2), 2^{11} : M_{24}\}$$

is not unique, even when the character is prescribed.

Within the classification of the finite simple groups J_4 falls into the most dramatic class of the *quasi-thin groups* (Mason 1980; Aschbacher and Smith 2003). In Berlin in 1990 Geoff Mason suggested me that the simple connectedness proof for $\mathcal{G}(J_4)$ can be extended to the uniqueness proof for J_4. This was fully accomplished in (Ivanov 1992). An independent uniqueness proof along similar lines by M. Aschbacher and Y. Segev appeared in 1991.

9.7 Lempken's construction

In (Lempken 1993) J_4 was constructed as a group of 1333×1333-matrices over the field $GF(11)$. The construction is similar to our's in Chapter 6. First a representation of $G^{[0]} \cong 2^{10} : L_5(2)$ was constructed, then it was restricted to the subgroup $G^{[0]} \cap G^{[1]} \cong 2^{10} : 2^4 : L_4(2)$ and finally a matrix t was found, which extends $G^{[0]} \cap G^{[1]}$ to the whole of $G^{[1]}$. To justify the fact that the group generated is indeed J_4, the existence of J_4 was needed.

9.8 Computer-free construction

In July 1990 during the Durham Symposium U. Meierfrankenfeld and the author agreed that the simple connectedness proof for the 2-local geometries of J_4 can be turned into a computer-free construction of J_4. It took almost 10 years before the project was completed and published in (Ivanov and Meierfrankenfeld 1999). The present volume is essentially based on this project.

9.9 The maximal subgroups

The maximal p-local subgroups of a group of J_4-type were determined in (Lempken 1989) while the non-local subgroups were determined in (Kleidman and Wilson 1988) with considerable use of computer.

Theorem 9.9.1 *The group J_4 contains exactly* 13 *classes of maximal subgroups. If X is a maximal subgroup in J_4, then one of the following holds:*

(i) *X is 2-local and is isomorphic to one of the following: $2^{11} : M_{24}$, $2^{10} : L_5(2)$; $2^{1+12}_+ \cdot 3 \cdot \mathrm{Aut}(M_{22})$, $2^{3+12} \cdot (L_3(2) \times \mathrm{Sym}_5)$;*

(ii) *X is a Sylow p-normalizer for some $p \in \{11, 29, 37, 43\}$ and is isomorphic to the respective groups $11^{1+2}_+ : (GL_2(3) \times 5)$, $29 : 28$, $37 : 12$, $43 : 14$;*

(iii) *X is a non-local subgroup isomorphic to one of the following: $U_3(11) : 2$, $M_{22} : 2$, $L_2(32) : 5$, $L_2(23) : 2$, $U_3(3)$.*

9.10 Rowley–Walker diagram

In (Rowley and Walker 1994) the following suborbit diagram of the action of J_4 on the cosets of $2^{11} : M_{24}$ was computed.

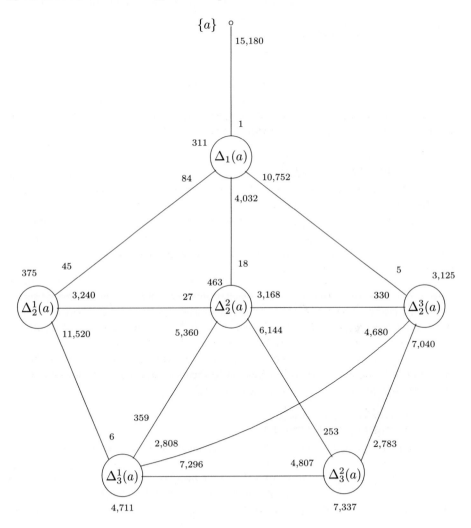

9.11 Locally projective graphs

Let $n \geq 2$ be an integer and q be a prime power. Recall that a locally finite, vertex-transitive action of a group F on a graph Φ is said to be *locally projective* of type (n, q) if for a vertex u of Φ the subconstituent $F(u)^{\Phi(u)}$ contains a normal subgroup isomorphic to the special linear group $L_n(q)$ acting on the set of 1-dimensional subspaces in its natural module (which is an n-dimensional $GF(q)$-space), so that the valency of Φ is $(q^n - 1)/(q - 1)$,

$$L_n(q) \trianglelefteq F(u)^{\Phi(u)} \leq P\Gamma L(n, q),$$

and the action of F on Φ is 2-arc-transitive.

The most interesting case among the locally projective actions is when $q = 2$ and the following condition holds: whenever $\{u, v\}$ is an edge of Φ, there is an element $t \in F\{u, v\} \setminus F(u, v)$ which centralizes $F(u, v)/O_2(F(u, v))$.

Under these assumptions the amalgam

$$\mathcal{F} = \mathcal{A}(F, \Phi) = \{F^{[0]}, F^{[1]}\},$$

where $F^{[0]} = F(u)$ and $F^{[1]} = F\{u, v\}$ is said to be the amalgam determined by the locally projective action or simply a *locally projective amalgam*. The locally projective amalgams can be characterized as abstract rank 2 amalgams subject to the following conditions:

(A1) $F^{[0]}/O_2(F^{[0]}) \cong L_n(2)$ for some $n \geq 2$;

(A2) $F^{[01]} := F^{[0]} \cap F^{[1]}$ is the stabilizer of a 1-space U_1 in the natural module U of $F^{[0]}/O_2(F^{[0]})$;

(A3) $[F^{[1]} : F^{[01]}] = 2$;

(A4) $F^{[1]}/O_2(F^{[01]}) \cong L_{n-1}(2) \times 2$ or, equivalently $[t, F^{[1]}] \leq O_2(F^{[01]})$ for some $t \in F^{[1]} \setminus F^{[0]}$;

(A5) no non-identity subgroup in $F^{[01]}$ is normal in both $F^{[0]}$ and $F^{[1]}$.

D. Ž. Djoković and G. L. Miller (1980) used the classical Tutte's theorem (Tutte 1947), to show that there are exactly seven locally projective amalgams for $n = 2$. In Ivanov and Shpectorov (2004) we used Trofimov's theorem (Trofimov 2003) to extend the classification to the case $n \geq 3$ by proving the following.

Theorem 9.11.1 *Let $\mathcal{F} = \{F^{[0]}, F^{[1]}\}$ be a locally projective amalgam, that is an amalgam satisfying* (A1) *to* (A5) *for some $n \geq 3$. Then exactly one of the following holds:*

(i) $\mathcal{F} \cong \mathcal{A}(AGL_n(2), K_{2^n})$;

(ii) $\mathcal{F} \cong \mathcal{A}(O_{2n}^+(2), D^+(2n, 2))$;

(iii) $\mathcal{F}$ *is isomorphic to one of twelve exceptional amalgams in Table 4.*

In order to explain the data in Table 4 we need to recall some definitions. By (A2) the subgroup $F^{[01]}$ is the stabilizer in $F^{[0]}$ of a 1-space U_1 in the natural module U of $F^{[0]}/O_2(F^{[0]}) \cong L_n(2)$. Let

$$0 < U_1 < U_2 < \cdots < U_{n-1} < U$$

be a maximal flag in U containing U_1. For $2 \leq i \leq n-1$ let $F^{[0i]}$ be the stabilizer of U_i in $F^{[0]}$, $F^{[01i]} = F^{[0i]} \cap F^{[1]}$ and let $F^{[1i]}$ be the normalizer of $F^{[01i]}$ in $F^{[1]}$. Let $K^{[i]}$ be the largest subgroup in $F^{[01i]}$ which is normal in both $F^{[0i]}$ and $F^{[1i]}$. Let $\widehat{F^{[i]}}$ be the subgroup in the outer automorphism group Out $(K^{[i]})$ of $K^{[i]}$ generated by the natural images in Out $(K^{[i]})$ of $F^{[0i]}$ and $F^{[1i]}$.

TABLE 4. Exceptional amalgams

n	$\mathcal{F}$	$\dfrac{F^{[0]}/O_2(F^{[0]})}{O_2(F^{[0]})}$	$\dfrac{\widehat{F}^{[2]}}{K^{[2]}}$	$\dfrac{\widehat{F}^{[3]}}{K^{[3]}}$	$\dfrac{\widehat{F}^{[4]}}{K^{[4]}}$	Completions constrained at level 2
3	$\mathcal{A}_3^{(1)}$	$\dfrac{L_3(2)}{2^3}$	$\dfrac{O_4^+(2)}{2^4}$			—
	$\mathcal{A}_3^{(2)}$	$\dfrac{L_3(2)}{2^3}$	$\dfrac{\mathrm{Sym}_5}{2^4}$			M_{22}
	$\mathcal{A}_3^{(3)}$	$\dfrac{L_3(2)}{2^3}$	$\dfrac{\mathrm{Sym}_5}{2^4}$			—
	$\mathcal{A}_3^{(4)}$	$\dfrac{L_3(2)}{2^3\times 2}$	$\dfrac{2^4{:}O_4^+(2)}{2^5}$			$(\mathrm{Sym}_8 \wr 2)^+$
	$\mathcal{A}_3^{(5)}$	$\dfrac{L_3(2)}{2^3\times 2}$	$\dfrac{\mathrm{Sym}_5}{2^5}$			$\mathrm{Aut}\,(M_{22})$
4	$\mathcal{A}_4^{(1)}$	$\dfrac{L_4(2)}{1}$	$\dfrac{\mathrm{Sym}_5}{2^4{:}3}$	$\dfrac{1}{1}$		M_{23}
	$\mathcal{A}_4^{(2)}$	$\dfrac{L_4(2)}{2^6}$	$\dfrac{\mathrm{Sym}_5\times 2}{2^{1+8}_+{:}\mathrm{Sym}_3}$	$\dfrac{L_6(2)}{2^6}$		Alt_{64}
	$\mathcal{A}_4^{(3)}$	$\dfrac{L_4(2)}{2^{1+4+6}}$	$\dfrac{\mathrm{Sym}_5}{2^{4+10}.\mathrm{Sym}_3}$	$\dfrac{\mathrm{Aut}\,(M_{22})}{2^{10}}$		Co_2
	$\mathcal{A}_4^{(4)}$	$\dfrac{L_4(2)}{2^{4+4+6}}$	$\dfrac{\mathrm{Sym}_5}{2^{3+12+2}.\mathrm{Sym}_3}$	$\dfrac{3\cdot\mathrm{Aut}\,(M_{22})}{2^{1+12}_+}$		J_4
	$\mathcal{A}_4^{(5)}$	$\dfrac{L_4(2)}{2^{4+4+6}}$	$\dfrac{\mathrm{Sym}_5\times 2}{2^{3+12+2}.\mathrm{Sym}_3}$	$\dfrac{L_6(2){:}2}{2^{1+12}_+}$		Alt_{256}
5	$\mathcal{A}_5^{(1)}$	$\dfrac{L_5(2)}{2^{10}}$	$\dfrac{\mathrm{Sym}_5}{2^{3+12}.L_3(2)}$	$\dfrac{\mathrm{Aut}\,(M_{22})}{2^{1+12}_+{:}3}$	$\dfrac{1}{1}$	J_4
	$\mathcal{A}_5^{(2)}$	$\dfrac{L_5(2)}{2^{5+5+10+10}}$	$\dfrac{\mathrm{Sym}_5}{2^3.[2^{32}].L_3(2)}$	$\dfrac{\mathrm{Aut}\,(M_{22})}{2^{2+10+20}.\mathrm{Sym}_3}$	$\dfrac{Co_2}{2^{1+22}_+}$	BM

Let $\phi : \mathcal{F} \to F$ be a faithful generating completion (which might or might not be the universal completion). Let us identify $\mathcal{F}$ with its image under ϕ. For $2 \le i \le n-1$ let $F^{[i]}$ be the subgroup in F generated by $F^{[0i]}$ and $F^{[1i]}$. Then

$$\widehat{F^{[i]}} = F^{[i]}/K^{[i]}C_{F^{[i]}}(K^{[i]})$$

and the completion $\phi : \mathcal{F} \to F$ is said to be *constrained at level i* if

$$C_{F^{[i]}}(K^{[i]}) \le K^{[i]}.$$

The amalgam $\mathcal{H}$ defined in Section 2.3 is $\mathcal{A}(O_{10}^+(2), D^+(10,2))$, the amalgam $\mathcal{G}$ defined in Section 3.5 is $\mathcal{A}_5^{(1)}$ while the amalgam $\mathcal{G}^{[4]} = \{G^{[04]}, G^{[14]}\}$ as in Section 5.4 is $\mathcal{A}_4^{(4)}$. The Main Theorem of the present work states that J_4 is the only completion of $\mathcal{A}_5^{(1)}$ which is constrained at level 2.

Because of the classification of the flag-transitive Petersen type geometries accomplished in (Ivanov 1999) and (Ivanov and Shpectorov 2002) we know that the Baby Monster sporadic simple group BM and the non-split extension of an

TABLE 5. Dimensions of minimal representations

$\mathcal{F}$	$\mathcal{A}_3^{(1)}$	$\mathcal{A}_3^{(2)}$	$\mathcal{A}_3^{(3)}$	$\mathcal{A}_3^{(4)}$	$\mathcal{A}_3^{(5)}$	$\mathcal{A}_4^{(1)}$
$m(\mathcal{F})$	7	20	20	14	20	20
$\mathcal{F}$	$\mathcal{A}_4^{(2)}$	$\mathcal{A}_4^{(3)}$	$\mathcal{A}_4^{(4)}$	$\mathcal{A}_4^{(5)}$	$\mathcal{A}_5^{(1)}$	$\mathcal{A}_5^{(2)}$
$m(\mathcal{F})$	63	23	1333	255	1333	4371

elementary abelian group of order 3^{4371} by BM are the only completions of $\mathcal{A}_5^{(2)}$ which are constrained at level 2. It is desirable to prove this result starting from the basic principles (as we did here for J_4).

In (Ivanov and Pasechnik 2004) the following result was established.

Proposition 9.11.2 *Let $\mathcal{F}$ be a locally projective amalgam and $m = m(\mathcal{F})$ be the smallest positive integer such that $\mathcal{F}$ possesses $GL_m(\mathbf{C})$ as a faithful completion group (which might or might not be generating). Then*

(i) *if $\mathcal{F} = \mathcal{A}(AGL_n(2), K_{2^n})$, then $m(\mathcal{F}) = 3$, 7 and $2^n - 2$ for $n = 3$, $n = 4$ and $n \geq 5$, respectively;*

(ii) *if $\mathcal{F} = \mathcal{A}(O_{2n}^+, D^+(2n, 2))$ then $m(\mathcal{F}) = 7$, 28 and $(2^n - 1)(2^{n-1} - 1)/3$ for $n = 3$, $n = 4$ and $n \geq 5$, respectively;*

(iii) *if $\mathcal{F}$ is an exceptional amalgam then $m(\mathcal{F})$ is given in Table 5.*

The fact that $m(\mathcal{A}_4^{(4)}) = m(\mathcal{A}_5^{(1)}) = 1333$ is the dimension of the minimal representation of J_4 has played a crucial role in the present work. Notice that $m(\mathcal{A}_5^{(2)}) = 4371$ is the dimension of the minimal representation of the Baby Monster.

9.12 On the 112-dimensional module

The information on the 112-dimensional $GF(2)$-module B for J_4 given in (Conway *et al.* 1985) suggests that B is a quotient of the so-called *universal derived module* C of the Petersen geometry $\mathcal{G}(J_4)$ (cf. Ivanov and Shpectorov (2002) for the relevant definitions). The module C can be defined as follows:

(a) take one involutory generator $c(e)$ for every edge e of the locally projective graph $\Gamma = \Gamma(J_4)$;

(b) declare the $c(e)$'s to commute pairwise;

(c) whenever edges e, f and g are contained in a common geometric Petersen subgraph Σ in Γ and constitute an antipodal triple in Σ, impose the relation $c(e)c(f)c(g) = 1$.

Conjecture 9.12.1 *The above defined J_4-module C over $GF(2)$ is 112-dimensional.*

We can also define the non-abelian version D of C simply by dropping the condition (b). Then C is the quotient of D over the commutator subgroup of D.

Conjecture 9.12.2 *The group D is abelian.*

9.13 Miscellaneous

Further results on construction, characterization and properties of J_4 can be found in Cooperman et al. (1997), Finkelstein (1977), Ganief and Moori (1999), Green (1993), Guloglu (1981), Ivanov and Shpectorov (1989), Lempken (1978b), Mason (1977), Michler and Weller (2001), Sitnikov (1990), Smith (1980), Yoshiara (2000).

Exercises

1. Prove (or disprove) that the 112-dimensional $GF(2)$-module for J_4 is a quotient of the module C from (9.12.1).
2. Describe the group $G^{[0]}$ from the amalgam $\mathcal{A}_5^{(2)}$ in Table 4 in terms independent of the Baby Monster group.

10

APPENDIX: TERMINOLOGY AND NOTATION

In this appendix we indicate our main terminology and notation concerning groups, graphs, amalgams, and diagram geometries. As is common we could say that our notation and terminology are standard, but this would sound more like a declaration of what 'standard' is.

10.1 Groups

Let G be a group, H be a subset of G which might or might not be a subgroup and let $g \in G$ and let $h \in H$. Then $h^g = g^{-1}hg$, $H^g = \{h^g \mid h \in H\}$ and $\langle H \rangle$ denotes the subgroup in G generated by the elements of H. The subset (subgroup) H is said to be *normal* if $H^g = H$ for every $g \in G$. The *centralizer* and the *normalizer* of H in G are defined as follows:

$$C_G(H) = \{g \mid g \in G, h^g = h \text{ for every } h \in H\};$$
$$N_G(H) = \{g \mid g \in G, h^g \in H \text{ for every } h \in H\}.$$

The centralizer and normalizer are subgroups in G and

$$C_G(H) \leq N_G(H).$$

For a group G we write Aut (G) and Inn (G) for the automorphism group of G and for the group of inner automorphisms, respectively. An automorphism $\sigma \in$ Aut $(G) \setminus$ Inn (G) is called an *outer automorphism* of G; it should not be confused with the coset of Inn (G) in Aut (G) containing σ. The latter coset is an element of the *outer automorphism group* Out $(G) =$ Aut (G) /Inn (G) of G.

An *involution* in a group is an element of order 2.

If N and H are groups and $\sigma : H \to$ Aut N is a homomorphism, then $(H \times N, *)$, where $(h_1, n_1) * (h_2, n_2) = (h_1 h_2, n_1^{\sigma(h_2)} n_2)$ is a group called the *semidirect product* of N and H with respect to σ and is denoted by $S(N, H, \sigma)$ (or simply by $N : H$ when σ is clear from the context or irrelevant). If σ is trivial (in the sense that its image is the identity) the product is *direct*.

For a finite group G the symbol $O(G)$ denotes the largest odd order (solvable) normal subgroup of G. If p is a prime then $O_p(G)$ denotes the largest normal p-subgroup in G and $O^p(G)$ the smallest normal subgroup of G such that the corresponding factor group is a p-group. The centre of G is denoted by $Z(G)$ and G' is the commutator subgroup of G.

We mostly follow the Atlas notation (Conway et al. 1985) for groups. In particular p^n denotes the elementary abelian group of that order; 2_ε^{1+2n} denotes the extraspecial group of order 2^{2n+1} of type $\varepsilon \in \{+, -\}$. The symmetric and alternating groups of degree n are denoted by Sym_n and Alt_n, respectively. By $L_n(q)$ we denote the projective special linear group in dimension n over the field $GF(q)$ of q elements. By F_a^b we denote a Frobenius group with kernel of order a and complement of order b.

If H is a subgroup of a group G then there is a natural homomorphism of $N_G(H)$ into Aut (H); if this homomorphism is surjective then is said that H is *fully normalized in* G.

Suppose that a group G acts on a set Ω by permutation (which means there is a homomorphism of G into the symmetric group of Ω). If a is an element of Ω then the stabilizer of a in G is denoted either by $G(a)$ or by G_a. If $X \subset \Omega$ then $G\{X\}$ denotes the stabilizer in G of X as a whole and $G(X)$ denotes the elements wise stabilizer (then $G(X)$ is the intersection of the subgroups $G(x) = G_x$ taken for all $x \in X$). If $H \leq G\{X\}$ then by H^X we denote the permutation group induced by H on X (in this case clearly $H^X \cong H/H(X)$ and $H = H\{X\}$).

Let G be a group, let H be a subgroup of G and $G/H = \{gH | g \in G\}$ be the sets of left cosets of H in G. The action of G on G/H via

$$f : gH \longmapsto fgH$$

for $f \in G$ defines a homomorphism of G into the symmetric group of the set G/H. The image of this homomorphism is a transitive subgroup of the symmetric group; if $H = 1H$ is considered as an element of G/H then $G(H) = H$ and the kernel of the homomorphism is the largest normal subgroup of G contained in H. The latter normal subgroup is known as the *core* of H in G:

$$core(H, G) = \bigcap_{g \in G} H^g.$$

10.2 Amalgams

An amalgam of rank m is a collection

$$\mathcal{A} = \{(G^{[i]}, *_i) \mid 0 \leq i \leq m - 1\}$$

of m groups $(G^{[i]}, *_i)$, $0 \leq i \leq m - 1$ such that for all $0 \leq i < j \leq m - 1$ the intersection $G^{[ij]} := G^{[i]} \cap G^{[j]}$ of the element sets is non-empty and the group operations $*_i$ and $*_j$ coincide, when restricted to $G^{[ij]}$. If $(G, *)$ is a group and $G^{[0]}, G^{[1]}, ..., G^{[m-1]}$ are subgroups in G, then $\{(G^{[i]}, *|_{G^{[i]}}) \mid 0 \leq i \leq m - 1\}$ is an amalgam.

Let $\mathcal{A} = \{(G^{[i]}, *_i) \mid 0 \leq 0 \leq m-1\}$ be an amalgam, let $(G, *)$ be a group and let φ be a mapping of the union of the element sets of the groups constituting $\mathcal{A}$ into G such that for every $0 \leq i \leq m - 1$ and all $g, h \in G^{[i]}$ the equality

$$\varphi(g *_i h) = \varphi(g) * \varphi(h)$$

holds (which means that the restriction of φ to each $G^{[i]}$ is a homomorphism). Then the pair (G, φ) is called a *completion* of $\mathcal{A}$ (here G is the *completion group* and φ is the *completion map*). The completion is said to be *faithful* if φ is injective and *generating* if G is generated by the image of φ. A completion $(\widetilde{G}, \widetilde{\varphi})$ is said to be *universal* if for every completion (G, φ) there is a homomorphism ψ of $\widetilde{G}$ into G, such that φ is the composition of $\widetilde{\varphi}$ and ψ. A universal completion is always generating and the universal completion group is unique up to isomorphism. Furthermore, an amalgam possesses a faithful completion (which is not always the case) if and only if its universal completion is faithful. It is natural to define two completions $(G, \varphi^{(1)})$ and $(G, \varphi^{(2)})$ to be *equivalent* whenever there is $l \in G$, such that

$$\varphi^{(1)}(a) = l^{-1}\varphi^{(2)}(a)l$$

for every $a \in \mathcal{A} = \bigcup_{i=0}^{m} G^{[i]}$. Clearly equivalent *completions* generate in G conjugate subgroups.

Whenever the group operations are clear from the context or irrelevant, we simply write

$$\mathcal{A} = \{G^{[i]} \mid 0 \le i \le m - 1\}$$

for an amalgam of rank m. We will also drop the explicit reference to the completion maps whenever it does not cause confusion (in this case by 'completion' we mean the completion group). For an amalgam $\mathcal{A} = \{G^{[0]}, G^{[1]}\}$ of rank 2, the universal completion is faithful and the universal completion group is isomorphic to the free product of $G^{[0]}$ and $G^{[1]}$ amalgamated over the common subgroup $G^{[01]}$. The free amalgamated product is infinite whenever $G^{[01]}$ is proper in both $G^{[0]}$ and $G^{[1]}$.

10.3 Graphs

An undirected graph Γ without loops is a pair of sets $V(\Gamma)$ (the set of vertices or simply the *vertexset*) and $E(\Gamma)$ (the set of edges or simply the *edgeset*) together with an incidence relation with respect to which every edge is incident to exactly two distinct vertices. For such a graph we will write $\Gamma = (V(\Gamma), E(\Gamma))$ assuming that the incidence relation is clear from the context. Two vertices are called adjacent if they are incident to a common edge. A graph is said to contain no multiple edges if every pair of vertices is incident to at most one common edge.

In this case every edge is identified with the pair of vertices it is incident to, so that the incidence relation is via inclusion. For the remainder of the section we assume that $\Gamma = (V(\Gamma), E(\Gamma))$ contains no multiple edges, that $E(\Gamma)$ is simply a set of 2-element subsets of $V(\Gamma)$ and the incidence relation is via inclusion. For a vertex $x \in V(\Gamma)$ the number of edges incident to x is called the *valency* of x in Γ. Since there are no multiple edges the valency of x is equal to the number of vertices adjacent to x in Γ. If the valency is independent of the choice of the vertex x, it is called the *valency of the graph* Γ in which case the graph Γ is said to be *regular* of that valency.

A sequence $\pi = (x_0, x_1, ..., x_s)$ of vertices in a graph Γ is said to be an *s-arc* if $\{x_i, x_{i+1}\} \in E(\Gamma)$ for all $0 \le i \le s - 1$ and $x_i \ne x_{i+2}$ for all $0 \le i \le s - 2$. This arc is said to join x_0 with x_s. If in addition x_i and x_{i+2} are *not* adjacent in Γ then π is said to be a *path* (of length s). A graph is said to be connected if any two of its vertices are joint by a path. The length of a shortest path joining vertices x and y of Γ is called the *distance* between x and y in Γ and is denoted by $d_\Gamma(x, y)$. The *diameter* of a connected graph Γ is the maximum among distances between its vertices.

If $\Gamma = (V(\Gamma), E(\Gamma))$ and $\Sigma = (V(\Sigma), E(\Sigma))$ are graphs then Σ is said to be a *subgraph* of Γ if

$$V(\Sigma) \subseteq V(\Gamma) \text{ and } E(\Sigma) \subseteq E(\Gamma).$$

If in addition $\{x, y\} \in E(\Sigma)$ whenever both $x, y \in V(\Sigma)$ and $\{x, y\} \in E(\Gamma)$, then Σ is said to be the subgraph of Γ *induced* by $V(\Sigma)$. A subgraph Σ in Γ is said to be *convex* or *geodetically closed* if the following condition holds:

whenever $x, y \in V(\Sigma)$, Σ contains every shortest path (that is a path of length $d_\Gamma(x, y)$) which joins x and y in Γ.

Let G be a group of automorphisms of Γ, and suppose that the action of G on Γ is 2-arc-transitive, that is transitive on the set $\{(y, x, z) \mid y, x, z \in V(\Gamma), \{y, x\}, \{x, z\} \in E(\Gamma), y \neq z\}$ of 2-arcs in Γ. For $x \in V(\Gamma)$ let

$$\Gamma(x) = \{y \mid y \in V(\Gamma), \{x, y\} \in E(\Gamma)\}$$

be the set of neighbours of x in Γ and let

$$G(x) = \{g \mid g \in G, \ g(x) = x\}$$

be the stabilizer (subgroup) of x in G. We always assume that the action is *locally finite* so that $G(x)$ is of finite order. Then, because of the 2-arc-transitivity, the permutation group $G(x)^{\Gamma(x)}$ (known as the subconstituent) is doubly transitive.

The action of a group $G \leq \operatorname{Aut} \Gamma$ on Γ is *distance-transitive* if for every $0 \leq i \leq d$ (where d is the *diameter* of Γ) the group G acts transitively on the set

$$\Gamma_i = \{(x, y) \mid x, y \in \Gamma, d(x, y) = i\}.$$

A graph which possesses a distance-transitive action is called a *distance-transitive graph*. If Γ is distance-transitive then for every i, $0 \leq i \leq d$, the parameters

$$c_i = |\Gamma_{i-1}(y) \cap \Gamma(x)|, \quad a_i = |\Gamma_i(y) \cap \Gamma(x)|, \quad b_i = |\Gamma_{i+1}(y) \cap \Gamma(x)|$$

are independent of the choice of the pair $x, y \in \Gamma$ satisfying $d(x, y) = i$. Clearly in this case Γ is regular and $c_i + a_i + b_i = |\Gamma(x)| = k$ is the valency of Γ. The sequence

$$i(\Gamma) = \{b_0 = k, b_1, \ldots, b_{d-1}; c_1 = 1, c_2, \ldots, c_d\}$$

is called the *intersection array* of the distance-transitive graph Γ. If we put $k_i = |\Gamma_i(x)|$ for $1 \leq i \leq d$ then

$$k_i = \frac{k \cdot b_1 \cdot \ldots \cdot b_{i-1}}{c_1 \cdot c_2 \cdot \ldots \cdot c_i}.$$

To represent the decomposition of a distance-transitive graph with respect to a vertex we draw the following distance diagram:

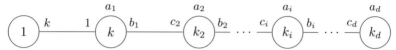

We draw similar diagrams for non-distance-transitive actions. Let G be a group acting on a graph Γ and x be a (basic) vertex. The *suborbit diagram* (with respect to x) consists of ovals (or circles) joined by curves (or lines). The ovals represent the orbits of $G(x)$ on the vertex set of Γ. Inside the oval which represents an orbit Σ_i (the Σ_i-oval) we show the size of Σ_i or place its name explained in the context. Next to the Σ_i-oval we show the number n_i (if non-zero) of vertices in Σ_i and adjacent to a given vertex $y_i \in \Sigma_i$. On the curve joining the Σ_i- and Σ_j-ovals we put the numbers n_{ij} and n_{ji} (called *valencies*.) Here n_{ij} (appearing closer to the Σ_i-oval) is the number of vertices in Σ_j adjacent to y_i. Clearly

$$|\Sigma_i| \cdot n_{ij} = |\Sigma_j| \cdot n_{ji}$$

and we draw no curve if $n_{ij} = n_{ji} = 0$. Normally we present the valencies n_i and n_{ij} as sums of lengths of orbits of $G(x, y_i)$ on the vertices in Σ_i and Σ_j adjacent to y_i. When the orbit lengths are unknown or irrelevant, we put the valencies into square brackets. Generally the suborbit diagram depends on the orbit of G on Γ from which the basic vertex x is taken. Even if a graph is not necessarily distance-transitive (but still generates an association scheme), we use the notation c_i, a_i, b_i if the corresponding parameters are independent of the choice of a pair of vertices at distance i.

If Γ and Δ are graphs, then a surjective mapping $\nu : V(\Gamma) \to V(\Delta)$ is said to be a *covering* if for every $x \in V(\Gamma)$ the restriction of ν to $\Gamma(x)$ is a bijection onto $\Delta(\nu(x))$.

Let Γ be a graph and G be a group of automorphisms of Γ. Suppose that the action is 2-arc transitive and for an integer $n \geq 2$ and a prime power q the following holds:

there is a group $SL(n, q) \leq H \leq \Gamma L(n, q)$, such that for every $x \in V(\Gamma)$ the action of $G(x)$ on $\Gamma(x)$ is permutation isomorphic to the permutation action of H on the set of 1-dimensional subspaces in its natural module (the latter being an n-dimensional $GF(q)$-space).

Then the action of G on Γ is said to be *locally projective* of type (n, q).

A very special role in our project is played by the famous *Petersen graph* Π. The standard way to define Π is the following. Take a set Ω of size 5. Then the vertices of Π are the 2-element subsets of Ω (so that there are 10 vertices) and two vertices are adjacent if they are disjoint (as subsets of Ω). Taking $\Omega = \{1, 2, 3, 4, 5\}$ we obtain the familiar picture of the Petersen graph presented below.

The automorphism group G of the Petersen graph is the symmetric group Sym_5 of Ω. The vertices of Π are naturally identified with the transpositions of G. In these terms two transpositions are adjacent if and only if they commute. Therefore x is the unique non-identity element of $G_1(x)$.

Due to the isomorphism $\mathrm{Sym}_5 \cong O_4^-(2)$ we obtain another description of Π. Let $V_4^- = (V, f, q)$ be a non-singular orthogonal space of minus type of dimension 4 over $GF(2)$. Then the vertices of Π are the vectors of V which are non-singular with respect to the quadratic form q; two such vectors are adjacent if they are perpendicular.

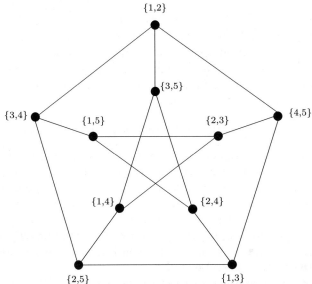

Let $l(\Pi)$ be the line graph of Π. This means that the vertices of $l(\Pi)$ are the edges of Π and two of them are adjacent if they are incident to a common vertex of Π. The graph $l(\Pi)$ is distance-transitive with the following intersection diagram:

The graph $l(\Pi)$ is *antipodal* in the following sense: the relation with respect to which two edges are related if they are either equal or are at distance three in the line graph is an equivalence relation (with classes of size 3). The classes with respect of this *antipodality* relation will be called antipodality classes. When the vertices of Π are treated as transpositions of the underlying 5-element set Ω, two distinct edges $\{x, y\}$ and $\{u, v\}$ are antipodal if and only if xy and uv are even permutations of Ω fixing the same element. Therefore the antipodality classes are indexed by the elements of Ω and the setwise stabilizer of an antipodality class in G is Sym_4.

In terms of the orthogonal space the antipodality classes are indexed by the non-zero singular vectors in V; a vertex is contained in an antipodality class if and only if the corresponding non-singular and singular vectors are perpendicular.

10.4   Diagram geometries

A *geometry* is a quadruple $(\mathcal{F}, I, t, \Omega)$, where $\mathcal{F}$ is a non-empty set (the element-set of the geometry), I is a symmetric binary relation on $\mathcal{F}$ (the incidence relation) and $t : \mathcal{F} \to \Omega$ is the type function which assigns to every element of $\mathcal{F}$ its *type*, the latter being an element of the set Ω of possible types. It is convenient to treat I as a subset of the set $\binom{\mathcal{F}}{2}$ of all 2-element subsets of $\mathcal{F}$. A *flag* in a geometry $(\mathcal{F}, I, t, \Omega)$ is a set of pairwise incident elements of the geometry. Thus $\Phi \subseteq \mathcal{F}$ is a flag whenever $\{a, b\} \in I$ for all distinct $a, b \in \Phi$. Clearly a subset of a flag is also a flag. A flag is said to be *maximal* if it is not a subset of a larger flag. The following conditions are assumed to hold in a geometry

(G1) $t(a) \neq t(b)$ for any two distinct elements a, b of a flag Φ;

(G2) if Φ is a maximal flag then $t(\Phi) = t(\mathcal{F}) = \Omega$.

In order to simplify the notation, instead of $(\mathcal{F}, I, t, \Omega)$ we simply write $\mathcal{F}$. The number of types of elements in $\mathcal{F}$ (that is the size of Ω) is called the *rank* of $\mathcal{F}$. For $i \in \Omega$ let $\mathcal{F}^{[i]}$ denote the set of elements of type i in $\mathcal{F}$ (so that $\mathcal{F}^{[i]} = t^{-1}(i)$). Let $\Gamma = \Gamma(\mathcal{F})$ be the *incidence graph* of $\mathcal{F}$ which is a graph on the element-set of $\mathcal{F}$ in which I is the set of edges. Then (G1) and (G2) imply that

(G3) Γ is an r-partite graph (where $r = |\Omega|$ is the rank of $\mathcal{F}$) with parts $\mathcal{F}^{[i]}$, $i \in \Omega$ and every maximal clique in Γ (that is a maximal complete subgraph) intersects all the parts.

It is clear that every graph which satisfies (G3) is the incidence graph of a geometry of rank r. Therefore it is common to identify a geometry with its incidence graph. A *path* in a geometry is a path in its incidence graph.

Important information about a geometry $\mathcal{F}$ is carried by its residual geometries often called *residues* and defined as follows. Let Φ be a flag in a geometry $\mathcal{F}$. The residue of Φ in $\mathcal{F}$ is the quadruple $(\mathcal{F}_\Phi, I_\Phi, t_\Phi, \Omega_\Phi)$, where

$$\mathcal{F}_\Phi = \{a \mid a \in \mathcal{F} \setminus \Phi, \{a\} \cup \Phi \text{ is a flag in } \mathcal{F}\}, \quad I_\Phi = I \cap \binom{\mathcal{F}_\Phi}{2},$$

t_Φ is the restriction of t to $\mathcal{F}_\Phi$, $\Omega_\Phi = \Omega \setminus t(\Phi)$.

It is immediate that the residue of a non-maximal flag is a geometry. Suppose that Φ is a premaximal flag of cotype $i \in \Omega$ (this means that $t(\Phi) = \Omega \setminus \{i\}$). Then $\mathcal{F}_\Phi$ is a geometry of rank 1 whose set of types is $\{i\}$. This geometry is determined by the number of its elements. If this number, say $q_i + 1$, is independent on the choice of the flag Φ of cotype i then q_i it is called the ith *index* of $\mathcal{F}$ (in what follows we assume that all the indexes exist). If $\mathcal{F}$ is the projective plane over $GF(q)$ then the point- and line-indexes are equal to q.

Suppose now that i and j are two different types in $\mathcal{F}$ and Φ is a flag of cotype $\{i,j\}$ in $\mathcal{F}$ (this means that $t(\Phi) = \Omega \setminus \{i,j\}$). Then the residue $\mathcal{F}_\Phi$ of Φ in $\mathcal{F}$ is a geometry of rank 2 over the set $\{i,j\}$ of types. Every bipartite graph without isolated vertices is the incidence graph of a rank 2 geometry. Since the indexes are assumed to exist, the vertices in one part of $\Gamma(\mathcal{F}_\Phi)$ have valency $q_i + 1$ while those from the other part have valency $q_j + 1$.

Some particularly nice geometries are characterized by their residual geometries of rank 2. It is common to present information about rank 2 residues in the form of a diagram. The diagram consists of nodes indexed by the types of elements in $\mathcal{F}$ (that is by the elements of Ω) and the specific edge which joins the nodes i and j symbolizes the class of rank 2 geometries which arise as residues in $\mathcal{F}$ of flags whose cotype is $\{i,j\}$. The meaning of some (labelled) edges will be explain below. On a diagram above the node we place the corresponding type $i \in \Omega$ (sometimes the types are suppressed) and below the node we put the index q_i. Having a diagram of $\mathcal{F}$ it is easy to see the diagrams of its residual geometries. In order to obtain the diagram of the residue of an element of type i (which is a flag of size 1) all we have to do is to remove from the diagram the node of type i along with all the edges incident to this node. In order to obtain the diagrams of residues of larger flags we proceed by induction.

The diagram business is probably best seen on the following example. Let $V = V_{r+1}(q)$ be a vector space of dimension $r + 1$ over the field $GF(q)$ of q elements. Let

$$\mathcal{P} = (\mathcal{P}, I, t, \Omega)$$

be the projective geometry of V. This means that $\mathcal{P}$ is the set of proper subspaces of V, I is the set of pairs $\{U, W\}$ of such subspaces where either $U < W$ or $W < U$ (in particular $U \neq W$). The type $t(U)$ of a subspace is the dimension of U, so that $\Omega = \{1, 2, \ldots, r\}$. Of course the particular names for the types are irrelevant, for instance, we could have defined $t(U)$ to be the *projective dimension* of U (which is dim $U - 1$). A flag in $\mathcal{F}$ is a chain

$$0 < V_{i_1} < V_{i_2} < \cdots < V_{i_s} < V$$

and the diagram of $\mathcal{P}$ is

$$A_r : \quad \overset{1}{\underset{q}{\circ}} \!\!-\!\!-\!\!-\!\! \overset{2}{\underset{q}{\circ}} \quad \cdots \quad \overset{r-1}{\underset{q}{\circ}} \!\!-\!\!-\!\!-\!\! \overset{r}{\underset{q}{\circ}}.$$

This can be illustrated as follows. Suppose that $r \geq 3$ and that Φ is a flag

$$0 < V_2 < V_3 < \cdots < V_{r-1} < V$$

of cotype $\{1, r\}$, then the collinearity graph of $\mathcal{P}_\Phi$ is the complete bipartite graph $K_{q+1,q+1}$ whose vertex-set consists of the 1-dimensional subspaces contained in V_2 and the r-dimensional subspaces containing V_{r-1}. Any geometry whose collinearity graph is complete bipartite is said to be a *generalized digon* and on a diagram it is symbolized by the empty edge. On the other hand, if Φ is a flag

$$0 < V_1 < \cdots < V_{i-2} < V_{i+1} < \cdots < V_r < V$$

of cotype $\{i-1, i\}$ then $\mathcal{P}_\Phi$ is the projective plane of order q associated with the 3-dimensional $GF(q)$-space V_{i+1}/V_{i-2}. On diagrams the projective planes are symbolized by simple edges.

Let $\mathcal{F} = (\mathcal{F}, I, t, \Omega)$ be a geometry. A permutation of the element-set is said to be an *automorphism* of the geometry if it preserves both the incidence relation and the type function. A permutation of the elements which preserves the incidence relation and permutes the types is called a *diagram automorphism* of the geometry. For instance the automorphism group of the above defined projective geometry $\mathcal{P}$ is the group $P\Gamma L(r+1, q)$ of projective semilinear transformations of V. On the other hand, the isomorphism between V and its dual space V^* induces a diagram automorphism of $\mathcal{P}$ which maps the elements of type i onto the elements of type $r+1-i$ for every $1 \le i \le \left[\frac{r}{2}\right]$.

Let $\mathcal{F} = (\mathcal{F}, I, t, \Omega)$ be a geometry and F be an automorphism group of $\mathcal{F}$. Then F is said to be *flag-transitive* if it acts transitively on the set of maximal flags in $\mathcal{F}$. In this case the whole of $\mathcal{F}$ can be described in terms of the group F and certain of its subgroups. This procedure works as follows. Let $\Phi = \{u_i \mid i \in \Omega\}$ be a maximal flag in $\mathcal{F}$, where $t(u_i) = i$. Let $F^{[i]} = F(u_i)$ be the stabilizer of u_i in F and let

$$\mathcal{A} = \{F^{[i]} \mid i \in \Omega\}$$

be the subamalgam in F forms by these stabilizers. Then the $F^{[i]}$'s are the *maximal parabolic subgroups* and $\mathcal{A}$ is the amalgam of maximal parabolic subgroups associated with the action of F on $\mathcal{F}$ (a parabolic subgroup in F is the stabilizer of a flag in $\mathcal{F}$). Because of the flag-transitivity the isomorphism type of $\mathcal{A}$ is independent on the choice of the maximal flag Φ. Furthermore for every element v_i of type i in $\mathcal{F}$ there is an element in F which maps u_i onto v_i; the set of all such elements is a left coset of $F^{[i]}$ in F which determines v_i uniquely. This observation enables one to show that the geometry $\mathcal{F}$ is isomorphic to the *coset incidence system* $\mathcal{F}(F, \mathcal{A})$ defined as follows:

(C1) the set of types is Ω;
(C2) the set of elements is $\{fF^{[i]} \mid f \in F, i \in \Omega\}$;
(C3) the type function is $fF^{[i]} \mapsto i$;
(C4) the incidence relation is $\{\{fF^{[i]}, hF^{[j]}\} \mid i \ne j, fF^{[i]} \cap hF^{[j]} \ne \emptyset\}$.

In the case of the projective geometry $\mathcal{P}$ a maximal flag Φ is a chain

$$0 < V_1 < V_2 < \cdots < V_r < V$$

the action of the automorphism group $L = P\Gamma L(r+1, q)$ is flag-transitive (the result is known as the main theorem of projective geometry). The maximal parabolic subgroup $L^{[i]}$ is the stabilizer in L of the i-dimensional subspace V_i from V. The subamalgam $\mathcal{L} = \{L^{[i]} \mid 1 \le i \le r\}$ can be specified as the one formed by the maximal p-local subgroups in L containing the normalizer of a given Sylow p-subgroup (here p is the characteristic of $GF(q)$ and the normalizer is the stabilizer of Φ in L). In this case the isomorphism $\mathcal{P} \cong \mathcal{F}(L, \mathcal{L})$ provides us with a purely group-theoretical definition of the projective geometry.

An important feature of the above description is that the proper residues in $\mathcal{F}$ are described in terms of the amalgam of maximal parabolics only (that is independently of whole group F). For instance the residue $\mathcal{F}_{\{u_i\}}$ is isomorphic to $\mathcal{F}(F^{[i]}, \mathcal{A}^{[i]})$, where

$$\mathcal{A}^{[i]} = \{ F^{[i]} \cap F^{[j]} \mid j \in \Omega \setminus \{i\} \}.$$

Now suppose that Ω is a set of size r, let $\mathcal{A} = \{ F^{[i]} \mid i \in \Omega \}$ be an abstract amalgam of rank r and $\varphi : \mathcal{A} \to F$ is a faithful completion of $\mathcal{A}$ (we identify $\mathcal{A}$ with its image under φ). Then the cosets incidence system $\mathcal{F}(F, \mathcal{A})$ might be a geometry but it also might not (in that case it contains so-called non-standard flags). The conditions for $\mathcal{F}(F, \mathcal{A})$ to be a geometry are known (cf. sections 10.1.3 and 10.1.4 in (Pasini 1994) and Lemma 1.4.1 in (Ivanov 1999)). These conditions are formulated in terms of the amalgam $\mathcal{A}$ and are independent on the completion group F. These conditions are satisfied in all the cases which occur in the present work. Therefore in the remaining of the section we assume that $\mathcal{F}(F, \mathcal{A})$ is a geometry.

The diagram of $\mathcal{F}(F, \mathcal{A})$ carries in a very compact form a lot of information about the structure of $\mathcal{A}$ (recall that the diagram is independent on the completion F) and we use this language extensively throughout the volume. The completion F is generating if and only if the geometry $\mathcal{F}(F, \mathcal{A})$ is connected (the latter means that the incidence graph of $\mathcal{F}(F, \mathcal{A})$ is connected in the usual sense).

Let $\varphi^{(1)} : \mathcal{A} \to F^{(1)}$ and $\varphi^{(2)} : \mathcal{A} \to F^{(2)}$ be two completions of $\mathcal{A}$ and $\psi : F^{(1)} \to F^{(2)}$ be a homomorphism of completions (so that $\varphi^{(2)}(a) = \psi(\varphi^{(1)}(a))$ for every $a \in \mathcal{A}$). Then the mapping χ of $\mathcal{F}(F^{(1)}, \mathcal{A})$ onto $\mathcal{F}(F^{(2)}, \mathcal{A})$ defined by

$$\chi : f\varphi^{(1)}(F^{[i]}) \mapsto \psi(f)\varphi^{(2)}(F^{[i]})$$

is a *covering* of geometries in the sense that the restriction of χ to every proper residue is an isomorphism (in fact, the residues are determined by the amalgam $\mathcal{A}$ which stays the same). Certainly we can introduce the notion of a covering for any geometry (which might or might not be flag-transitive) and to define the universal covering in the usual way (cf. Pasini 1994; Ivanov 1999; Ivanov and Shpectorov 2002). A geometry which is isomorphic to its universal cover is said to be *simply connected*. The following principle independently established by A. Pasini, S.V. Shpectorov and J. Tits is of a fundamental importance for the theory of flag-transitive diagram geometries.

Theorem 10.4.1 *If $\widetilde{\varphi} : \mathcal{A} \to \widetilde{F}$ is the universal completion of $\mathcal{A}$ then $\chi : \mathcal{F}(\widetilde{F}, \mathcal{A}) \to \mathcal{F}(F, \mathcal{A})$ is the universal covering.*

It is not so difficult to check that the projective geometry $\mathcal{P}$ of rank at least 3 is simply connected. Therefore $L = P\Gamma L(r+1, q)$ is the universal completion of the amalgam $\mathcal{L}$ of maximal parabolic subgroups.

Below we list the diagrams of some geometries we came across within the present volume.

$\underset{p}{\circ} \quad \underset{q}{\circ}$ – the generalized digon, whose incidence graph is complete bipartite $K_{p+1,q+1}$;

$\underset{2}{\circ}\!\!-\!\!-\!\!\underset{2}{\circ}$ – the projective plane of order 2 formed by the 1- and 2-dimensional subspaces in a 3-dimensional $GF(2)$-space, where the incidence is via inclusion;

$\underset{2}{\circ}\!\!=\!\!\underset{2}{\circ}$ – the generalized quadrangle of order $(2,2)$ formed by the 1- and 2-dimensional totally isotropic subspaces in a non-singular 4-dimensional symplectic $GF(2)$-space, the incidence is via inclusion;

$\underset{2}{\circ}\!\!\overset{\sim}{=}\!\!\underset{2}{\circ}$ – the rank 2 *tilde* geometry which is a triple cover of the generalized quadrangle of order $(2,2)$ associated with the non-split extension $3 \times S_4(2) \cong 3 \times \mathrm{Sym}_6$;

$\underset{2}{\circ}\!\!\overset{d}{-\!\!-}\!\!\underset{5}{\circ}$ – this geometry possesses the following description in terms of the generalized quadrangle of order $(2,2)$: take one clone of every totally isotropic 1-space and two clones of every totally isotropic 2-space, a 1-space is incident to the both clones of a 2-space containing the 1-space;

$\underset{1}{\circ}\!\!=\!\!\underset{2}{\circ}$ – the generalized quadrangle of order $(1,2)$ which is the geometry of vertices and edges of the complete bipartite graph $K_{3,3}$ with the natural incidence relation;

$\underset{1}{\circ}\!\!\overset{P}{-\!\!-}\!\!\underset{2}{\circ}$ – the geometry of vertices and edges of the Petersen graph with the natural incidence relation;

$\underset{5}{\circ}\!\!\overset{t}{-\!\!-}\!\!\underset{2}{\circ}$ – the geometry of vertices and the antipodal triples of edges of the Petersen graph with the natural incidence relation;

$\underset{2}{\circ}\!\!-\!\!-\!\!\underset{2}{\circ}\!\!-\!\!-\!\!\square$ – the geometry of 1- and 2-dimensional subspaces in a 4-dimensional $GF(2)$-space (this is a truncation of the rank 3 projective $GF(2)$-geometry).

A geometry of rank r with the digram

$$P_r: \quad \underset{1}{\circ}\overset{P}{-\!\!-}\underset{2}{\circ}-\!\!-\underset{2}{\circ}\cdots\underset{2}{\circ}-\!\!-\underset{2}{\circ}$$

is said to be a *P-geometry* (or Petersen-type geometry) of rank r.

A geometry of rank r with the diagram

$$T_r: \quad \underset{2}{\circ}\overset{\sim}{=\!\!=}\underset{2}{\circ}-\!\!-\underset{2}{\circ}\cdots\underset{2}{\circ}-\!\!-\underset{2}{\circ}$$

is said to be a *tilde geometry* of rank r.

11

APPENDIX: MATHIEU GROUPS AND THEIR GEOMETRIES

In this chapter we summarize some properties of the Mathieu groups M_{24} and M_{22}, and of the associated combinatorial structures. These properties are used throughout the second half of the volume. In order to fully appreciate Chapter 8 one should be fluent in Mathieu groups and their Steiner systems. In (Ivanov 1999) all the needed properties were deduced from the very basic principles. It would not be appropriate to reproduce all the proofs here. Most of the information is well known, not so hard to deduce and can be found in the extensive literature on the Mathieu groups including Aschbacher (1994), Brouwer et al. (1989), Conway et al. (1985), Curtis (1976), Griess (1998), James (1973), Mathieu (1860, 1861), Mazet (1979), Syskin (1980), Todd (1966), Witt (1938). Less commonly known (but crucial for our strategy) are the diagrams for $\mathcal{H}(M_{24})$ and $\mathcal{H}(M_{22})$ presented in Sections 11.5 and 11.6. These diagram were first introduced in (Ivanov and Meierfrankenfeld 1999) exactly for the purpose they are used here. The proofs for the diagrams are given in sections 3.7 and 3.9 in (Ivanov 1999). These diagram extend the calculations of the orbits of one maximal subgroup in a Mathieu group on the coset of another maximal subgroup pioneered in (Curtis 1976).

11.1 Witt design $S(5, 8, 24)$

Let P_{24} be a set of 24 elements. Then there exists a collection $\mathcal{B}$ of 8-element subsets of P_{24} such that every 5-element subset of P_{24} is contained in exactly one subset from $\mathcal{B}$. The pair $(P_{24}, \mathcal{B})$ is the unique Steiner system $S(5, 8, 24)$ also known as the Witt design (Witt 1938). In (Todd 1966) the name *octads* for the subsets in $\mathcal{B}$ was introduced. An elementary combinatorial calculation shows that there are exactly

$$\binom{24}{5} \bigg/ \binom{8}{5} = 759$$

octads.

The automorphism group M of the Witt design $(P_{24}, \mathcal{B})$ is the Mathieu sporadic simple group M_{24} discovered by E. Mathieu as early as in 1860 (cf. Mathieu (1860, 1861)). The order of M_{24} is

$$|M_{24}| = 2^{10} \cdot 3^3 \cdot 5 \cdot 7 \cdot 11 \cdot 23,$$

and the action of M_{24} on P_{24} is 5-fold transitive, that is transitive on the set of ordered 5-element subsets. Let M_{24-i} denote the elementwise stabilizer in M_{24} of an i-element subset of P_{24} and let M_{24-i}^* denote the stabilizer of such a subset as a whole. Then M_{23} and M_{22} are two further Mathieu sporadic simple groups of orders

$$|M_{23}| = |M_{24}|/24 = 2^7 \cdot 3^2 \cdot 5 \cdot 7 \cdot 11 \cdot 23$$

and

$$|M_{22}| = |M_{23}|/23 = 2^7 \cdot 3^2 \cdot 5 \cdot 11,$$

while $M_{21} \cong L_3(4)$. We have $M_{22}^*/M_{22} \cong 2$, $M_{21}^*/M_{21} \cong \mathrm{Sym}_3$, so that

$$M_{22}^* \cong \mathrm{Aut}\,(M_{22}) \quad \text{and} \quad M_{21}^* \cong P\Gamma L_3(4),$$

respectively.

Let $B \in \mathcal{B}$ be an octad. The stabilizer M_b of B in $M \cong M_{24}$ is the semidirect product of a subgroup $Q_b \cong 2^4$ which fixes B elementwise acting regularly on $P_{24} \setminus B$ and a subgroup $K_b \cong L_4(2)$ which fixes an element outside B. The subgroup K_b acts faithfully on B as the alternating group of degree eight and $K_b \cong L_4(2)$ acts on Q_b by conjugation as on its natural 4-dimensional $GF(2)$-module. Therefore

$$M_b = Q_b : K_b \cong 2^4 : L_4(2) \cong AGL_4(2).$$

In order to simplify the terminology we will identify B with the partition of P_{24} into B and the complement of B.

The *octad graph* $\Gamma(M_{24})$ is the graph on the set $\mathcal{B}$ of 759 octads in which two octads are adjacent if they are disjoint. The octad graph is distance-transitive with respect to the action of M_{24}. The intersection diagram of $\Gamma(M_{24})$ is the following

A *trio* is a partition $T = \{B_1, B_2, B_3\}$ of P_{24} into three pairwise disjoint octads. By the definition the trios are in a natural correspondence with the triangles in the octad graph $\Gamma(M_{24})$ and it is immediate from the intersection diagram of $\Gamma(M_{24})$ that every octad appears in exactly 15 trios, so that the total number of trios is $3795 = (759 \cdot 15)/3$. The group M_{24} permutes the trios transitively and the stabilizer M_t of a trio is a semidirect product of a subgroup $Q_t \cong 2^6$ and a subgroup $K_t \cong L_3(2) \times \mathrm{Sym}_3$. The subgroup K_t acts on Q_t by conjugation as it acts on the tensor product of a natural module U_3 of $L_3(2)$ and a natural module U_2 of $L_2(2) \cong \mathrm{Sym}_3$. Thus

$$M_t = Q_t : K_t \cong 2^6 : (L_3(2) \times \mathrm{Sym}_3).$$

A 4-element subset of P_{24} is called a *tetrad*. Every tetrad S is contained in a unique *sextet* $\Sigma = \{S_1 = S, S_2, ..., S_6\}$ which is a partition of P_{24} into six tetrads such that the union of any two distinct ones is an octad. The octads refined by a given sextet induce in the octad graph $\Gamma(M_{24})$ a subgraph known as a *quad* and isomorphic to the collinearity graph of the generalized quadrangle of order $(2, 2)$ (cf. Shult and Yanushka (1980) and Brouwer et al. (1989) for further properties of the octad graph, which is also an example of a near hexagon with three point per a line). There are

$$1771 = \binom{24}{4} \Big/ 6$$

sextets transitively permuted by M and the stabilizer M_s of a sextet is a semi-direct product of a subgroup $Q_s \cong 2^6$ and a subgroup $K_s \cong 3 \cdot \mathrm{Sym}_6$. The latter group is a non-split non-central extension of a group of order 3 by Sym_6. The group K_s acts on Q_s as on the *hexacode module*, in particular $O_3(K_s)$ acts on Q_s fixed-point freely. The group M_s induces the symmetric group on the set of tetrads in the corresponding sextet.

11.2   Geometries of M_{24}

Every octad from $\mathcal{B}$ can be identified with the partition of P_{24} into two subsets, one of them being the octad itself. Let $\mathcal{T}$ and $\mathcal{S}$ denote, respectively the set of trios and sextets in $(P_{24}, \mathcal{B})$. The Ronan–Smith geometry $\mathcal{H}(M_{24})$ is the geometry on

$$\mathcal{B} \cup \mathcal{T} \cup \mathcal{S}$$

(so that every element of $\mathcal{H}(M_{24})$ is a partition of P_{24}). Two elements of $\mathcal{H}(M_{24})$ are incident if one of them refines the other one. The geometry $\mathcal{H}(M_{24})$ belongs to the truncated C_4-diagram

$$C_4^t : \quad \begin{array}{ccccc} \circ & = & \circ & \rule{1cm}{0.4pt} & \circ \rule{1cm}{0.4pt} \square \\ 2 & & 2 & & 2 \end{array} .$$

The natural action of M_{24} on $\mathcal{H}(M_{24})$ is flag-transitive and

$$\mathcal{B}(M_{24}) = \{M_b, M_t, M_s\}$$

is the amalgam of maximal parabolic subgroups associated with this action (here the octad, the trio and the sextet stabilized by M_b, M_t, and M_s, respectively are assumed to be pairwise incident).

The following is a slightly weakened version of the main result of (Ronan 1982).

Theorem 11.2.1 *Let* $\mathcal{X} = \{X_b, X_t, X_s\}$ *be an amalgam of rank 3, which corresponds to the locally truncated diagram* C_4^t *and suppose that* $X_b \cong M_b \cong 2^4 : L_4(2)$. *Then the following assertions hold:*

(i) $\mathcal{X} \cong \mathcal{B}(M_{24})$;

(ii) *the geometry* $\mathcal{H}(M_{24})$ *is simply connected;*

(iii) M_{24} *is the only faithful completion of* $\mathcal{B}(M_{24})$.

Proof The assertion (i) is proved in section 12.4 of (Ivanov and Shpectorov 2002), the assertion (ii) is proposition 3.2.4 in (Ivanov 1999). Finally (iii) is an immediate consequence of (i), (ii), and the simplicity of M_{24}. ∎

As above let $\mathcal{B}(M_{24})$ denote the amalgam of maximal parabolic subgroups associated with the action of $M \cong M_{24}$ on $\mathcal{H}(M_{24})$. Then $Q_b \cap Q_t$ is elementary abelian of order 2^3 and $M_b \cap M_t \cong 2^{3+3+1} : L_3(2)$ induces on $Q_b \cap Q_t$ an action of $L_3(2)$ as on the natural module. Let $\tau \in Q_b \cap Q_t$, let $R \cong 2^2$ be the subgroup in Q_t generated by the images of τ under $O_{2,3}(M_t)$ and consider the amalgam

$$\mathcal{A}(M_{24}) = \{C_{M_b}(\tau), N_{M_t}(R), M_s\}.$$

Then the isomorphism type of $\mathcal{A}(M_{24})$ is independent on the choice of such τ;

$$C_{M_b}(\tau) = C_M(\tau) \cong 2^{1+6}_+ : L_3(2)$$

is isomorphic to the centralizer of a central involution in $L_5(2)$ and the coset geometry corresponding to the embedding of $\mathcal{A}(M_{24})$ into M_{24} is described by the tilde diagram

$$T_3 : \quad \overset{\sim}{\underset{2}{\circ}\!\!=\!\!=\!\!=\!\!\underset{2}{\circ}\!\!-\!\!-\!\!\underset{2}{\circ}.}$$

Theorem 11.2.2 *Let* $\mathcal{Y} = \{Y_b, Y_t, Y_s\}$ *be an amalgam of rank 3 which corresponds to the diagram* T_3 *and suppose that*

$$|Y_s| = |M_s| = |2^6 : 3 \cdot \mathrm{Sym}_6|.$$

Then

(i) *$\mathcal{Y}$ is isomorphic either to $\mathcal{A}(M_{24})$ or to one further amalgam $\mathcal{A}(He)$;*
(ii) *M_{24} is the only faithful completion of $\mathcal{A}(M_{24})$;*
(iii) *the sporadic simple group He of Held is the only faithful completion of $\mathcal{A}(He)$.*

Proof The assertion (i) is proved in section 12.3 in (Ivanov and Shpectorov 2002). The assertions (ii) and (iii) have been established by coset enumeration on a computer (cf. proposition 12.3.6 in Ivanov and Shpectorov (2002) and Heiss (1991)). ∎

Below we reproduce the presentations for the universal completions of $\mathcal{A}(M_{24})$ and $\mathcal{A}(He)$ obtained by S. V. Shpectorov and the author in 1989. The generators are 13 involutions a_i, $1 \le i \le 13$ such that the first 12 generate

$$Y_b \cong C_{M_b}(\tau) \cong 2^{1+6}_+ : L_3(2)$$

which is also the centralizer of a central involution in $L_5(2)$. These 12 generators correspond to elementary 5×5 matrices over $GF(2)$ as shown on the matrix below. It should be understood that the generator a_i for $1 \le i \le 12$ corresponds to the matrix with 0s everywhere except the diagonal and the position where a_i stands. This position is of course 1. It is easy to see that a_7 generates the centre of

Y_b, while $a_1, a_2, a_4, a_7, a_8, a_9, a_{10}$ generate $O_2(Y_b) \cong 2^{1+6}_+$ and $a_3, a_5, a_6, a_{11}, a_{12}$ generate an $L_3(2)$-complement.

$$\begin{pmatrix} 1 & 0 & 0 & 0 & 0 \\ a_1 & 1 & a_{11} & 0 & 0 \\ a_2 & a_3 & 1 & a_{12} & 0 \\ a_4 & a_5 & a_6 & 1 & 0 \\ a_7 & a_8 & a_9 & a_{10} & 1 \end{pmatrix}$$

Using the above matrix it is easy to calculate the commutators $[a_i, a_j]$ for $1 \leq i, j \leq 10$ as given below (the missing commutators equal to the identity):

$$\begin{array}{llll} [a_1, a_3] = a_2, & [a_1, a_5] = a_4, & [a_1, a_8] = a_7, & [a_2, a_6] = a_4, \\ [a_2, a_9] = a_7, & [a_2, a_{11}] = a_1, & [a_3, a_6] = a_5, & [a_3, a_9] = a_8, \\ [a_4, a_{10}] = a_7, & [a_4, a_{12}] = a_2, & [a_5, a_{10}] = a_8, & [a_5, a_{11}] = a_6, \\ [a_5, a_{12}] = a_3, & [a_6, a_{10}] = a_9, & [a_8, a_{11}] = a_9, & [a_9, a_{12}] = a_{10}. \end{array}$$

These commutator relations together with the relations

$$(a_3 a_{11})^3 = (a_6 a_{12})^3 = (a_{11} a_{12})^4 = 1.$$

provide us with a presentation for Y_b (this is basically a fragment of the Steinberg presentation for $L_5(2)$).

The additional generator a_{13} extends

$$\langle a_i \mid 1 \leq i \leq 11 \rangle \text{ and } \langle a_i \mid 1 \leq i \leq 12, i \neq 11 \rangle$$

(both isomorphic to $2^6 : (\mathrm{Sym}_4 \times \mathrm{Sym}_2)$) to $Y_t \cong 2^6 : (\mathrm{Sym}_4 \times \mathrm{Sym}_3)$ and to $Y_s \cong 2^6 : 3 \times \mathrm{Sym}_6$, respectively (where $O_2(Y_t) = \langle a_4, a_5, a_6, a_7, a_8, a_9 \rangle$ and $O_2(Y_s) = \langle a_2, a_3, a_4, a_5, a_7, a_8 \rangle$). The following relations are common for both amalgams

$$[a_2, a_{13}] = a_4 a_5, \ [a_3, a_{13}] = a_4, \ [a_7, a_{13}] = a_4, \ [a_8, a_{13}] = a_5.$$

The extra relations for $\mathcal{A}(M_{24})$ are

$$[a_1, a_{13}] = a_6, \ [a_9, a_{13}] = a_6, \ (a_{10} a_{13})^3 = 1,$$

$$[a_{11}, a_{13}] = a_6 a_4, \ (a_{12} a_{13})^5 = (a_6 a_{12} a_{13})^5 = (a_{10} a_{13} a_{12})^5 = 1.$$

For $\mathcal{A}(He)$ they are

$$[a_1, a_{13}] = a_6 a_7, \ [a_6, a_{13}] = a_4, \ [a_9, a_{13}] = a_4 a_6 a_7, \ [a_{11}, a_{13}] = a_1 a_9 a_4 a_7,$$

$$(a_2 a_7 a_{10} a_{13})^3 = (a_7 a_{12} a_{13})^5 = (a_6 a_{12} a_{13})^5 = (a_2 a_7 a_{10} a_{13} a_{12} a_7)^5 = 1.$$

Enumeration of the cosets of Y_s gives

$$1,771 = [M_{24} : Y_s]$$

in the case of $\mathcal{A}(M_{24})$ and

$$29,155 = [He : Y_s]$$

in the case of $\mathcal{A}(He)$.

Proposition 11.2.3 *The following assertions hold:*

(i) *each of the subgroups $M_{22}^* \cong \mathrm{Aut}\ (M_{22})$, $M_{21}^* \cong P\Gamma L_3(4)$, $M_b \cong 2^4 :$ $L_4(2)$, $M_t \cong 2^6 : (L_3(2) \times \mathrm{Sym}_3)$ and $M_s \cong 2^6 : 3 \cdot \mathrm{Sym}_6$ is maximal in $M \cong M_{24}$;*

(ii) *if $C_{M_b}(\tau) < F < M$ then $F = M_b$;*

(iii) *if $N_{M_t}(R) < F < M$ then $F = M_t$.*

11.3   Golay code and Todd modules

Let $\mathcal{P}_{24}$ denote the $GF(2)$-permutation module of M_{24} acting on P_{24} defined as in Section 3.2 and let $\mathcal{C}_{12}$ be the subspace in $\mathcal{P}_{24}$ generated by the octads (considered as elements of $\mathcal{P}_{24}$). Then $\mathcal{C}_{12}$ is 12-dimensional, known as the *Golay code module* for M_{24}. Besides the 759 octads and the empty set, $\mathcal{C}_{12}$ contains the set P_{24} itself and the 759 complements to the octads; the remaining 2576 elements of $\mathcal{C}_{12}$ are the so-called *dodecads*. A dodecad is a 12-element subset of P_{24} which can be represented as the symmetric difference of two octads (intersecting in two elements). The dodecads are transitively permuted by M_{24} and the stabilizer M_d of a dodecad is another Mathieu sporadic simple group M_{12} of order

$$|M_{12}| = 2^6 \cdot 3^3 \cdot 5 \cdot 11.$$

The complement of a dodecad is again a dodecad and the stabilizer M_{12}^* of a complementary pair of dodecads is $\mathrm{Aut}\ (M_{12})$. The smallest Mathieu sporadic simple group M_{11} is the stabilizer in M_d of an element of P_{24}, its order is

$$|M_{11}| = 2^4 \cdot 3^2 \cdot 5 \cdot 11.$$

The only composition series of $\mathcal{P}_{24}$ is

$$0 < \mathcal{P}_{24}^c < \mathcal{C}_{12} < \mathcal{P}_{24}^e < \mathcal{P}_{24}$$

and $\mathcal{C}_{11} = \mathcal{C}_{12}/\mathcal{P}_{24}^c$ is known as the *irreducible Golay code module*. In that module M_{24} has two orbits on the set of non-zero vectors indexed by the octads and pairs of complementary dodecads, respectively. The module

$$\bar{\mathcal{C}}_{12} = \mathcal{P}_{24}/\mathcal{C}_{12}$$

is called the *Todd module*. The non-zero vectors in the Todd module are indexed by the 1-, 2-, 3-element subsets of P_{24} and by the sextets. The submodule $\bar{\mathcal{C}}_{11} = \mathcal{P}_{24}^e/\mathcal{C}_{12}$ of codimension 1 in the Todd module is called the *irreducible Todd module*. The non-zero vectors of $\bar{\mathcal{C}}_{11}$ are indexed by 2-element subsets of P_{24} and by the sextets. It is clear that $(\mathcal{C}_{12}, \bar{\mathcal{C}}_{12})$ and $(\mathcal{C}_{11}, \bar{\mathcal{C}}_{11})$ are dual pairs.

For the proof of the following lemma see section 3.8 in (Ivanov 1999).

Lemma 11.3.1 *Let $Y = \bar{\mathcal{C}}_{11}$ and $x = b$, t or s. Then*

$$1 < C_Y(Q_x) < [Y, Q_x] < Y$$

is the only composition series of Y as a module for M_x. Furthermore,

(i) $C_Y(Q_b) \cong \bigwedge^2 Q_b$, $[Y, Q_b]/C_Y(Q_b) \cong Q_b$, $Y/[Y, Q_b] \cong 2$;

(ii) $C_Y(Q_t) \cong U_3^*$, $[Y, Q_t]/C_Y(Q_t) \cong Q_t$, $Y/[Y, Q_t] \cong U_2$;

(iii) $C_Y(Q_s) \cong 2$, $[Y, Q_s]/C_Y(Q_s) \cong Q_s^*$, $Y/[Y, Q_s]$ *is the natural symplectic module of $M_s/O_{2,3}(M_s) \cong S_4(2)$.*

11.4 Shpectorov's characterization of M_{22}

In terms of the previous section let $\Psi = (B, T, \Sigma)$ be a flag in $\mathcal{H}(M_{24})$, where $B \in \mathcal{B}$ is an octad, $T = \{B = B_1, B_2, B_3\}$ is a trio and $\Sigma = \{S_1, ..., S_6\}$ is a sextet and suppose that $B_1 = S_1 \cup S_2$, $B_2 = S_3 \cup S_4$, $B_3 = S_5 \cup S_6$. Let $\mathcal{A}(M_{24}) = \{M_b, M_t, M_s\}$ be the amalgam of maximal parabolic subgroups in $M = M_{24}$ associated with the flag Ψ. Let $x, y \in S_6$ and let K be the setwise stabilizer of $\{x, y\}$ in $M \cong M_{24}$. Then

$$K \cong M_{22}^* \cong \text{Aut } (M_{22}).$$

Let $K_o = K \cap M_b$, $K_t = K \cap M_t$, $K_p = K \cap M_s$ and put

$$\mathcal{A}(\text{Aut } (M_{22})) = \{K_o, K_t, K_p\}.$$

Then $K_o \cong 2^3 : L_3(2) \times 2$, $K_t \cong (2_+^{1+4} \times 2) : (\text{Sym}_3 \times 2)$ and $K_p = 2^5 : S_5$, in particular $[K_t : K_o \cap K_t] = 2$.

Let $\Gamma(M_{22})$ be the graph whose vertices and edges are the cosets in K of K_o and K_t, respectively; a vertex and an edge are incident if the corresponding cosets have non-empty intersection (in terms of (2.3.2) $\Gamma(M_{22})$ is $\Lambda(\{K_o, K_t\}, \iota, K)$, where ι is the identity mapping). Alternatively $\Gamma(M_{22})$ is the subgraph in the octad graph $\Gamma(M_{24})$ induced by the set of images of $B = B_1$ under K. This set of images consists of the 330 octads disjoint from $\{x, y\}$ (these octads are called *octets* and therefore $\Gamma(M_{22})$ is called the *octet graph*). Notice that two octets are adjacent in the octet graph $\Gamma(M_{22})$ if and only if they are disjoint (for instance $\{B_1, B_2\}$ is an edge). The octet graph is distance-transitive with the following intersection diagram.

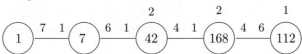

The action of K (and also of M_{22}) is distance-transitive and locally projective of type $(3, 2)$. The geometric cubic subgraphs in the octet graph are isomorphic to the Petersen graph and K_p is the stabilizer in K of one of these subgraphs. The geometry of vertices, edges, and Petersen subgraphs in $\Gamma(M_{22})$ is the Petersen type geometry $\mathcal{G}(M_{22})$ with the diagram

$$P_3 : \quad \underset{1}{\circ} \overset{\text{P}}{\underline{\hspace{2cm}}} \underset{2}{\circ} \underline{\hspace{2cm}} \underset{2}{\circ}$$

and $\mathcal{A}(\text{Aut } (M_{22}))$ is the amalgam of maximal parabolic subgroups associated with the flag-transitive action of K on $\mathcal{G}(M_{22})$.

The octet graph possesses a 3-fold antipodal cover $\Gamma(3 \cdot M_{22})$ known as *Ivanov–Ivanov–Faradjev graph* (Brouwer et al. 1989) whose intersection diagram is the following:

The automorphism group $\widetilde{K}$ of $\Gamma(3 \cdot M_{22})$ is the non-split non-central extension by K of a group of order 3, so that

$$\widetilde{K} \cong 3 \cdot \mathrm{Aut}\ M_{22}.$$

The action of $\widetilde{K}$ on $\Gamma(3 \cdot M_{22})$ is also locally projective of type $(3, 2)$, the geometric cubic subgraphs are still isomorphic to the Petersen graph; the geometry $\mathcal{G}(3 \cdot M_{22})$ of vertices, edges, and Petersen subgraphs in $\mathcal{G}(3 \cdot M_{22})$ is a cover of $\mathcal{G}(M_{22})$, in particular it also belongs to the diagram P_3. Furthermore the amalgam of maximal parabolic subgroups of the flag-transitive action of $\widetilde{K}$ on $\mathcal{G}(3 \cdot M_{22})$ is isomorphic to $\mathcal{A}(\mathrm{Aut}\ (M_{22}))$.

The following beautiful combinatorial result was established almost 20 years ago (cf. Shpectorov (1985) and proposition 3.5.5 in Ivanov (1999)). This result lies in the foundation of the geometric approach to the sporadic simple groups.

Theorem 11.4.1 *Let $\mathcal{G}$ be a geometry with the diagram P_3 (which might or might not be flag-transitive). Then the number of elements corresponding to the leftmost node on the diagram is at most 1898.*

Theorem 11.4.2 *Let $\mathcal{Z} = \{Z_o, Z_t, Z_p\}$ be an amalgam of rank 3 which corresponds to the diagram P_3 and suppose that $Z_o \cong K_o \cong 2^3 : L_3(2) \times 2$. Then*

(i) $\mathcal{Z} \cong \mathcal{A}(\mathrm{Aut}\ (M_{22}))$;
(ii) $K \cong \mathrm{Aut}\ (M_{22})$ *and* $\widetilde{K} \cong 3 \cdot \mathrm{Aut}\ (M_{22})$ *are the only faithful completions of* $\mathcal{A}(\mathrm{Aut}\ (M_{22}))$ *(with the latter completion being the universal one).*

Proof The assertion (i) was originally proved in Shpectorov (1985) and a proof is given in section 11.2 in Ivanov and Shpectorov (2002). Now (ii) is immediate from (11.4.1) since $\Gamma(3 \cdot M_{22})$ contains 990 vertices (which is more than half of the upper bound). ∎

The classification of the flag-transitive geometries with diagram P_3 obtained in Shpectorov (1985) (see also Ivanov and Shpectorov (2002)) implies the following.

Theorem 11.4.3 *Let Θ be a graph of valency 7 and L be a group of automorphisms of Θ whose action on Θ is locally projective of type $(3, 2)$. Suppose further that a geometric cubic subgraph in Θ is isomorphic to the Petersen graph.*

Then one of the following holds:

(i) Θ *is the octet graph* $\Gamma(M_{22})$ *and* L *is either* M_{22} *or* Aut M_{22};

(ii) Θ *is the Ivanov–Ivanov–Faradjev graph* $\Gamma(3 \cdot M_{22})$ *and* L *is either* $3 \cdot M_{22}$ *or* $3 \cdot$ Aut (M_{22}).

In terms introduced at the beginning of the section put $P_{22} = P_{24} \setminus \{x, y\}$, $H = B_3 \setminus \{x, y\}$ and let $\mathcal{H}$ be the set of images of H under K. Then $(P_{22}, \mathcal{H})$ is the unique Steiner system $S(3, 6, 22)$ whose blocks are called *hexads*. Let K_h be the stabilizer in K of the hexad H. Then

$$K_h \cong 2^4 : \mathrm{Sym}_6$$

and K_h contains K_t with index 15.

Let us say that two edges of the octet graph $\Gamma(M_{22})$ are *locally antipodal* if they are contained in a common geometric cubic subgraph Σ (isomorphic to the Petersen graph) and if they are antipodal in the line graph of Σ. Let $\Phi_{\Gamma(M_{22})}$ be the graph (called *local antipodality* graph) on the set of edges of $\Gamma(M_{22})$ in which two edges are adjacent if they are locally antipodal. The following result can be read from lemmas 3.4.4 and 7.1.7 in Ivanov (1999).

Lemma 11.4.4 *Let* $e = \{B_1^e, B_2^e\}$ *and* $f = \{B_1^f, B_2^f\}$ *be edges of the octet graph* $\Gamma(M_{22})$. *Then* e *and* f *are contained in the same connected component of the local antipodality graph* $\Phi_{\Gamma(M_{22})}$ *if and only if*

$$P_{22} \setminus (B_1^e \cup B_2^e) \quad \text{and} \quad P_{22} \setminus (B_1^f \cup B_2^f)$$

is the same hexad of $(P_{22}, \mathcal{H})$.

In the local antipodality graph $\Phi_{\Gamma(3 \cdot M_{22})}$ every connected component is isomorphic to the triple cover (associated with $3 \cdot \mathrm{Sym}_6$) of the point graph of the generalized quadrangle of order $(2,2)$.

Thus there is a bijection between the hexads in $(P_{22}, \mathcal{H})$ and the connected components of the local antipodality graph $\Phi = \Phi_{\Gamma(M_{22})}$. In these terms the component Φ^c containing the edge $\{B_1, B_2\}$ corresponds to the hexad $H = B_3 \setminus \{x, y\}$. Then Φ^c is the point graph of the unique generalized quadrangle of order $(2, 2)$ (the vertices are the 2-subsets of a 6-sets in which two of subsets are adjacent if they are disjoint). The triangles of Φ^c correspond to the 15 geometric cubic subgraphs in $\Gamma(M_{22})$ whose edge-sets intersect the vertex-set of Φ^c. These subgraphs naturally correspond the 2-element subsets of P_{22} contained in H. The group K_h (which is the stabilizer of H in K) induces the full automorphism group Sym_6 of Φ^c; the kernel $O_2(K_h)$ stabilizes as a whole (but not vertexwisely) every edge of $\Gamma(M_{22})$ contained in Φ^c.

The amalgam

$$\mathcal{B}(\mathrm{Aut}\ (M_{22})) = \{K_p, K_o, K_h\}$$

is the amalgam of maximal parabolic subgroups associated with the flag-transitive action of K on the Ronan–Smith geometry $\mathcal{H}(M_{22})$ (Ronan and

Smith 1980). The elements of $\mathcal{H}(M_{22})$ are the geometric cubic subgraphs in $\mathcal{G}(M_{22})$, the vertices of $\Gamma(M_{22})$ and the connected components of the local anti-podality graph $\Phi_{\Gamma(M_{22})}$. Alternatively elements are the *pairs* (2-element subsets of P_{22}), the octets and the hexads and the incidence relation is such that

$$(S_6 \setminus \{x,y\}, B_1, B_3 \setminus \{x,y\})$$

is a maximal flag (recall that S_6 is the last tetrad in the sextet discussed in the beginning of the section). The geometry $\mathcal{H}(M_{22})$ belongs to the following diagram

$\mathcal{H}(Mat_{22})$:

where the nodes from left to right correspond to pairs, octets and hexads, respectively.

The geometry $\mathcal{H}(M_{22})$ is convenient to view through the *hexad graph*. This is a graph Ξ on the set of 77 hexads in the Steiner system $(P_{22}, \mathcal{H})$ of type $S(3,6,22)$ in which two hexads are adjacent if they intersect in a *pair* of elements of P_{22}. The graph is strongly regular (which means distance-regular of diameter 2) with the following intersection diagram:

The action on Ξ of the group $K \cong \mathrm{Aut}\ (M_{22})$ is of rank 3, which means the action is transitive both on the pairs of adjacent and on the pairs of non-adjacent vertices. The stabilizer of a vertex h is $K_h \cong 2^4 : \mathrm{Sym}_6$. Thus K_h is transitive on $\Xi(h)$ consisting of the hexads intersecting h in pairs and on the set $\Xi_2(h)$ consisting of the hexads disjoint from h. We will need the following result which is well known and easy to check.

Lemma 11.4.5 *In the above terms $O_2(K_h)$ acts on $\Xi(h)$ with orbits of length 4 and its action on $\Xi_2(h)$ is transitive.*

There are five hexads containing a given pair p. Clearly these five hexads form a complete subgraph in the hexad graph. This subgraph we denote by $\Xi(p)$. If $p \subset h$ then $\Xi(p) \setminus \{h\}$ is an orbit of $O_2(K_h)$ on $\Xi(h)$. The subgraphs $\Xi(p)$ containing h are in the natural bijection with $\binom{h}{2}$. Since $K_h/O_2(K_h)$ induces the symmetric group of h, the following lemma can be justified just by mere looking at the above intersection diagram of Ξ. Alternatively it can be deduced from the fact that any two hexads are either disjoint or intersect in a pair.

Lemma 11.4.6 *Let* $p_1, p_2 \in \binom{h}{2}$. *Then*

(i) *if* $|p_1 \cap p_2| = 1$ *then every vertex from* $\Xi(p_1) \setminus \{h\}$ *is adjacent to every vertex in* $\Xi(p_2) \setminus \{h\}$;

(ii) *if* $|p_1 \cap p_2| = 0$ *then every vertex from* $\Xi(p_1) \setminus \{h\}$ *is adjacent to a half of the vertices in* $\Xi(p_2) \setminus \{h\}$.

The following final result in this section is well known and easy to deduce.

Proposition 11.4.7 *Each of the subgroups* $K_p \cong 2^5 : \mathrm{Sym}_5$, $K_o \cong 2^3 : L_3(2) \times 2$ *and* $K_h \cong 2^4 : \mathrm{Sym}_6$ *is maximal in* $K \cong \mathrm{Aut}\ (M_{22})$.

11.5 Diagrams of $\mathcal{H}(M_{24})$

In this section we present the diagrams which describe the action of the Mathieu group $M \cong M_{24}$ on its Ronan–Smith geometry $\mathcal{H}(M_{24})$. Let us explain the information given on the diagrams. The proof of the diagrams can be found in section 3.7 of (Ivanov 1999).

For $x = b$, t and s the diagram $D_x(M_{24})$ describes the orbits of the maximal parabolic subgroup M_x on the elements of $\mathcal{H}(M_{24})$. An orbit is represented by a circle; the number inside the circle is the length of the orbit and the index at this number (which is b, t, or s) indicates the type of elements in the orbit (octads, trios, or sextets, respectively). The length is given in the form $2^l k$ which means that $Q_x = O_2(M_x)$ acts with k orbits of length 2^l each (when $l = 0$ the factor 1 is suppressed).

Let $O^{(1)}$ and $O^{(2)}$ be orbits of M_x. Then the line joining the circles corresponding to these orbits indicates that an element $\alpha \in O^{(1)}$ is incident to a non-zero number m of elements in $O^{(2)}$. The number m is shown next to the line and closer to the circle of $O^{(1)}$. If the number m is shown as a sum of two numbers, say m_1 and m_2 then the stabilizer of α in M_x acting on the set of elements in $O^{(2)}$ incident to α has two orbits with lengths m_1 and m_2; otherwise the action is transitive. Notice that on each diagram every orbit is uniquely determined by its length and type, which enables us to refer to an orbit simply by giving its length and type.

For instance consider the diagram $D_t(M_{24})$. It shows that the subgroup $M_t \cong 2^6 : (L_3(2) \times \mathrm{Sym}_3)$ acting on the set of octads has three orbits with lengths 3, 84, and 672. The subgroup $Q_t = O_2(M_t)$ acts trivially on the first of the orbits, has 21 orbits of length 4 on the second orbit and 21 orbits of length 32 on the third orbit. Let α be an octad from the third orbit. Then among the 35 sextets incident to α one is in the M_x-orbit of length $84 = 2^2 \cdot 21$, six are in the orbit of length $336 = 2^3 \cdot 42$, and twenty-eight are in the orbit of length $1344 = 2^6 \cdot 21$. Finally the stabilizer of α in M_x acts on the set of sextets incident to α with four orbits of lengths 1, 6, 12, and 16.

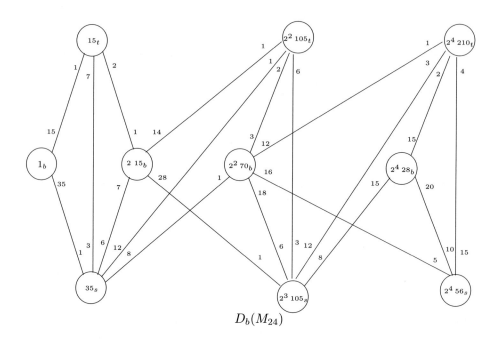

$D_b(M_{24})$

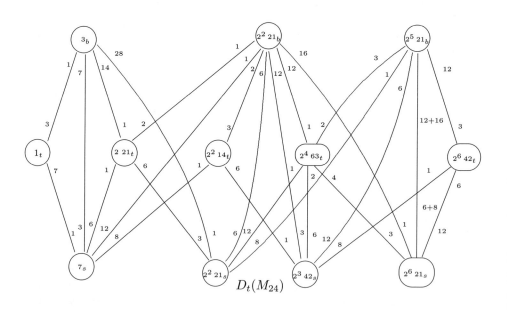

$D_t(M_{24})$

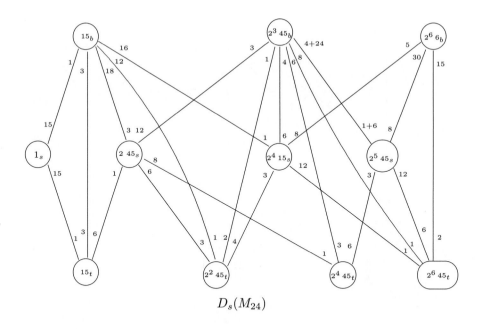

$$D_s(M_{24})$$

11.6   Diagrams of $\mathcal{H}(M_{22})$

In this section we present the diagrams describing the action of $K \cong \mathrm{Aut}\ (M_{22})$ on the Ronan–Smith geometry $\mathcal{H}(M_{22})$. The notation is similar to that in the previous section. The proofs of the diagrams can be found in section 3.9 of (Ivanov 1999).

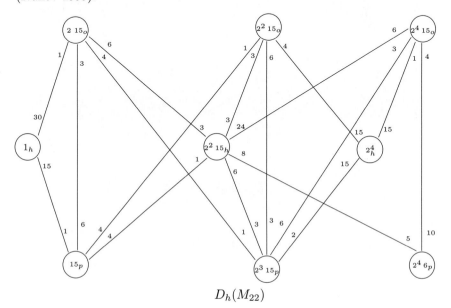

$$D_h(M_{22})$$

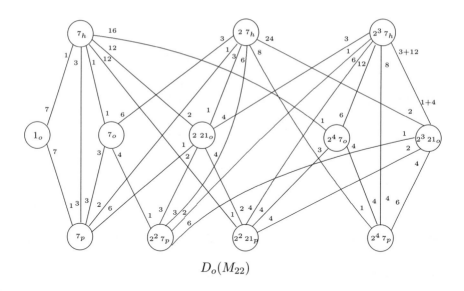

$$D_o(M_{22})$$

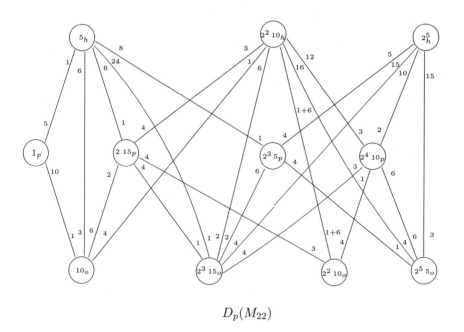

$$D_p(M_{22})$$

Exercises

1. Deduce from (11.4.2) that the geometry $\mathcal{H}(M_{22})$ is simply connected.

REFERENCES

Aschbacher, M., *Finite Group Theory*, Cambridge University Press, Cambridge, 1986.

Aschbacher, M., *Sporadic Groups*, Cambridge University Press, Cambridge, 1994.

Aschbacher, M., *3-Transposition Groups*, Cambridge University Press, Cambridge, 1997.

Aschbacher, M. and Segev, Y., The uniqueness of groups of type J_4, *Invent. Math.* **105** (1991), 589–607.

Aschbacher, M. and Smith, S. D., A classification of quasithin groups, Preprint, 2003.

Azad, H., Barry, M., and Seitz, G., On the structure of parabolic subgroups, *Comm. Algebra* **18** (1990), 551–562.

Bell, G., On the cohomology of finite special linear groups I and II, *J. Algebra* **54** (1978), 216–238 and 239–259.

Benson, D. J., The Simple Group J_4, Ph.D. Thesis, Trinity College, Cambridge, 1980.

Brouwer, A. E., Cohen, A. M., and Neumaier, A., *Distance-Regular Graphs*. Springer Verlag, Berlin, 1989.

Cameron, P. J. and Praeger, C. E., Graphs and permutation groups with projective subconstituents, *J. London Math. Soc.* **25** (1982), 62–74.

Conway, J. H., A group of order $8, 315, 553, 613, 086, 720, 000$, *Bull. London Math. Soc.* **1** (1969), 79–88.

Conway, J. H., The hunting of J_4, *Eureka* **41** (1981), 46–54.

Djoković, D. Ž. and Miller, G. L., Regular groups of automorphisms of cubic graphs, *J. Cobin. Theory (B)* **29** (1980), 195–230.

Conway, J. H., Curtis, R. T., Norton, S. P., Parker, R. A., and Wilson, R. A., *Atlas of Finite Groups*, Clarendon Press, Oxford, 1985.

Cooperman, G. D., Lempken, W., Michler, G. O., and Weller, M., *A New Existence Proof of Janko's Simple Group J_4*, Preprint No. 16 IEM, Essen, 1997.

Curtis, R. T., A new combinatorial approach to M_{24}, *Math. Proc. Camb. Phil. Soc.* **79** (1976), 25–42.

Finkelstein, L., Finite groups with a standard component of type J_4, *Pasific. J. Math.* **71** (1977), 41–55.

Frenkel, I., Lepowsky, J., and Meurman, A., *Vertex Operator Algebras and the Monster*, Acad. Press, Boston, 1988.

Ganief, S. and Moori, J., 2-Generations of the fourth Janko group J_4, *J. Algebra* **212** (1999), 305–322.

Goldschmidt, D., Automorphisms of trivalent graphs, *Annals Math.* **111** (1980), 377–406.

Gorenstein, D., *Finite Groups*, Harper & Row, New York, 1968.

Green, D. J., On the cohomology of the sporadic simple group J_4, *Math. Proc. Camb. Phil. Soc.* **113** (1993), 253–266.

Griess, R. L., Automorphisms of extra special groups and non-vanishing degree 2 cohomology, *Pasific. J. Math.* **48** (1973) 403–422.

Griess, R. L., *Twelve Sporadic Groups*, Springer-Verlag, Berlin, 1998.

Guloglu, I. S., A characterization of the simple group J_4, *Osaka J. Math.* **18** (1981), 13–24.

Harada, K., *Monster Group Expansion*, Iwanami Shoten, 1999. [In Japanese]

Heiss, St., On a parabolic system of type M_{24}, *J. Algebra* **142** (1991), 188–200.

Ivanov, A. A., On 2-transitive graphs of girth 5, *Europ. J. Combin.* **8** (1987), 393–420.

Ivanov, A. A., A geometric approach to the uniqueness problem for sporadic simple groups, *Dokl. Akad. Nauk SSSR* **316** (1991), 1043–1046. [In Russian]

Ivanov, A. A., A presentation for J_4, *Proc. London Math. Soc.* **64** (1992), 369–396.

Ivanov, A. A., *Geometry of Sporadic Groups I. Petersen and Tilde Geometries*, Cambridge University Press, Cambridge, 1999.

Ivanov, A. A., From $O_{10}^+(2)$ to J_4, *Algebraic Combinatorics, RIMS Kokyuroku* **1327** (2003), 66–75. [In Japanese]

Ivanov, A. A. and Meierfrankenfeld, U., A computer free construction of J_4, *J. Algebra* **219** (1999), 113–172.

Ivanov, A. A. and Pasechnik, D. V., Minimal representations of locally projective amalgams, *J. London Math. Soc.* **70** (2004).

Ivanov, A. A. and Shpectorov, S.V., Geometries for sporadic groups related to the Petersen graph. II, *Europ. J. Combin.* **10** (1989), 347–362.

Ivanov, A. A. and Shpectorov, S. V., P-geometries of J_4-type have no natural representations, *Bull. Soc. Math. Belgique* (A) **42** (1990), 547–560.

Ivanov, A. A. and Shpectorov, S. V., The universal non-abelian representation of the Petersen type geometry related to J_4, *J. Algebra*, **191** (1997), 541–567.

Ivanov, A. A. and Shpectorov, S. V., *Geometry of Sporadic Groups II. Representations and Amalgams*, Cambridge University Press, Cambridge, 2002.

Ivanov, A. A. and Shpectorov, S. V., Amalgams determined by locally projective actions, *Nagoya J. Math.* **176** (2004).

Ivanov, A. A. and Shpectorov, S. V., Tri-Extraspecial groups, *J. Group Theory* (submitted).

James, G. D., The modular characters of the Mathieu groups, *J. Algebra* **27** (1973), 57–111.

Janko, Z., A new finite simple group of order 86, 775, 571, 046, 077, 562, 880 which possesses M_{24} and the full covering group of M_{22} as subgroups, *J. Algebra* **42** (1976), 564–596.

Kleidman, P. B. and Wilson, R. A., The maximal subgroups of J_4, *Proc. London Math. Soc.* **56** (1988), 484–510.

Lempken, W., A 2-local characterization of Janko's group J_4, *J. Algebra* **55** (1978a), 403–445.

Lempken, W., The Schur multiplier of J_4 is trivial, *Archiv Math.* **30** (1978b), 267–270.

Lempken, W., On local and maximal subgroups of Janko's simple group J_4, *Rendic. Accad. Naz. Sci., Mem. Matem.* 107° **XIII** (1989), 47–103.

Lempken, W., Constructing J_4 in $GL(1333, 11)$, *Comm. Algebra.* **21** (1993), 4311–4351.

Mason, G., Some remarks on groups of type J_4, *Archiv Math.* **29** (1977), 574–582.

Mason, G., Quasithin groups. In: *Finite Simple Groups II* (Collins, M., ed.), Academic Press, London, 1980, pp. 181–197.

Mathieu, E., Mémoire sur le nombre de valeurs que peut acquérir une fonction quand on y permut ses variables de toutes les manières possibles, *J. de Math. Pure et App.* **5** (1860), 9–42.

Mathieu, E., Mémoire sur l'étude des fonctions de plusieres quantités, sur la manière des formes et sur les substitutions qui les laissent invariables, *J. de Math. et App.* **6** (1861), 241–323.

Mazet, P., Sur le multiplicateur de Schur du groupe de Mathieu M_{22}, *C. R. Acad. Sci. Paris*, Serie A–B, **289** (1979), 659–661.

Michler, G. O. and Weller, M., A new computer construction of the irreducible 112-dimensional 2-modular representation of Janko's group J_4, *Comm. Algebra* **29** (2001), 1773–1806.

Norton, S. P., The construction of J_4. In: *Proc. Symp. Pure Math.* No. 37 (Cooperstein, B. and Mason, G., eds.), AMS, Providence, RI, 1980, pp. 271–278.

Parrott, D., Characterization of the Fischer groups, I, II, III, *Trans. AMS* **265** (1981), 303–347.

Pasini, A., *Diagram Geometries*, Clarendon Press, Oxford, 1994.

Reifart, A., Some simple groups related to M_{24}, *J. Algebra* **45** (1977a), 199–209.

Reifart, A., Another characterization of Janko's simple group J_4, *J. Algebra* **49** (1977b), 621–627.

Reifart, A., A 2-local characterization of the simple groups $M(24)'$, .1, and J_4, *J. Algebra* **50** (1978), 213–227.

Ronan, M. A., Locally truncated building and M_{24}, *Math. Z.* **180** (1982), 489–501.

Ronan, M. A. and Smith, S., 2-Local geometries for some sporadic groups. In: *Proc. Symp. Pure Math.* No. 37 (Cooperstein, B. and Mason, G., eds.), AMS, Providence, RI, 1980, pp. 283–289.

Rowley, P. and Walker, L., The maximal 2-local geometry for J_4, I, Preprint, UMIST, No. 11, 1994.

Serre, J.-P., Arbres, amalgams, SL_2, *Astérisque* **46**, 1977.

Shpectorov, S. V., A geometric characterization of the group M_{22}, *Investigations in Algebraic Theory of Combinatorial Objects*, VNIISI, Moscow, 1985, pp. 112–123. [In Russian, English translation by Kluwer Academic Publishers, Dordrecht, 1994]

Shpectorov, S. V., On geometries with diagram P^n, Preprint, 1988. [In Russian]

Shult, E. E. and Yanushka, A., Near n-gons and line systems, *Geom. Dedic.* **9** (1980), 1–72.

Sitnikov, V. M., The minimal permutational representation of the finite simple group J_4 of Janko, *Mat. Zametki.* **47** (1990), 137–146.

Smith, S. D., The classification of finite groups with large extraspecial 2-subgroups. In: *Proc. Symp. Pure Math.* No. 37 (Cooperstein, B. and Mason, G., eds.), AMS, Providence, RI, 1980, pp. 111–120.

Staffort, R. M., A characterization of Janko's simple group J_4 by centralizer of elements of order 3, *J. Algebra* **57** (1979), 555–566.

Stroth, G., An odd characterization of J_4, *Israel J. Math.* **31** (1978), 189–192.

Stroth, G. and Weiss, R., Modified Steinberg relations for the group J_4, *Geom. Dedic.* **25** (1988), 513–525.

Syskin, S. A., Abstract properties of the simple sporadic groups, *Russ. Math. Survey.* **35** (1980), 209–246.

Taylor, D. E., *The Geometry of the Classical Groups*. Heldermann Verlag, Berlin, 1992.

Thompson, J. G., Finite-dimensional representations of free products with an amalgamated subgroup, *J. Algebra* **69** (1981), 146–149.

Todd, J., A representation of the Mathieu group M_{24} as a collineation group, *Annali di Math. Pura ed Applicata* **71** (1966), 199–238.

Trofimov, V. I., Vertex stabilizers of locally projective groups of automorphisms of graphs. A summary. In: *Groups, combinatorics and geometry, Durham 2001* (Ivanov, A. A., Liebeck, M. W., and Saxl, J., eds.), World Scientific Publishing, River Edge, NJ, 2003, pp. 313–326.

Van Trung, T., Eine Kennzeichnung der endlichen Gruppe J_4 durch eine 2-local Untergruppe, *em Rend. Cont. Sem. Mat. Padova* **62** (1980), 35–45.

Tutte, W., A family of cubical graphs, *Proc. Camb. Phil. Soc.* **43** (1947), 459–474.

Venkatesh, A., *Graph Coverings and Group Liftings*, Unpublished manuscript, 1998.

Vick, J. W., *Homology Theory: An Introduction to Algebraic Topology*, Springer Verlag, New York, 1994.

Wilson, R. A., Vector stabilizers and subgroups of the Leech lattice groups, *J. Algebra* **127** (1989), 387–408.

Witt, E., Über Steinersche Systeme, *Abh. Math. Sem. Univ. Hamburg* **12** (1938), 325–275.

Yoshiara, S., The radical 2-subgroups of the sporadic simple groups J_4, Co_2, and Th, *J. Algebra* **233** (2000), 309–341.

INDEX